HISTORY IN THE MAKING

EUROPE 1870-1966

RICHARD FOGARTY

THE EDUCATIONAL COMPANY

First published 1994
The Educational Company of Ireland
Ballymount Road
Walkinstown
Dublin 12

A trading unit of Smurfit Services Ltd

© R. Fogarty

AHI 7311 S
6789

Editor: Ursula Ní Dhálaigh
Design: Peanntrónaic Tta
Disc conversion and layout: Phototype-Set Ltd.
The publishers are deeply indebted to Michael Lynam and Noel Murphy for their unfailing thoroughness and graciousness.
Cover design: Identikit
Printed in the Republic of Ireland by
Smurfit Web Press, Dublin 9

Cover shows *Café Deutschland* by Jörg Immendorff, courtesy Galerie Michael Werner, Cologne & New York.

Acknowledgements

The publishers wish to thank the following for permission to reproduce illustrations in this book:
Hulton Deutsch Collection; Topham Picture Library; Imperial War Museum; The Mansell Collection; Popperphoto; Keystone Press; Photosource; Camera Press; Mary Evans Picture Library; The National Portrait Gallery; Associated Press; UPI; Snark International; Penguin Ltd.; Odham's Press; Victoria and Albert Museum; John Hillelson Agency, United Nations Headquarters; Novosti Press Agency; Staatsbibliothek, Berlin; Ullstein Bilderdienst, Berlin; Bilderdienst Süddeutscher Verlag; Bundesarchiv, Coblenz; Photo Deutsches Museum, München; Bibliothèque Nationale, Paris; La Documentation Francaise Photothèque; H. Roger-Viollet, Paris; *London Evening Standard*; *Punch*; Sgt. White of Military Archives, Cathal Brugha Barracks, Dublin, for the photograph of the Irish Army serving with the UN in Cyprus.

In some cases the publishers have failed to trace copyright holders. However, they will be happy to come to a suitable arrangement with them at the earliest opportunity.

CONTENTS

SECTION ONE

1870-1914: NATIONALISM, IMPERIALISM & INDUSTRIAL EXPANSION

SECTION TWO

1914-1924: WAR AND PEACE

SECTION THREE
1919-1939: The Rise of Fascism

SECTION FOUR
1939-c. 1970:
The Shaping of Modern Europe

SECTION FIVE
1870-c. 1970: Aspects of Science, Technology and Culture

Preface

History in the Making: EUROPE is intended for use by senior post-primary students. It is written as a companion volume to *History in the Making:* IRELAND by M.E. Collins.

Special care has been taken in this book to present the information in a clear and simple manner. Long and complex sentences are avoided and difficult concepts or ideas are explained. Chapters are divided into manageable sections and sub-sections. Lavish use is made of photographs, political cartoons and clearly-defined maps. In the later chapters a link is established between the events described and current affairs. A list of examination questions is included after each section. The select bibliography at the end of Chapter 30 recommends a number of books which students may wish to use for additional reading.

I am grateful to M.E. Collins, Geraldine Grogan and Timothy Nyhan for advice on the text and to Anne Murphy for help with the maps. I owe thanks to all at The Educational Company, especially to the editor, Ursula Ní Dhálaigh, who has provided valuable help, inspiration and guidance. Finally, I would like to thank my colleagues at St. Fintan's High School for their encouragement and support.

Richard Fogarty
St. Fintan's High School, Sutton,
Dublin 13.
March 1994

Introduction: Europe in the Nineteenth Century

From hence, let fierce contending nations know
What dire effects from civil discord flow.

(Joseph Addison, *The Campaign*)

The Congress of Vienna 1815

After the collapse of the Napoleonic Empire in 1815, statesmen representing the major powers of Europe met at Vienna. They wanted to secure a lasting peace and bring to an end the very idea that there should be 'contending nations'. At the same time they were concerned with the maintenance of order within states and with the prevention of revolution.

The agreements made in Vienna proved successful in forming a framework which ensured that there was no major conflict between the powers between 1815 and 1870 with the exception of the Crimean War, 1854-56. This was achieved by setting up a **'balance of power'** in Europe. The term meant that no country should be strong enough to cause disruption to other countries, as Napoleon had done. Such a country, with territorial ambitions, would face the combined forces of other powers.

The Vienna Settlement

At Vienna, France was reduced to her pre-1789 borders and the Bourbon monarchy was restored. In Germany a loose union of 38 states known as the **German Confederation** was established. One of the strongest German states, Prussia, was further strengthened when she was given the Rhineland. This placed her in a position where she would have to take the lead in any future war against France.

Austria remained the most powerful German state. She was given the Presidency of the Confederation. Austria also had a large empire and was granted more territory in Northern Italy. The position of the Tsar of the Russian Empire was strengthened when he was confirmed as ruler of Poland. In South-East Europe, the rule of the Turkish (Ottoman) Emperor was not disturbed.

Weaknesses of the Settlement

The settlement held firm during the Crimean War when Russia failed to extend her influence over territories within the Turkish Empire. However, the statesmen of 1815 were very conservative and did not allow for the problems of a changing world.

They based their settlement on the old principle of the **'divine right'** of monarchs who, they believed, received their authority from God. This meant that they could rule without any reference to their subjects.

Monarchs who maintained all power in their own hands were known as **autocrats.** They resisted the dangerous ideas that emerged during the French Revolution, namely liberty, equality and fraternity.

Other theories and ideals emerged in nineteenth-century Europe. Many of these theories challenged the very principle of 'divine right'. Chief among them were **liberalism, nationalism, socialism** and **communism**.

Liberalism

The term 'liberal' became a widespread political description in the nineteenth century. It was associated with human rights and liberties: **the right to property, freedom of speech, of the press, of worship and of political assembly**. Government was to be elected by the people, have its powers limited and be responsible to the people.

Liberals also demanded **economic freedom.** They believed in the principle of *laissez-faire*, that is **freedom from government interference or regulation**. Liberalism appealed mainly to the middle classes. Liberals felt that they had obligations to the lower classes, but were not anxious to grant them equality.

Nationalism

The term 'nationalism' came into common usage in the 1830s. It is defined as the desire of people of common descent, language, culture or historical tradition to be united under a single government. The area ruled by such a government came to be known as a **state** or a **nation state**.

The emergence of nationalism in the nineteenth century led to the unification of Italy which had been a collection of smaller independent states. The Kingdom of Italy was proclaimed in 1861 and the process of unification was completed in 1870 when the nationalists captured Rome. During the 1860s nationalism also grew stronger in the German Confederation.

Socialism

Socialism is the theory which advocates that **the means of production, distribution and exchange should be owned and controlled by the state**. Some people who called themselves socialists were merely concerned that there should be a greater measure of justice and equality in the way society was ordered. They were concerned, in particular, with the welfare of the lower classes. A more extreme interpretation of socialism was provided by Karl Marx.

Communism

The basis of Marxism is explained by the **German Socialist, Karl Marx**, in his two great books, *The Communist Manifesto* (1848) and *Das Kapital* (1867). Marx believed that economic factors explained all changes that occurred in the history of the world. He predicted that one day the working-class (proletariat) would overthrow the capitalist system of private ownership. Communism developed from the Marxist philosophy. It aimed at **the abolition of private property and the creation of a truly classless society**.

Europe in 1870

The structures laid down by the statesmen of Europe in 1815 were beginning to break down by 1870. The various theories and 'isms' had caused **'civil strife'** in 1848 and were threatening internal order in a number of states. **The balance of power between states was looking unsteady.** The first great conflict among 'contending nations' would be between France and Prussia.

SECTION ONE

1870-1914: NATIONALISM, IMPERIALISM & INDUSTRIAL EXPANSION

Chapter 1

The Franco-Prussian War

Background to the War

German Unification

The Franco-Prussian War should be seen as the final part of the process of German unification. The Congress of Vienna (see introduction) in 1815 created a German Confederation of 39 states. The transformation of these 39 independent states into the German Empire in 1871 was due to four main factors:

- German nationalists throughout the 19th century sought the unification of their country. They hoped to create a *Grossdeutsch* or 'Greater German' state and looked to Austria for leadership. However, this scheme was abandoned because the Habsburg rulers of Austria insisted on retaining their non-German territory and people. German patriots then turned to Prussia to lead a *Kleindeutsch* or 'Lesser German' solution to unification. This solution would exclude Austria, a major Germanic state.
- There was a powerful economic impetus towards unification. Rapid industrial growth, improved communications and commercial co-operation between parts of the country showed the potential of a united Germany.
- The success of the movement for Italian unification was an example which inspired many Germans.
- The emergence of a politician willing to accept Prussia's role as leader was the final and successful spur to the process of unification. This politician was **Otto von Bismarck**.

Opposition to Unification

The unification of Germany was opposed by other European powers, especially France and Austria. The French felt that it would upset the balance of power in Europe. A united Germany would be very strong and would be a threat to peace. Austria and Prussia were the two major Germanic powers and were traditional rivals. Austria was opposed to any movement which would increase Prussia's growing power and prestige.

In 1862, **Otto von Bismarck**, a wealthy **junker** (landowner), became Prime Minister of Prussia. He was appointed by **William I**, the king of Prussia. Bismarck's aim was to create a united Germany which would be under the control of Prussia. He believed that the unification of Germany would be achieved by war if necessary. In his first speech to the Prussian National Assembly he stated that:

> the position of Prussia in Germany will be determined by its power... not through speeches and majority decisions are the great questions of the day decided, but through blood and iron.

1864: war with Denmark

Austria believed that she should be the dominant state in a united Germany. At first, Bismarck decided to deceive her about his real intentions.

In 1864 he persuaded Austria to join Prussia in a war against Denmark. Bismarck wanted the Danish provinces of **Schleswig** and **Holstein** which had large German populations. He promised Austria a share of the spoils. The Danes were quickly defeated, thanks to the the efficient Prussian army. Prussia took over Schleswig and Austria took Holstein.

June-July 1866: the Seven Weeks War

Bismarck now turned against Austria, his former 'ally'. France had promised Bismarck that she would remain neutral in any war between Prussia and Austria. But France believed that Austria would easily defeat Prussia if war broke out and would put an end to Bismarck's ambitions. In 1866 the Prussian army again invaded Denmark and took back Holstein from Austrian control. As a result, Austria declared war on Prussia. Within seven weeks the Austrians were defeated. At the decisive battle of **Sadowa**, 40,000 Austrians were killed while only 10,000 Germans lost their lives.

The North German Confederation

A year later Bismarck formed the states north of the river Main into the **North German Confederation** under Prussia's control. The remaining states, south of the river Main, became independent and free of Austrian influence. After that, Austria played no further role in German unification.

Bismarck, however, still had to unite the south German states with the North German Confederation. He needed a good reason for them to support German unification. He believed that the south German states would fight on his side if France declared war on Germany.

The Spanish throne

Bismarck was looking for an excuse to provoke France into war. A chance came in 1868 when a revolt took place in Spain. **Queen Isabella** was deposed and she and her family were sent into exile.

Bismarck now proposed that **Prince Leopold von Hohenzollern** should become king of Spain. Prince Leopold was related to the Prussian royal family, a fact which

GERMANY BEFORE THE FRANCO-PRUSSIAN WAR

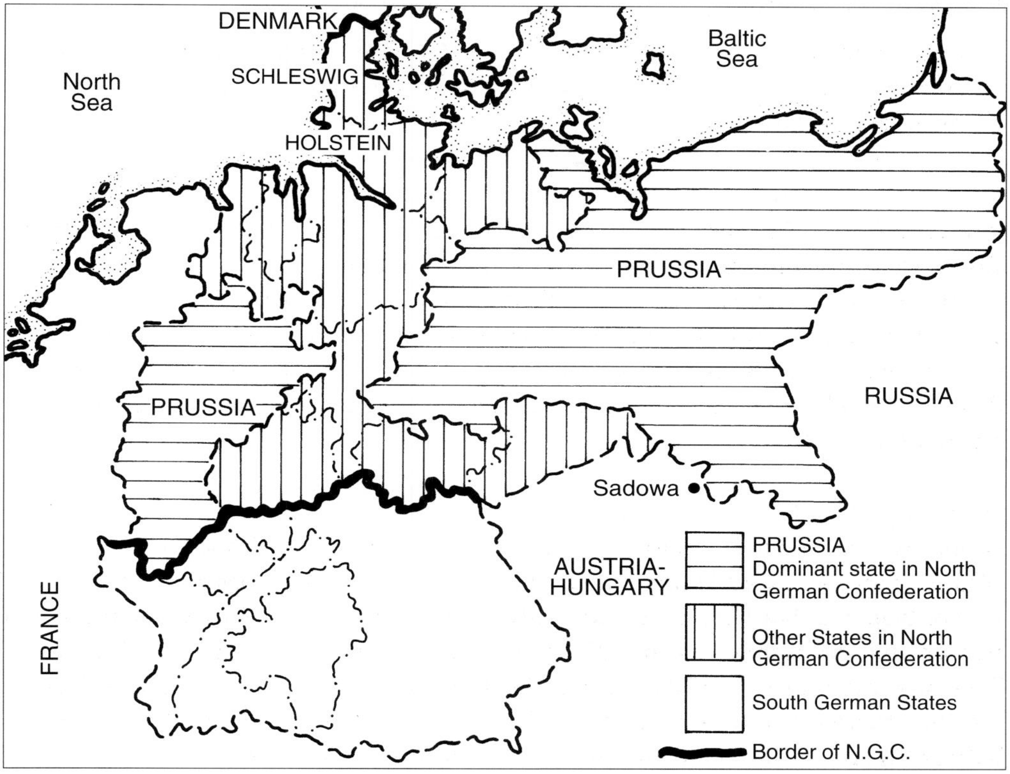

Bismarck knew would annoy France. He even sent a secret envoy to Spain to bribe the Spanish parliament to support Leopold's claim. However, the French were determined that no Prussian should sit on the Spanish throne. If a Prussian did become king of Spain, France would be surrounded by hostile states. At the last minute, Leopold withdrew his claim. For the moment, war between France and Germany had been averted.

The Ems telegram

Napoleon III, Emperor of France, was urged on by his government, the French press and his wife, **Empress Eugenie**, to stand up to Bismarck. Advised by his foreign minister, the **Duc de Gramont**, Napoleon ordered his ambassador to Prussia to meet with King William I. He wanted the Prussian king to promise that he would never again support a Hohenzollern claim to the Spanish throne.

In July 1870 King William was staying at a health spa in **Ems**, taking the waters. He received the French ambassador with courtesy and gave him no cause for offence. He refused, however, to give any guarantee for the future. The king then sent a telegram to Bismarck in Berlin informing him of the meeting.

Bismarck seized this opportunity to upset the French. He shortened the telegram and also reworded it in such a way as to suggest that the French ambassador had been insulted by the Prussian king. The edited version of the telegram was published in the Berlin papers. It was also telegraphed to Prussian embassies throughout Europe.

War is declared

France fell into the trap set by Bismarck. Excited crowds gathered in city streets and village squares. Everywhere there was talk of national honour and revenge. Empress Eugenie and her circle of friends were determined that the 'insult' to France should be immediately avenged. Napoleon III, despite his personal reservations, found it impossible to resist public opinion. On 14 July 1870 the mobilisation of French troops began. On 19 July 1870 France declared war on Prussia.

Analysis

The Ems telegram was the excuse for war, not the cause of it. In the past, historians believed that Bismarck deliberately used the telegram to force France to declare war.

Napoleon III, Emperor of France

Modern historians, however, have argued that Bismarck could not have foreseen the war hysteria that the telegram would cause. Yet when France declared war, Bismarck did not find it unwelcome. For their own selfish reasons other European powers, including Britain, Russia, Italy and Austria agreed to remain neutral.

Course of the War

The War begins

Almost the entire Franco-Prussian War was fought on French soil. Despite an initial victory by the French at **Saarbrucken**, the war was a disaster for France. On 4 August 1870 a major battle took place at **Worth**. The French were outnumbered two to one. They were driven back to **Metz** where they took refuge in a huge fortress. Two days later, on 6 August, the Prussians drove the French back at **Spickeren** in Lorraine. They also took Alsace.

Metz and Sedan

France now faced a major invasion. Napoleon seemed to lose his nerve. The fortress at Metz, under the command of **General Bazaine**, was now besieged by the Germans. **Marshal MacMahon** joined forces with Napoleon (who had decided to take personal command of the French army) and tried to raise the siege of Metz. Outflanked by the Germans, they were forced to retreat.

THE FRANCO-PRUSSIAN WAR

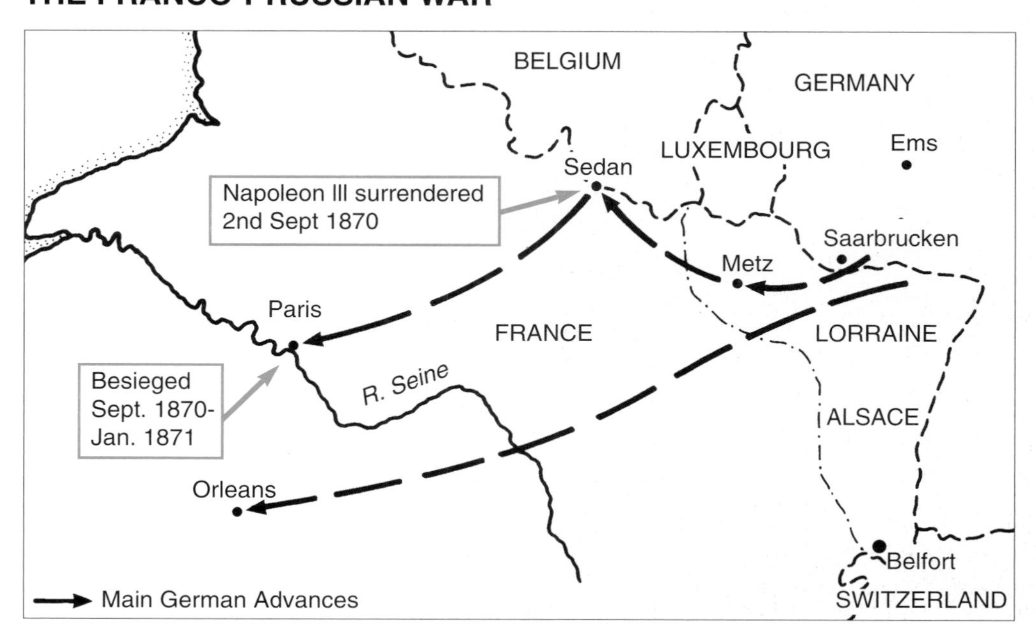

On 1 September the greatest battle of the war began at **Sedan**. King William and Bismarck were among the observers on a nearby hill. Under constant bombardment, the charges of the French cavalry were repulsed by the Prussian guns. On 2 September Napoleon surrendered in order to prevent further slaughter.

Reasons for the German victory

The Germans had an army of over 400,000 soldiers while the French had about half that number. Although the French had a new rifle (*chassepot*) and a new machine gun (*mitrailleuse*), in the long run they did not compare well with German weaponry, particularly the **Krupps** heavy artillery. Military historians believe that the new French machine gun was never used to proper effect in the war. The French were also short of ammunition.

In addition, the French were badly prepared for war. Reservists had to report to their home barracks rather than to those near where they were stationed. Therefore mobilisation was slowed up. Civilians also continued to use the railway system. This held up the transportation of men and war materials. The French generals, Bazaine and MacMahon, were experienced and brave commanders, but their tactics and war strategies were badly thought out. Napoleon's decision to take personal command of

Count von Moltke, Commander-in-chief of the German army.

the French army was disastrous. He was a sick man, suffering from stomach cancer.

The Germans, on the other hand, mobilised their forces quickly. In fact, they reached the border before the French had fully mobilised. They made full use of the rail system to transport men, food and ammunition. All German reservists reported to the regiment nearest to where they were stationed. The commander-in-chief of the German army, **Helmuth von Moltke**, was an expert tactician. Together with his Minister for War, **von Roon**, he had planned the war in great detail and they were confident of victory.

The end of the Second Empire

Two days after the French surrender, the National Assembly met in Paris. Napoleon was deposed. The Second Empire was over and was replaced with the **Third French Republic**. **Leon Gambetta** became leader of a new revolutionary government called the **Government of National Defence**.

When he learned that Prussia wanted to keep the provinces of Alsace and Lorraine, Gambetta decided that the French army must fight on. Partisan units attacked German troops who retaliated by burning villages and shooting captured

Shooting an elephant for food during the Siege of Paris

soldiers. General Bazaine held out in Metz until the end of October. Meanwhile, the German armies marched on Paris.

The first Siege of Paris

Late in September the siege of Paris began. Prussian troops encircled the city and cut off its food supplies. Bismarck wanted to starve it into submission. Early in October Gambetta made a dramatic escape from the city in a balloon. He tried to raise an army of resistance in the Loire. Meanwhile, the Parisians continued their resistance against the Prussian troops. Winter set in and food and fuel became desperately short. The animals in the zoo were slaughtered to feed the hungry. Even dogs and rats were eaten.

In January 1871 the Prussians began a bombardment of the city. Four hundred shells a day were fired from their Krupps guns. On 18 January, even before the siege had ended, King William of Prussia was declared **Emperor of Germany**. The ceremony took place in the **Hall of Mirrors** at Versailles on the outskirts of Paris. Finally, on 28 January, Paris surrendered. An armistice was signed between the government of National Defence and the Prussians. The Franco-Prussian War was over.

Consequences of the War

1871: the Treaty of Frankfurt

Bismarck would not negotiate the terms of the peace treaty with the Government of National Defence. He insisted on elections being held for a democratically-elected National Assembly. In February 1871 the French voted for a conservative assembly. The veteran statesman, **Adolphe Thiers**, became Prime Minister of a Republican government which formally signed the Treaty of Frankfurt on 10 May 1871.

The terms of the treaty were considered at the time to be extremely harsh.

- France lost the provinces of Alsace and Lorraine to the German Empire.
- France had to pay an indemnity or compensation of five billion francs (£200 million) to Germany within three years.
- A German army of occupation would remain in the North and East of France until the indemnity was paid.
- The French were allowed to keep the town of Belfort in Alsace but only after agreeing to allow a triumphant march by German soldiers through the streets of Paris.

Resentment

Most French people were shocked by the harsh terms of the treaty. The provinces of Alsace and Lorraine were rich in minerals and had a population of some 1,600,000.

William I of Germany proclaimed Emperor at Versailles, 1871

France had been humiliated and Germany united in a new German Empire. Men like **Gambetta**, who resigned as Minister for War, voted against the Treaty. They made revenge *(revanche)* and the recovery of the lost provinces part of their new political objectives. These objectives dominated French foreign policy up to World War I.

Bismarck was well aware of the smouldering French resentment. As Chancellor of the new German Empire he was determined that France would not be able to pursue her policy of revenge against Germany.

Chapter 2

Bismarck's Germany, 1871-1890

Bismarck: a profile

Bismarck's early life

Otto von Bismarck (1815-1898) was born on 1 April 1815. He was the son of a Prussian **Junker** or landowner. The Junkers were the conservative, landed aristocracy of East Prussia. After his student days and the death of his father, Bismarck spent eight years as a squire and farmer. Later he claimed that this was the life he liked best.

Bismarck entered parliament in 1847 and remained in politics for the next forty-three years. His attitude to politics is summed up in a statement he made on his own ambitions:

I want to play only the music I myself like or no music at all.

Otto von Bismarck, First Chancellor of the German Empire

Political aims

His policies as a member of the **Diet** (parliament) were to achieve German unity under Prussian leadership and to seek what he termed 'equality with Austria'. He was appointed Prussian Ambassador to Austria in 1852, to Russia in 1859 and to France in 1862. During the year 1862, William I of Prussia faced a deadlock with the National Assembly over army reforms. Bismarck was summoned home from Paris and appointed Prime Minister of Prussia.

One of Bismarck's political opponents saw his time as Prime Minister as 'government without budget, rule by the sword in home affairs and war in foreign affairs'. Through wars with Denmark, Austria and France (see chapter 1) he succeeded in his aim of creating a German Empire under Prussian influence.

Bismarck's Domestic Policy

The Constitution

It was Bismarck who created the constitution or system of government of the new German Empire. This became law on 20 April 1871. In drawing up the constitution, Bismarck was primarily concerned that the government would be controlled by the executive, i.e. the Emperor and Prime Minister.

This would enable him to dominate Parliament. He could secure Prussia's position as leading state in the Empire and keep a check on militant army officers. Bismarck considered it important that the new Empire should be stabilised and strengthened through strong central Government.

The role of the Kaiser

The head of the empire was always to be the king of Prussia. The ancient title of **Kaiser** (Emperor) was adopted to please the Bavarians. Only the Kaiser could appoint or dismiss his Prime Minister, the Imperial Chancellor. This office was held by Bismarck from 1871 to 1890. The Kaiser was also in sole charge of foreign affairs, and was Commander-in-Chief of the army. He assembled or dismissed the two houses of parliament and along with the Chancellor, signed all laws. While William I was Kaiser, he was happy that Bismarck should exercise the powers of both Emperor and Chancellor.

The houses of parliament

The Upper House of parliament or **Bundesrat** was made up of delegates from the twenty-five states. The number of representatives from each state was in proportion to the state's importance in the empire. Prussia held seventeen of the fifty-eight seats and in practice could block any laws she did not favour.

The Lower House, the **Reichstag**, was elected by the votes of all men over the age of twenty-five. As members of parliament were not paid until 1906, few working-

Kaiser William I

class men were in a position to stand for election. The Reichstag had limited powers and had no control over the army or foreign affairs, although it did control taxes.

The German Empire was a federal state with power divided between the Central Government and the twenty-five states from which it was formed. These states also had their own rulers and parliaments and they were responsible for their own internal affairs.

Political parties

The parties that competed for seats in the Reichstag represented the sectional interests of groups in the Empire. The **German Conservative Party** drew most of its support from Protestant Prussian Junkers. They opposed any concessions to liberals and socialists. Also on the right was the **Reichspartei** which represented landowners and industrialists and supported the Chancellor's policies.

The **Zentrum** (Central Party) looked after the interests of Catholics. Although conservative in most of its policies, it was sympathetic to moderate social reforms. The **National Liberals** were mainly Protestant wealthy middle-class supporters of Bismarck. They wanted no state interference in business and favoured free trade. In later years they opposed concessions to the **Socialists**. Another party, the **Progressives**, won support from intellectuals, small businessmen and skilled workers, while the **Social Democrats** were elected by the industrial working-class.

The Kulturkampf

The new German Empire was dominated by Lutheran Prussia. However, about one third of the total population was Catholic. In 1870 the **First Vatican Council** issued a document on the infallibility of the Pope. This stated that the Pope could not be wrong when making official announcements on matters of faith and morals.

Bismarck feared this would mean that Catholics would be more loyal to the Pope than to the new German Empire which he was trying to consolidate. By 1871 the Centre Party, representing German Catholics, had 57 out of 382 deputies in the Reichstag. Catholics felt that they needed their own party because German Liberals had put forward a programme which would reduce the role of the Churches in society, especially their role in education.

Bismarck thought that it was time for action. He attacked the power of the Catholic Church in Germany:

> It is not a matter of a struggle between belief and unbelief. It is a matter of a conflict between monarchy and priesthood. What is at stake is the defence of the state.

Kulturkampf means 'struggle for civilisation'. It resulted in a battle between Bismarck and the Church for control of education. In 1871 the Jesuits were forbidden to mention politics while preaching. From 1872 they were not allowed to preach or to enter schools and eventually most Jesuits were expelled from Germany. In 1873 the anti-clerical **Aldabert Falk** was appointed Prussian Minister for Ecclesiastical (Church) affairs.

The May Laws

Falk issued the **May Laws**. These laws decreed that priests had to graduate from a German university and pass a state examination. The state was made responsible for the discipline of the Church over its members. Education was placed under state control. Civil marriage was made compulsory.

The Church responded with a policy of passive resistance and gained strength from persecution. By 1876 most of the Prussian bishops were in prison or in exile. Support for the Centre Party grew and it increased its numbers in the Reichstag. Bismarck began to weary of the struggle. He wanted the support of the Catholics for his economic policies and for other challenges ahead. The royal family, the Liberal Party and the Protestant Church were against religious persecution. Instead of uniting Germany, the Chancellor came to realise that he was dividing Germans.

Bismarck's opportunity for a change of course came with the death of **Pope Pius IX**. The new Pope, **Leo XIII**, wrote to the Kaiser expressing a wish for better relations. The May Laws were repealed and Falk was dismissed in 1879, as Bismarck cleverly shifted the blame away from himself.

Bishops were allowed back to their dioceses and all religious orders, except the Jesuits, could now carry out their religious duties. The Church again became

Bismarck offers two options to the Socialists – limited Socialism, or close the Reichstag!

responsible for the education of its own clergy. State supervision of schools and civil marriage remained. Bismarck looked forward to co-operating with the Centre Party in the upcoming fight against socialism.

The growth of Socialism

In 1869 various socialist groups in Germany joined together to form the **Social Democratic Party**. At an important conference in Gotha in 1875, the party drew up its programme. Private business was condemned and a state takeover of industry was advocated. The Social Democrats demanded that all profits should be shared among workers and that social injustices in Germany be removed.

Germany's rapid industrialisation led to the growth of an urban workforce. Conditions in the factories were very bad. The law offered little protection to the workers and discontent grew. By 1877 the Social Democrats had twelve seats in the Reichstag. Bismarck was convinced that socialism would have to be destroyed in order to preserve the unity of the Empire. In 1878 two attempts on the life of Kaiser William I gave Bismarck an excuse to attack the socialists, even though they were not involved in the incidents.

Bismarck and the socialists

An anti-socialist bill was passed by the Reichstag. Socialism was banned in Germany.

Some 1,500 socialists were sent to prison in the period up to 1890 and many were deported. Their newspapers and magazines were suppressed.

No measures were taken to improve the working or living conditions of the poor and support for socialist policies continued to grow. Bismarck realised that these policies could not be overcome by oppression alone. He introduced a programme of **'State Socialism'.** Medical insurance and sick pay for workers was provided. In 1889 old-age pensions were paid to people who reached the age of seventy. This was two decades before old-age pensions were introduced in Britain. The laws against socialists were allowed to lapse in 1890 and by 1912 the Social Democratic Party was the largest in the Reichstag.

Industry

While the foundations of German economic development had been laid before 1870, it was in the twenty years up to 1890 that the enormous expansion of German industries began. The size of the population, which increased from 41 million in 1871 to nearly 50 million in 1899, meant that more workers were available to work in the new factories. German technical schools and universities provided a constant stream of highly-educated recruits into science-based industries. German unity provided one large market.

The coal and iron deposits of Lorraine and the textile industry of Alsace contributed to the country's prosperity. There were enormous quantities of raw materials in the Ruhr and Saar basins and in Silesia. Bismarck encouraged the

The first electric train, designed by Werner von Siemens, 1879.

development of heavy industry and, during his term as Chancellor, Germany became one of the world's great economic powers.

One of the most famous industrial firms was **Alfred Krupp's** steel factory at Essen. It became famous for the production of arms and railway equipment. The electrical factories of **Werner von Siemens** also became known throughout the world. The chemical industry prospered and Germany became a major producer of synthetic dyes, potassium, potash and ammonia. The formation of 'cartels' helped profits by avoiding the dangers of competition.

Agriculture

Agriculture also prospered in Bismarck's Germany. The Chancellor looked after his own class, the East German Junkers. He introduced high tariffs on goods imported into Germany. This favoured the production of cereals on the large estates owned by the Junkers. Agriculture became more mechanised. The new sugar beet industry grew. Co-operatives were established and banks provided credit for farmers.

Bismarck's policy of protecting both industry and agriculture led to a clash with the National Liberals, who favoured free trade.The Chancellor won the argument. Agricultural production and industrial output continued to grow and Germany soon became one of the most powerful countries in Europe.

Bismarck's foreign Policy

Bismarck's aims

Bismarck had used wars to unite Germany. After 1871 he realised that another major war might shatter the German Empire. Germany was open to attack from two sides because of its central position. France was bitter at losing the provinces of Alsace and Lorraine and the French press constantly referred to a war of revenge. Bismarck knew that France could do very little without an ally. His foreign policy after 1871 was built around the principle of keeping France isolated.

The Dreikaiserbund

In 1873 Bismarck formed the **Dreikaiserbund**, or **League of the Three Emperors**. The three empires involved were Germany, Austria-Hungary and Russia. The main aim of the League was to maintain existing borders in Europe. The three countries would unite to fight socialism and republicanism. The Dreikaiserbund was not at this stage a military alliance, as Russia and Austria-Hungary were rivals in the Balkans.

The Congress of Berlin

The Dreikaiserbund lasted from 1873 until 1878. Russia defeated the Turks in a war over areas of influence in the Balkans in south-east Europe. The **Treaty of San Stefano** meant that Russian dominance in the region was extended. Austria-Hungary was not willing to accept this arrangement. Bismarck did not wish to see a war between his two allies and called the **Congress of Berlin** in 1878.

Bismarck tried to act as 'honest broker' in the dispute. He was supported by the British Prime Minister, **Disraeli**, who was also suspicious of Russian motives in the Balkans. Russia's plan to create the large state of Bulgaria was dropped. She felt that Bismarck had cheated her of the prize she had gained at San Stefano. Immediate war was avoided, but long-term problems remained.

The Dual Alliance

The Tsar was very annoyed with Bismarck's role in the Congress of Berlin. His verdict on the agreements reached at the Congress was harsh. He believed them to be part of:

> a coalition of Europe against Russia under the chairmanship of Prince Bismarck.

Relations quickly worsened and the Russians moved large numbers of troops to the German frontier. Bismarck saw that he had to choose between Austria-Hungary and Russia. He signed the **Dual Alliance** with the former in 1879. Germany and Austria-Hungary agreed to help each other if attacked by Russia. If one was attacked by a country other than Russia, the other would remain neutral. It seems that Bismarck wanted to scare Russia but not to break with her and the details of the Dual Alliance were kept secret.

Bismarck's plan was successful. For a while Russia turned her colonial attentions to the East. However, she was afraid of isolation and agreed to renew the Dreikaiserbund in 1881.

The Triple Alliance

Italy asked to join the Dual Alliance in 1882. She was in dispute with France over the colonisation of Tunis. Bismarck did not think much of Italy's radical government or her military strength. But he allowed her to join the alliance, because this deprived France of a possible ally. The Dual Alliance now became the **Triple Alliance**.

The Reinsurance Treaty

Tensions between Russia and Austria-Hungary over who would control the Balkans remained. In the Bulgarian Crisis of 1885 (see chapter 7), Russia supported Bulgaria and Austria-Hungary supported Serbia. The Dreikaiserbund broke up and was not renewed in 1887.

Bismarck was now worried that Russia might enter into an alliance with France. This would undo all his work in foreign affairs. To prevent this, he signed the **Reinsurance Treaty** with Russia in 1887. Each would stay neutral in a war in which the other was involved, unless Russia attacked Austria-Hungary, or France attacked Germany.

Colonialism

Bismarck said publicly on several occasions that he thought colonies were expensive luxuries. However, German industrialists and merchants were anxious that Germany should build an overseas empire. As a result, the **German Colonial Union** was founded in 1882.

This public pressure at home was the main reason for Bismarck's change of policy in 1883. In the next couple of years Germany acquired South-West Africa, the Cameroons, Togoland, German East Africa and some islands in the Pacific.

But Bismarck was never really committed to colonialism. He refused to use colonies as naval or military bases and made merchants pay the costs of governing them. After his resignation in 1890, Germany supported a more active colonial policy.

Bismarck's fall from power

A new Kaiser

Kaiser William II became Emperor of Germany in 1888. Bismarck hoped that he would be able to manage German affairs without interference, just as he had done since 1871. But William II had other ideas. A few weeks after becoming Emperor he declared:

> I shall let the old man [Bismarck] snuffle on for six months, then I shall rule myself.

Bismarck did not try to gain the friendship of the young Emperor and treated his suggestions with contempt. The Chancellor was not concerned with popular opinion, while William wanted to be loved and honoured by all classes of German society. The Kaiser was ambitious and was anxious to direct the affairs of his country. Serious differences occurred between the two men.

During the Great Strike at the Ruhr coalfields in 1889, Bismarck sided with the employers. William ordered the coal owners to raise miners' wages. The Kaiser also planned social reforms such as banning child labour, reducing working hours and ending Sunday working. These plans led to further tensions with the more conservative Chancellor.

Resignation

The final break came in 1890. During a violent row, Bismarck allowed William to see a letter in which Tsar Alexander III of Russia referred to him as 'an ill-bred youngster of bad faith'. The Kaiser felt insulted and was accused by Bismarck of being anti-Russian. William demanded and received his Chancellor's resignation.

Bismarck in retirement

Retirement

So at the age of seventy-five, Bismarck retired to his estates in Prussia. He wrote his memoirs and many newspaper articles criticising the government's new policies. He was deeply unhappy and he hoped that each new crisis would mean a recall to his old post as Chancellor. The summons never came and Bismarck died in 1898 at the age of eighty-three.

Bismarck: the Great Debate

With the exception of Napoleon Bonaparte, no nineteenth-century figure has attracted the controversy that surrounds the life and achievements of Otto von Bismarck.

Early German historians saw him as a great politician who sacrificed everything in the interests of the state. However, the picture of Bismarck as a Machiavellian who was prepared to use questionable methods to achieve his aims has also emerged, especially among twentieth-century non-German historians:

> the Prussian victories of the 1860s and the transformation which they wrought in the minds of most Germans laid the foundations to the defeats of 1918 and 1945.
>
> (H. Kohn, *Rethinking Recent German History*)

We have seen that Bismarck achieved German unity through a policy of 'blood and iron'. Through his moderation and skill after 1871, he contributed to a period of European peace and stability. Some commentators have talked about his foreign policy being in ruins by 1890. However, it is worth remembering that in 1890 the Russians were eager to renew the Reinsurance Treaty. Bismarck had controlled the ambitions of Germany's chief ally, Austria-Hungary, and French ambitions were cleverly directed towards colonialism in Africa. He never raised the issue of sea power and, therefore, did not antagonise Britain. It was short-sighted successors who threw away his legacy.

In domestic affairs, Bismarck's authoritarian personality led him into unnecessary conflicts with the Catholic Church and with the socialists. His low tolerance level was responsible for him completely over-reacting to the problems he encountered. However, Bismarck was extraordinarily clever and cunning. He knew when he was beaten and, like a good general, could retreat without suffering a humiliation. The main criticism of his policy must be his failure to allow the development of proper parliamentary democracy in Germany.

It is hard to get away from A.J.P. Taylor's portrayal of Bismarck. He painted the chancellor as a brilliant but unstable neurotic, driven less by great principles than by overwhelming personal ambition. Such a policy was a recipe for disaster when the German ship of state was guided by less capable hands.

Chapter 3

France, 1870-1914

End of the Empire

The Republic proclaimed

Since the French Revolution of 1789, the people of France were divided into two groups: those who supported a Republic and those who favoured a Monarchy or Empire. Louis Napoleon, a nephew of Napoleon I, ruled France from 1848 to 1870. In 1852 he adopted the title Napoleon III and the **Second Empire** was born.

This Empire collapsed during the crisis of 1870 (see chapter 1). France suffered a humiliating defeat at the Battle of Sedan on 1-2 September. Napoleon III abdicated and the National Assembly met in Paris. On 4 September 1870, the Assembly proclaimed a **Republic**. It also formed a **Government of National Defence** to continue the war against Germany.

New National Assembly

But many people in France still wanted a monarchy. In the elections of February 1871 a new National Assembly was elected. **Adolphe Thiers**, a liberal monarchist, became

Adolphe Thiers, Head of the National Assembly 1871

head of the National Assembly. Two-thirds of its members were conservatives and did not favour the creation of a republic. Thiers and the National Assembly accepted the harsh German peace terms at the Treaty of Frankfurt. This angered many Republicans.

Period of Instability

The period from 1870 to 1914 was one of major instability for France. This means that the Republic was often in danger of collapsing due to the many problems it encountered. The main problems can be examined under seven main headings:

- i The Paris Commune
- ii Monarchists v Republicans
- iii The Boulanger Crisis
- iv The Panama Scandal
- v The Dreyfus Affair
- vi Church-State conflicts
- vii Class divisions

The Commune

Reasons for Commune Rebellion

In 1871 the people of Paris were extremely discontented. The poor had suffered most during the First Siege of Paris the previous winter and were restless. Parisians had for generations resented any attempt by the government to move to the provinces. Adolphe Thiers, the leader of the National Assembly, had allowed the Germans to make a triumphal march through the French capital. Paris expected that the Assembly would return to the city from **Bordeaux**. However, it moved to **Versailles**, the seat of the Bourbons.

The government insisted that all commercial bills and rent arrears should be settled immediately. The people of Paris felt that they were under attack. The government ordered the army to take back over 400 cannons which the **National Guard** had captured in order to prevent them being seized by the Germans as war booty. At dawn on 18 March thousands of government troops were sent into Montmartre to take the guns, but they did not have enough horses to drag them away. Confusion reigned and angry crowds surrounded the soldiers. The two generals in charge were killed and the troops withdrew while the Parisians jeered.

New Town Council

A new town council of ninety members was elected to rule Paris. The council was known as the **Commune** and had its centre at the **Hôtel de Ville**. A minority of the **Communards**, especially the small shopkeepers, could be called moderates. They looked forward to minor reforms. Others were socialists and saw in the Commune a chance to introduce revolutionary ideas.

The Commune issued a manifesto which set forth its aims and stressed its independence. They called on other cities in France to follow suit. However, support from the provinces was weak and minor outbreaks of violence in Marseilles and Lyons were suppressed.

Second Siege of Paris

On 2 April 1871 the second siege of Paris began. Bismarck had offered German troops to Thiers to suppress the Communards. The offer was refused and the government troops besieged the city. Paris was bombarded, but it was not until 21 May that the troops broke through the gates of the city.

A week of bloodshed – *'la semaine sanglante'* – followed. The real horrors of civil war were revealed. Both sides shot hostages. The Archbishop of Paris died at the hands of the Communards. Some of the finest buildings in the city were destroyed. For a week the Communards fought against overwhelming odds. By noon on 28 May, the guns of Paris were silenced and the tricolour of France replaced the red flag of the revolutionaries.

Execution of Communards, who were regarded as traitors

Aftermath

Several French historians today agree that between 20,000 and 25,000 people died in the fighting. It is estimated that only about 1,000 government troops were killed. After the actual fighting ceased, the government rounded up the Communards. Thousands were shot without mercy. Many were transported overseas or imprisoned in France.

The Paris Commune had achieved nothing constructive. The harsh measures that the government took against the Communards caused a rift which slowed down the reform of workers' conditions for decades. They made the people of Paris and the people of rural areas of France very suspicious of each other.

Great myths grew up about the Commune. **Karl Marx** later called it 'essentially a working-class government'. Historians now know that the majority of the members of the Commune were intellectuals, doctors, teachers, lawyers and journalists. Radicals long remembered with bitterness the savage crushing of the Communards. Conservatives did not forget the terrifying disregard for authority shown by those they called extremists and terrorists.

The Establishment of the Third Republic

Monarchist splits

The elections for the National Assembly were held in February 1871. Two thirds of the members elected were Monarchists. The Assembly concluded peace with the Germans and dealt with the threat posed by the Commune. The next task would be to choose the form of government and to establish a constitution.

The Monarchists were divided. The **Bonapartists** hoped for the restoration of Napoleon III or of his son. **Henri, Comte de Chambord**, of the true Bourbon line, opposed the **Comte de Paris** of the Orléanist line. Each claimed that he was the lawful heir to the throne. Because of undue delay and disagreement, neither Monarchist acceded to the throne.

Vote for the Republic

The Monarchists continued to disagree and to delay. The Assembly could not prolong forever the task of drafting a new constitution. In 1875 serious debates began. On 30 January a vote was taken on a motion which included the words 'The President of the Republic is elected... for seven years'. It was carried by 353 votes to 352. France had become a Republic, almost by default.

The Monarchists failed to realise that they had not voted for a temporary seven-year president, but for a permanent office to be filled every seven years. Many insisted that they had not properly understood what they were voting for, but they were comforted by their belief that it would not last long.

Leon Gambetta, Republican statesman

The Constitution

The Constitution decreed that the **President** of the Republic should be elected by a majority of members of the **Senate** and **Chamber of Deputies**, united in a National Assembly. The Senate had 300 members. Each commune or local council, irrespective of size, elected one member. This meant that the conservative country areas dominated the Senate. The Chamber of Deputies contained 600 members, elected for a four-year term.

Electing the President in this way protected France from adventurers who might rise to power as Louis Napoleon had done by attracting the support of the masses. The President was elected for a period of seven years and was eligible for re-election.

The Third Republic was born in most unusual circumstances. It was always unstable and faced many crises and scandals. Despite the instability, it was to last until 1940.

Right and Left

Party divisions

The French invented the political terms 'Right' and 'Left'. At the start of the French Revolution in 1789, members of the National Assembly took their places in a big arc of seating. Those who supported royal power, mainly the privileged classes, sat to the right of the Speaker's Chair. Those who wanted the will of the people to prevail in any dispute sat to the left. Men who sat in the centre were uncommitted and regarded themselves as moderates.

For those on the Right, the root cause of France's problems after 1870 was the weak liberal attitude of the Left. What was needed, they believed, was strong authoritarian government. They called for an increase in the privileges and influence in public life of the aristocracy, the army and the Church. In economic affairs they were fully committed to capitalism with the emphasis on private enterprise. Socialism and communism were seen as the enemy within.

Those who were on the Left wished to continue the work of the Revolution of 1789. The **Radicals** wanted gradual change and were just to the left of centre. Their leader was **Georges Clemenceau.** Although he did not enter any cabinet before 1914, Clemenceau, who was anti-clerical, had a considerable influence. Most governments between 1885 and 1914 had some Radical ministers.

The **Socialists** supported workers' rights and wanted social welfare for ordinary people. The **Marxists** wanted revolutionary changes and demanded nationalisation of the means of production. The **Syndicalists** advocated change through trade union activity, especially using the strike weapon.

Social unrest

The Industrial Revolution of the nineteenth century led to a polarisation of French society into Right and Left. In 1905 **Jean Jaurès** became leader of the Socialists in the Chamber of Deputies. Though not ruling out strikes and peaceful revolution, he mainly worked for reforms from within the system. Between 1870 and 1914, socialist pressures led to the freedom of the press, legalisation of trade unions and the abolition of child labour.

In 1906 the Syndicalists called for the use of the general strike weapon to destroy capitalism. Great bitterness was caused by the actions of two Premiers, Clemenceau, a Radical, and **Briand,** a former Socialist. They used troops to break up the strikes. Clashes and unrest continued until 1914. When World War I broke out, a minority of French socialists did not want the French workers to take part in a war against German workers. However, the vast majority responded to the rallying cry for national unity. Differences between the government and Socialists were shelved until after the war.

Meanwhile, parties of the Right and of the Left played their parts in the various crises and scandals that threatened the very existence of the Third Republic, on a number of occasions between 1870 and 1914.

The Boulanger Crisis

Minister for war

After the elections of 1886, **General Georges Boulanger**, a Radical, became Minister for War. Boulanger was a dashing figure, rode a black circus horse and seemed to have an unerring flair for publicity. He won great popularity by his reforms in the army.

A Boulangist 'party' demanded a more powerful presidency and Boulanger spoke of the inefficiency of the Assembly. He was hailed as *'General Revanche'*, the man who would win back Alsace-Lorraine. The government became alarmed at his popularity and thought he might cause a war with Germany. He was relieved of his post as Minister for War.

The Assembly in danger

Boulanger now offered himself as a candidate for election to the Chamber of Deputies. He soon attracted many Socialists, Monarchists and Bonapartists. In 1888 he won six out of seven by-elections and in January 1889 he gained an overwhelming victory in a Paris election. The crowds urged him to march to the President's palace and seize power. However, Boulanger hesitated and the government planned to try him for treason against the state. Fearing arrest, he fled to Belgium where, in 1891, he committed suicide on his lover's grave.

Although the episode ended in farce, it had been a dangerous moment for the Republic. Not only had the old conservative Monarchists supported Boulanger, but so had the ordinary Parisians. The long-term effect of the crisis was to rally all parties

General Boulanger

to the defence of the Constitution of 1875. In the short term, other signs of instability emerged.

The Panama Scandal

Bankrupt company

No sooner had the Boulanger episode died down than a fresh problem appeared. **Ferdinand de Lesseps**, who had successfully planned the Suez Canal, tried to repeat his triumph with a canal linking the Atlantic and Pacific Oceans. He was joined by **Gustave Eiffel**, the builder of the Eiffel Tower. The reputation of these men persuaded many French investors to buy shares in the **Panama Canal Company**.

The construction of the canal proved difficult. Thousands of lives and millions of francs were lost in a vain attempt to cut through the rocky hills and mosquito-ridden swamps of Central America. In 1888 the company was given government permission to issue lottery bonds. However, the company's affairs were mishandled and in 1889 it was declared bankrupt.

Corruption

In 1892 an investigation revealed that the company had been involved with dishonest financiers in a system of corruption. As much as one third of its capital had been spent on bribery. Two German Jews had carried out much of the bribery. Proceedings were taken against ten politicians, but only one, the former Minister for Public Works, was convicted. Ferdinand de Lesseps and his son were given five years imprisonment.

Long-term results

A mood of cynicism against all politicians spread through France. The enemies of the Republic were loudest in their condemnations. Anti-Jewish feeling increased. The scandal gave an opportunity to anti-Semites to pour out their propaganda against the Jews in popular newspapers. **Clemenceau**, a future Prime Minister of France, was forced to withdraw temporarily from public life because his newspaper had been financed by one of the Jews involved in the bribery. The anti-Jewish hysteria continued and reached its high point during the **Dreyfus Affair**.

The Dreyfus Affair

The Case against Dreyfus

Captain Alfred Dreyfus was born in Alsace which had been under German rule since 1871. He was the son of a rich Jewish textile manufacturer. As a young man he joined the army. His fellow soldiers found him cold and rather aloof. His personality and his Jewish-Alsatian background made him unpopular. He became an easy target for suspicions.

Things came to a head in 1894, when Dreyfus was accused of passing secret information to the Germans. The evidence against Dreyfus was based on a *bordereau*, an official document which in this case contained some handwritten personal comments. Suspicion fell on Dreyfus because of the handwriting, despite disagreement among handwriting experts about the identity of the writer.

When Dreyfus appeared at his court-martial, he was virtually a condemned man. He was sentenced to life imprisonment on the dreaded Devil's Island, off the South American coast. In a public ceremony, his officer decorations were torn off and his

Artist's impression – Dreyfus at his trial 1899

sword was broken. When Dreyfus cried 'I am innocent, vive la France', the watching crowd shouted back 'Death to the Jew!'.

Picquart suspects Esterhazy

After his imprisonment, secret information continued to reach the Germans. A new figure, **Colonel Picquart**, became head of the French Intelligence Service. He carried out his own investigation and concluded that the guilty party was **Major Esterhazy**, a personal friend of **Colonel Henry**, the intelligence officer involved in the arrest of Dreyfus. Picquart's senior officers became alarmed at the implications of the investigation. They were unwilling to re-open the case and Picquart was transferred to Tunisia. Esterhazy was cleared of all suspicions.

Emile Zola

In January 1898 the newspaper *L'Aurore* published a letter which **Emile Zola** had written to the President of the French Republic. Zola was a famous novelist and his letter boldly headed *'J'Accuse'* was a violent attack on the army and the conservative institutions. He accused the army of a cover-up. Zola was tried and sentenced to one year's imprisonment. All this time Alfred Dreyfus remained in prison, unaware of what was going on.

Dreyfusards versus anti-Dreyfusards

France was now split between Dreyfusards and anti-Dreyfusards. The Dreyfusards included Republicans, intellectuals and socialists. They were opposed by the army, the Catholic Church, monarchists and right-wing nationalists. They looked on Dreyfus as a Jew and a traitor, while the army represented stability and was valiantly spreading the glory of France.

Dreyfus found innocent

A new War Minister, **General Cavaignac**, was determined to end the continuing Dreyfus agitation and produced some of the relevant documents to the French Assembly. Meanwhile, Colonel Henry, caught in a trap, confessed to forging the key document. He was arrested, but slit his own throat while awaiting trial. Major Esterhazy fled to London. Cavaignac resigned and the whole basis of the Dreyfus case seemed to be collapsing. The government agreed to a re-trial.

Dreyfus was brought back from Devil's Island in 1899, an 'old, old man of 39', as an observer noted. Over four years of solitary confinement had taken its toll.

The verdict of the judges, by five to two, was again guilty, but this time with special circumstances. His sentence was reduced to ten years. The President offered a pardon. It was not proof of innocence, but Dreyfus, in bad health, accepted. His family and friends continued to work to prove his innocence. Finally, in 1906 an Appeal Court, acting on new evidence which had been withheld from earlier trials, declared that Alfred Dreyfus was innocent. He was re-instated in the army and awarded the Legion of Honour. Both he and Picquart were promoted in rank. Dreyfus served his country well in the First World War and died quietly in retirement in 1935. Esterhazy, the guilty one, lived under a false name in England for the rest of his life.

The Dreyfus Affair had almost caused a civil war in France. The result was a defeat for the military and the anti-Jewish Right. It was a victory for the Left and the anti-clerical radicals. They were to dominate the Third Republic for the remainder of its existence.

Church and State

Background to the conflict

Of five million children in France, two million were educated in Catholic schools. Republicans looked on the Church as a right-wing institution. They were deeply concerned about control of education. The case for educational reform was reinforced by the belief that the Prussian education system had contributed as much to the victories of 1870 as had the well-organised army. In addition, the Church was seen as upholding the interests of the upper and middle classes. Many republicans said it was hostile to scientific advances and out of sympathy with the problems of the lower classes.

The Ferry Laws

The most prominent critics of the Catholic Church were the Radicals, who grouped around **Georges Clemenceau**. He demanded a severe reduction in the authority and educational role of the Church. In 1879 **Jules Ferry** had been appointed Education Minister. He believed that education should be compulsory, free and subject to the authority of the state. The Ferry Laws resulted in the expulsion of the Jesuits from the country in 1880. Other religious orders were only allowed to teach with permission of the government. Primary education was made free and the state controlled the granting of all degrees and diplomas.

Rapprochement

Relations between republicans and the Church were poor until the Vatican attempted to improve them in the late 1880s. Pope Leo XIII urged Catholics to accept the Republican Government. He asked them to participate fully in political life. However, members of some religious orders made their hostility to the Republic and to the Radicals clear during the Dreyfus Affair. The *rapprochement,* also known as *ralliement,* was not very successful.

Waldeck-Rousseau and Combes

The Waldeck-Rousseau government (1899–1902) further reduced the influence of the Church. The Assumptionist Order was dissolved and all religious orders had to be authorised by parliament. In 1902, the election was fought on the question of Church authority. The anti-clerical, **Emile Combes**, was Premier from 1902 until 1905. Over 1,000 religious houses with a membership of 20,000 were closed. Members of religious orders were forbidden to teach in schools and education thus came completely under state control.

Law of Separation

The **Law of Separation** ended the centuries-old connection between the Catholic Church and the French state. State payment of the clergy came to an end. Many clerics suffered extreme hardship. The Law of Separation, framed under the Combes government, was passed by **Aristide Briand**, the new Premier, in December 1905. All Church property was taken over by the state and leased back to the Church.

The Vatican condemned the Law of Separation in 1906 and Church-state relations remained poor. However, by 1914 the Catholic Church had come to see the benefits of separation. Freedom of religion was preserved and, in time, clericalism and anti-clericalism, which had divided the country since the Revolution, ceased to arouse the same degree of passion and bitterness.

The Economy

Industry

In 1870 France ranked second to Britain in European industrial production. It is true that France made great progress up to 1914. However, her progress in industry did not match that of the new united Germany, or even that of Russia.

The loss of Alsace meant that France lost a large number of her best cotton factories. The takeover of Lorraine deprived her of valuable coal and iron-ore deposits. Payment of the indemnity to Germany also left France short of much-needed capital immediately after 1870. The political instability, the scandals and the deep class divisions discouraged investment. Much of the capital accumulated by French banks and capitalists was lost abroad, especially to Russia.

However, France continued to reign supreme in the production of luxury goods like silk, fashionable clothes, cosmetics and artistic products. By 1914, stimulated by the growing demand for armaments, steel production had shown a great increase.

Agriculture

France retained her traditional image of a country of small peasant farmers with few landless labourers. The peasants did not have enough resources to purchase machinery and fertilisers. They were very slow in adopting modern methods of farming. Despite this, France was almost completely self-sufficient in agricultural produce. She was second only to the United States in the amount of wheat grown and to Italy in wine production.

Cultural Achievements

La Belle Epoque

During the era of the Third Republic, France gained an international reputation for outstanding artistic achievements. Paris became known as the 'City of Light' and attracted writers and artists from all over the world. The capital became a centre of fashion and luxury and in 1900 held a World Fair which was visited by over 50 million people. There on display were images of France's glory, her culture and her prosperity. This age of achievement was later remembered fondly as ***La Belle Époque***.

Painting and Sculpture

Paris became the home of a new school of **Impressionist** painters. These artists attempted to depict the initial momentary impression made upon them by the subject. They usually painted outdoor scenes and were particularly concerned with the effect of light and colour.

Claude **Monet**, a famous landscape painter, gave the movement its name. Other famous Impressionist painters were Edouard **Manet**, Auguste **Renoir**, Edgar **Dégas**, Paul **Cezanne** and Henri de **Toulouse-Lautrec**. The **Post Impressionists** attempted to make painting less objective and more symbolic through the use of vivid colours. Paul **Gaugin** and Georges **Seurat** were the best-known figures from this school.

Sculpture was also revitalised at this time. The movement was led by Auguste **Rodin,** who became known as the 'father of modern sculpture'.

Literature

During the era of the Third Republic, France also produced a number of famous literary figures. Guy **de Maupassant** was the master of the short story. Emile **Zola** and Anatole **France** originated the school of Naturalism. Their novels were noted for social criticism and the depiction of the ugly and seamy side of life. Gustave **Flaubert's** famous novel, ***Madame Bovary***, provided a marvellous picture of romantic love and of life in the provinces. Some readers were shocked because the author failed to pass judgement on the moral conduct of his characters.

Music

Claude **Debussy** applied Impressionism to achieve a special mood. George **Bizet** was famous for ***Carmen***, on opera of Spanish life and passion. Jacques **Offenbach**, although German-born, was a composer of French comic opera. He is well known for ***The Tales of Hoffman.*** Maurice **Ravel's** music is notable for its exotic flavour.

Science

France's tradition as a nation of scientists was upheld. Henri **Becquerel** discovered radioactivity in uranium salts. **Pierre and Marie Curie** discovered radium. Louis **Pasteur** introduced the germ theory of disease. He was the founder of modern bacteriology and immunisation.

Chapter 4

Tsarist Russia, 1855–1914

Alexander II, 1855-81

Personality and position

Since the seventeenth century, Russia had been ruled by the **Romanovs** as emperors or tsars. **Alexander II** succeeded his father, **Nicholas I**, in 1855. He was an autocrat; this meant that he was in complete control of government and did not answer to any parliament or council. Alexander has been known in history as the 'Tsar Liberator'. He gained this name by his decision to free the serfs and bring in other reforms.

Alexander II had one major weakness. He was not prepared to give up his autocratic position. He was deeply shocked by an attempt on his life in 1866 and this blunted his reforming zeal. Many modern historians now see his reign as an opportunity lost. He could have introduced a constitutional monarchy in Russia. However, he was not willing to share power with an elected parliament.

Emancipation of the serfs

Until 1861 the Russian peasants were serfs. Serfs were owned by landowners for whom they had to work three days a week as well as paying various fees and rents. Serfs' lives were controlled by their landowner. He could stop them leaving their village and sell or barter them. In 1861 the serfs were emancipated. This meant that they were free to marry, to own property and to set up in business. They were no longer owned by their masters and no longer owed their masters fees or services.

Other reforms

The emancipation of the serfs and the attempt to westernise Russia led to other reforms. Alexander established ***Zemstvos*** which were rural district and provincial assemblies. Their job was to look after primary education, health, local industry and road maintenance. Great work in the administration of local affairs was done by those who served on the *Zemstvos*. Many of them saw this system of elected local government as a step on the road to full parliamentary democracy.

Alexander also saw the need for reform in other areas. Law cases were now conducted in public, the law applied equally to all classes and a jury system was brought in. Many reforms were introduced in the army. Universities were given greater freedom and the number of students allowed to rise. Censorship was relaxed, both in education and in the press.

Discontent

Emancipation failed to conciliate the serfs. The land remained the property of the nobles and landowning class. Later a land purchase scheme was introduced by the state. The peasant could repay the cost of the land over a period of forty-nine years. Until payment was completed, the land was not owned by the peasant, but rather by the village commune, known as the ***Mir***. This commune controlled the land and often redistributed strips among peasants. This meant that a hard-working or progressive peasant was not likely to experiment with modern agricultural methods or improve the land in case it was taken away from him. Another problem with the land purchase scheme was that peasants who fell into debt lost their freedom of movement. Peasants resented making payments for what they had always considered their own property.

Many outbreaks of violence were recorded and rural unrest seemed to grow rather than die down in the decades after emancipation. Household serfs received no land. Some stayed on as free servants with their former masters. Others, together with landless labourers, drifted into the cities in search of work. Growth of this new proletarian class of industrial worker was to lead to the emergence of revolutionary groups.

An end to reforms

Alexander drew back from further reforms after the assassination attempt on his life in 1866. He explained his views to a member of the St Petersburg *Zemstvo*:

> I suppose you consider that I refuse to give up any of my powers for motives of petty ambition. I give you my imperial word that, this very minute, at this very table, I would sign any constitution you like if I felt it was good for Russia. But I know that, were I to do so today, tomorrow Russia would fall to pieces.

The Nihilists and the Anarchists

The Tsar's failure to grant further reforms led to the growth of revolutionary groups. The **Nihilists** (from the Latin word *Nihil*, meaning nothing) accepted nothing in society as it was then structured. The Nihilists tried to organise peasant revolts and to assassinate the Tsar. The **Anarchists**, led by **Mikhail Bakunin**, advocated terrorism and violence and worked towards the destruction of all state institutions.

The Narodniki

During the 1870s, thousands of young radicals went into the countryside to open the eyes of the people to their miserable plight. These radicals were called ***Narodniki***, meaning 'to the people'. Dressed as peasants, they lived in the villages and tried to teach their revolutionary ideas to the masses. The peasants, however, were not very interested and some handed the unwelcome visitors over to the police. Many were imprisoned or exiled to Siberia.

'The Peoples' Will'

An extreme breakaway group of young terrorists, **The Peoples' Will**, made life very difficult for those who held high office. In 1879 this group passed sentence of death on 'Alexander Romanov'. Several more attempts were made on the Tsar's life. This led to further repression.

Death of Alexander II

Alexander now realised that he must make some concessions to settle the situation. In 1881 plans were prepared for the calling of a national assembly. However, on 13 March 1881 Alexander was killed by the second of two bombs thrown at his sledge in a St Petersburg street. Plans for a new assembly died with Alexander II.

Alexander III, 1881-1894

The reactionary conservative

Alexander III was a giant of a man who could bend a horse-shoe in his hands. He succeeded his father as Tsar in 1881, and was determined to crush all opposition. The years of his reign are known as 'the reaction', the time when the government and its supporters did all they could to undo the advances made under Alexander II.

The greatest influence on the views of the new Tsar came from his former tutor and now his trusted advisor, **K.P. Pobedonostsev**. He, like the Tsar, was a firm believer in autocracy and said that 'parliamentarianism is the great lie of our time'.

The new Tsar was firmly opposed to liberal reforms. The murder of his father convinced him that previous policies were what he called 'a failure', and he firmly stated his belief in the 'power and right of autocratic government'. To further his policy, Alexander and Pobedonostsev planned a programme of repression, racial discrimination and police terror.

Russification

The policy of **Russification** was an attempt to suppress the local characteristics of the various peoples within the Empire and to spread Russian culture among all the Tsar's subjects. The Poles, Ukrainians, Tartars and Georgians were considered 'disloyal'. These peoples were subjected to Russian education and forced to learn Russian. The policy of 'Russification' was a failure and the various racial groups remained independent in outlook and language.

However, the severest penalties issued by the government were imposed on the Jews. Harsh treatment led many of them to emigrate to Western Europe and to America. Propaganda was used against them and '**pogroms**' were encouraged: mobs were urged to attack Jews and destroy their homes and businesses.

Forced labour: political prisoners at work in Siberia

Education, censorship and state police

Repression was felt above all in the field of education. The Minister for Education, **I.V. Delyanov**, was opposed to what he considered the 'dangerous' advances in education under the previous Tsar. Student organisations were suppressed and the independence of the universities was severely limited. The raising of fees in schools, which placed education out of the reach of many, was done for a specific reason by Delyanov:

> Children of coachmen, servants, cooks, washerwomen, small shopkeepers and persons of similar type...should not be brought out of the social environment to which they belong.

By as late as 1897, Russia's illiteracy rate was a staggering 79 per cent. Less than half the children of school-going age attended primary school.

There was also strict control of the press during Alexander III's reign. Newspapers and books were censored and freedom of speech and of assembly was severely limited. The ***Okhrana***, the Tsar's state police, were given additional powers. Terrorists and revolutionary groups were hunted down and many were executed. To ensure that the peasants and their village communes were kept under check, the hated **land captains** were appointed to each locality.

Industrialisation

Despite his repressive policies, Alexander III did not neglect economic growth. Russia was far behind Western Europe in industrial development. A movement to change this had been made during the reign of Alexander II. Trade was liberalised and tariffs lowered. The cotton industry grew and the beginnings of the Donets coal industry were seen. Government subsidies helped an emerging railway system.

Trans-Siberian Railway

Under Alexander III industrialisation continued. Factory legislation controlled the working hours of women and children. In 1891 the decision was taken to construct the **Trans-Siberian Railway**, the most ambitious scheme of its type in the world. Industrial and agricultural production grew, though slowly by western standards. The land payments from the peasants to the *Mirs* were lowered and the **Peasants' Land Bank** provided cheap loans to buy land. Agriculture was exploited as a source of export earnings. This helped the economy, but it led to a series of famines throughout Russia.

Nicholas II, 1894-1917

A weak Tsar

In 1894 Alexander III died unexpectedly and was succeeded by his twenty-nine-year-old son, Nicholas. The determined and dominant father had produced a shy and weak-willed son. He was a pleasant, charming man and very devoted to his family. However, he completely failed to understand the new forces at work in Russia and historians agree that he was not equal to the tasks that confronted him.

Like his father, Nicholas had been tutored by Pobedonostsev. Those who hoped for reforms were soon to be disappointed as Nicholas declared on ascending the throne that:

> I will uphold the principle of autocracy as firmly and unflinchingly as my late, unforgettable father.

Tsar Nicholas II and the Tsarina, in their coronation robes

His wife, the German-born **Princess Alexandra**, was also devoted to the principle of autocracy. A strong-willed woman, she firmly resisted the good advice of those who might have saved the monarchy. When the representatives of the *Zemstvos* met the new Tsar and expressed their hopes for an elected assembly, he told them that they were 'carried away by senseless dreams'.

Sergei Witte

Sergei Witte had become Minister for Finance in 1892 and held the office until 1903. He was a self-confident, dynamic man and hoped that his policies would save Russia from chaos and create a strong modern state. He was convinced that future greatness lay in industrialisation. Protective tariffs were strengthened in order to guard infant Russian industries against competition.

Investment funds within Russia were lacking and Witte concentrated on attracting foreign capital. He sold railway bonds abroad and the Trans-Siberian Railway, over seven thousand kilometres in length, was completed in 1904. The

THE RUSSO-JAPANESE WAR 1904-05

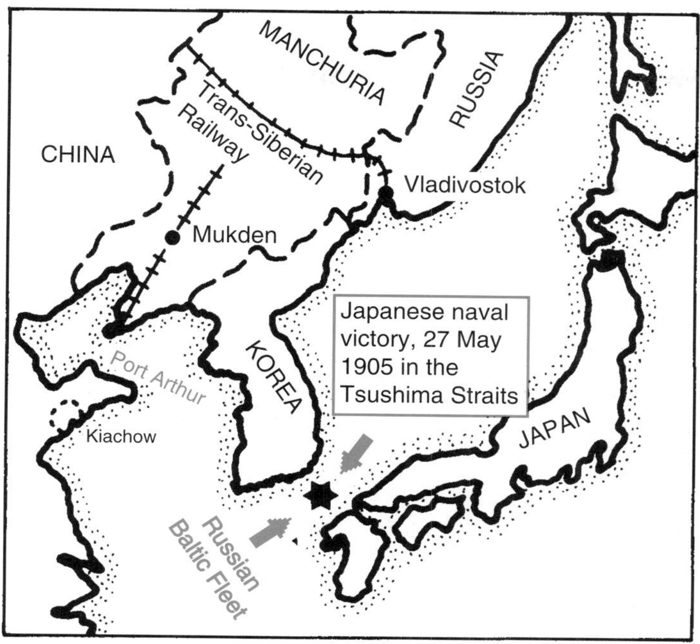

Russian currency was placed on the Gold Standard and a major export drive was encouraged. Figures for the production of coal and petroleum showed an impressive growth pattern and the textile industry flourished.

1904-05: The Russo-Japanese War

In chapter 2 we saw that Russia turned to expansion in the East after her diplomatic defeat at the Congress of Berlin in 1878. By 1904 the Trans-Siberian Railway linked Moscow with Vladivostok. Russia then gained a lease on the ice-free base of **Port Arthur** in the Chinese province of Manchuria. Japan watched Russian progress and was very indignant when Russia acquired territory in Manchuria.

Nicholas refused a Japanese request that Russian soldiers be withdrawn from Korea. He felt that victory in a foreign war would halt opposition at home and earn prestige abroad. On 8 February 1904 the Japanese launched a surprise naval attack on Port Arthur. The Russian Far Eastern Fleet was surprisingly and decisively defeated and the Japanese moved into Manchuria.

After a long siege, Port Arthur surrendered in January 1905. In February, the Russians suffered huge casualties at the land battle of **Mukden**. By May, the Russian Baltic Fleet had reached the Gulf of Korea. The Japanese fleet attacked the Russians in the **Straits of Tsushima.** The Baltic Fleet was decisively beaten, their vessels either destroyed or captured.

President **Theodore Roosevelt** acted as mediator between the two sides. In September they signed the **Treaty of Portsmouth.** The terms of this treaty included Russian recognition of Japanese influence in Korea and the handing over of the Liaotung peninsula to Japan.

The Russo-Japanese War was significant because it was the first victory of an Asian power over a European state in modern times. Defeated in the Far East, Russia now turned her gaze back to the Balkans. An immediate result was the Russian Revolution of 1905.

Background to the Revolution of 1905

Some two and a half million workers were employed in factories or in workshop production by 1900. Conditions of work were dreadful and fatal accidents in the factories were commonplace. The country witnessed regular student demonstrations, strikes by the industrial workers and peasant revolts. However, Nicholas II refused to offer a political solution to the problems. As a result, opposition groups became more active and offered their own solutions. Three main groups emerged – the **Liberals**, the **Social Revolutionaries** and the **Social Democrats**.

The attitudes of the Liberals

The Liberals advocated the end of autocracy and sought a western-type constitutional government. They were drawn mainly from the *Zemstvo* politicians, some enlightened gentry and sympathetic middle-class professional people. The Liberals hoped that they could achieve political freedom by peaceful means. Their programme included a constitutional monarchy with a parliament elected by universal male suffrage (right to vote). They wanted reforms to help the working-classes and an end to the policy of Russification. In 1904, the Liberals formed the **Constitutional Democratic Party (Kadets)**.

The Social Revolutionaries

The Social Revolutionaries grew from the Populist movement which had been driven underground by the persecutions of Alexander III. Many were forced into exile and they issued their manifestos from abroad. These called for a socialist society, political freedom and the complete separation of Church and state. The Social Revolutionaries appealed mainly to the peasants, to whom they promised land. They did not believe in too much theory, but favoured direct action and terrorist methods to achieve their aims.

The Social Democrats

The Social Democrats adopted Marxist policies. Among its leaders were Vladimir Ilyich Ulyanov, who took the name **Lenin**, and Leon Bronstein, later known as **Trotsky**. Other major figures in the party were **Georgii Plekhanov** and **Julii Martov**. Because of his revolutionary activities, Lenin had been exiled to Siberia and later went to England. In his newspaper ***Iskra*** *(The Spark)* and in pamphlets he advocated that the Social Democratic Party should be run by a small elite group of dedicated revolutionaries. Martov and Plekhanov disagreed and the party split. Lenin's followers called themselves **Bolsheviks** which means 'the majority'. The other group, which followed Plekhanov were called the **Mensheviks**, which means 'the minority'.

1905: Bloody Sunday

Discontent had reached a high point by 1905. The economic slump since 1900 led to great hardship. News of defeats in the Russo-Japanese War and of the sufferings of the poorly-equipped soldiers made matters worse. Acts of terrorism were common and food prices rose.

On Sunday 22 January a march by more than 150,000 demonstrators approached the Tsar's **Winter Palace,** in St. Petersburg. It was led by **Father Gapon**, a controversial figure who, some left-wing workers felt, was being used by the

Bloody Sunday, St Petersburg 1905. Troops guarding the Winter Palace open fire on demonstrators

government. The demonstrators were petitioning the Tsar in his traditional role of 'Little Father' of his people. The workers asked for improved working conditions and a national assembly. They sought basic civil liberties, a transfer of land to the people and an end to the war.

The crowds moved peaceably through the streets carrying icons and portraits of the Tsar and singing hymns and patriotic songs. Most of them thought that once the Tsar knew of their troubles, he would help them. Suddenly the marchers were charged by the cavalry. Troops who were guarding the palace opened fire. Estimates of up to 1,000 dead are now accepted. Thousands more were wounded.

Reaction to Bloody Sunday

Reaction to Bloody Sunday resulted in disorders throughout Russia. Work stopped in every large city as the workers came out on strike. In the countryside, the Social Revolutionaries led the peasants in acts of terrorism. Landlords were murdered and their property destroyed. The Tsar's uncle was assassinated. In June, the crew of the battleship ***Potemkin*** mutinied and sailed away from the port of Odessa after raising the red flag.

Despite these developments the Tsar continued to resist calls for an assembly and for other reforms from middle-class liberals such as *Zemstvo* leaders and the Kadets. Meanwhile, the Social Democrats had set up ***Soviets*** in St Petersburg, Moscow and other large cities. These *Soviets* were workers' councils that ran the strikes and set out workers' demands.

In September, news of the humiliating peace treaty with Japan led to more unrest and strikes. A general strike now brought the country to a standstill. The Tsar turned for advice to Witte who said that the only way to avoid a civil war was to grant an assembly.

The October Manifesto

Nicholas reluctantly agreed to follow Witte's advice. An Imperial Manifesto issued on 17 October announced that an elected parliament or ***Duma*** would be set up. The autocratic government of the Tsar was to be replaced by a **Council of Ministers** led by a Prime Minister.

The October Manifesto was received with great jubilation and a general return to work. This was advised by the St Petersburg Soviet which sensed that the people were tired of the strike. The Manifesto brought an end to the 1905 revolution, although it was some months before all strikes ceased and order was restored. Government troops put down all the peasant risings and defeated the people of Georgia who had attempted to break away from the Empire.

The First Duma

In the spring of 1906 elections took place for the First Duma. Even before it met, it was clear that the Tsar had broken his promise to share power with this first Russian parliament. He announced:

> To the Emperor of all Russians belongs supreme autocratic power. Submission to his power is commended by God himself.

The Bolsheviks and Mensheviks boycotted the elections as they believed the Duma would be a sham. In the end, most members of the Duma were either Social Revolutionaries or Kadets. At the first meeting they demanded a full share of government, the transfer of nobles' land to the peasants and the freeing of all political prisoners. The Tsar found all these demands 'inadmissible'. After seventy-three days, members arrived one morning to find the meeting-hall surrounded by troops and a notice on the door announcing that the Duma had been closed.

Stolypin

The Tsar and his wife disliked Witte, whom they thought too liberal, and replaced him with **Peter Stolypin** in 1906. He organised elections for the Second Duma. Lenin agreed that Social Democrats should stand but only sixty-five were elected. The assembly was again dominated by Kadets and Social Revolutionaries. Extremists from left and right used the parliament for their own propaganda purposes. Nicholas was annoyed by the ensuing uproar and had a good excuse for dissolving the Second Duma in June 1907.

Third and Fourth Dumas

Stolypin now had the electoral laws changed. This resulted in the election of the more conservative assemblies known as the Third Duma in 1907 and Fourth Duma in 1912.

These Dumas slowly introduced reforms. The hated land captains were replaced and compulsory health insurance for industrial workers was introduced. Much progress was made in primary, secondary and university education. Stolypin worked with the Dumas for these reforms. He was also responsible for a new scheme of land ownership. Peasants could buy their freedom from the Mir and own their own private farms These farms were in single units rather than in scattered strips. The farmers could sell their farms and move without permission.

Reform and repression

Stolypin also encouraged the development of industry and the production of iron, steel and coal rose sharply. But Stolypin combined reform with repression. Thousands of political offenders were exiled to the harsher areas of Siberia. So many people were executed that the gallows became known as 'Stolypin's necktie'. The population rose rapidly and it was difficult to satisfy restless industrial workers and land-hungry peasants. The Prime Minister had made many enemies and he was assassinated in 1911 in the Opera House in Kiev.

Grand-Duke Sergei Witte, Minister for Finance 1892-1903

Russia in 1914

The Duma had not become a truly representative assembly because the Tsar had never wanted it. However, by 1914 political parties were legally formed. While rebellion was severely punished, open political discussion was tolerated and allowed to appear in the press. All these factors represented advances scarcely dreamed of before 1905. The system could have developed into a true democracy. However, despite the wave of patriotism that greeted the outbreak of World War I, by the end of the war all hopes for a constitutional monarchy lay in tatters.

Chapter 5

Britain, 1870–1914

The Victorian Era

A powerful empire

Britain in 1870 was a constitutional monarchy. This means that the head of state was a monarch, but that the real power was held by parliament. The monarch in 1870 was **Queen Victoria**, who ruled from 1837 to 1901, the longest rule in Britain's history. The second half of the nineteenth century is often referred to as the Victorian era.

In 1870 Britain was still the most powerful nation in the world. As leader of the industrial revolution, Britain became known as the 'workshop of the world'. A large merchant fleet served this successful trading nation. Her huge overseas empire continued to expand. It provided a source of raw materials and a market for manufactured goods. The empire and the commercial fleet were protected by the **Royal Navy**, which was bigger than the combined strengths of any two of her competitors.

Queen Victoria, longest-serving monarch in Britain's history – 1838-1901

Political parties

By 1870 politics in Britain was dominated by two political groups, the Whigs, who formed the **Liberal Party** and the Tories, who formed the **Conservative Party**. Gilbert and Sullivan, in the opera *Iolanthe,* reflected the situation accurately:

> Every boy and every gal
> That's born into this world alive
> Is either a little Liberal
> Or else a little Conservative.

Traditionally, the Conservatives were supported by the aristocracy and the land-owning classes. They were opposed to rapid change in society. They favoured imperialism and the expansion of the empire. The Liberals were supported mainly by the middle classes. They were in favour of reforming government and society to meet the needs of the new industrial age.

Problems in Victorian Britain

Britain may have appeared successful, but she had many problems. The Industrial Revolution resulted in the movement of people into the large towns and cities, which in turn resulted in overcrowding and slums. Reforms were needed in the factories, in education and in health. However, the major problem was with the structure of society itself. Britain was divided rigidly into the aristocracy or upper class, the middle class and the working or lower class. The vast majority of the people lived in poverty and belonged to the lower class.

The rule of government was undemocratic. The story of the Victorian era is the attempt by reformers to introduce laws which would bring democracy to Britain. They had to wrest power from an unwilling upper class. Such concerns as universal suffrage, education for all and changes in the army and civil service were key issues.

The parliamentary system itself would have to be reformed before such matters could be dealt with. Parliamentary and social reforms are closely associated with Benjamin **Disraeli,** Conservative leader from 1868 to 1881, and with William **Gladstone,** who led the Liberals, almost without a break, from 1867 to 1894.

Disraeli and Gladstone

Benjamin Disraeli (1804-1881)

Benjamin Disraeli was the grandson of an immigrant Jew, but as a child was baptised a Christian. He first made a name for himself as a popular novelist. He was a fine speaker but opponents often scoffed at his fancy clothes and gaudy rings. Unlike his great opponent, Gladstone, Disraeli was not the leader of a great moral crusade. He wished to see the Conservatives in power and saw himself as leader of the party. He

Gladstone enviously watching through a window, as Disraeli serves a new measure of reform

achieved his goal despite the anti-semitic feelings in the country. Bismarck usually referred to him as 'the Jew' and to Gladstone as 'the professor'. When Disraeli took over as Prime Minister for 10 months in 1868 he said, 'I have climbed the top of the greasy pole'.

Pragmatic politician

Disraeli was, above all, a practical politician. When the Conservatives lost the 1868 General Election he saw that a change in direction was needed. His alternative to Gladstone's liberalism and policy of Home Rule for Ireland, was what he called 'imperialism and social reform'. His policies broadened the appeal of the Conservatives. He summed up his policies thus:

> the maintenance of our institutions, the preservation of our empire and the improvement of the conditions of the people.

Disraeli and the Conservatives returned to power in 1874 and remained until 1880. We shall see later how his policy of 'Tory democracy' worked. He was true to his word. In a letter to *The Times* in 1879, the leader of the Engineers Society summed up his achievements:

> The Conservative Party has done more for the working class in five years than the Liberals have done in fifty.

Disraeli was in poor health during the 1880 election and was unable to campaign effectively. The country was suffering from economic depression and Gladstone's Liberals had an overall majority. Disraeli did not long survive his party's election defeat and died in April 1881.

William Ewart Gladstone (1809-98)

William E. Gladstone, the son of a rich Liverpool merchant who had made his money from the corn and slave trades, was a brilliant student at both Eton and Oxford. He was an Anglican, but showed toleration for all religions. He felt that religion and politics were related and during his political life he tried to do what he believed to be the correct and moral thing. When, aged fifty-nine, he was appointed Prime Minister, his attitude differed from that of Disraeli:

> I ascend a steepening path, with a burden ever gathering weight. The Almighty seems to sustain and spare me for some purpose of His own, deeply unworthy as I know myself to be.

Gladstone's 'equality of opportunity'

Gladstone was Prime Minister on four separate occasions. He wanted to 'pacify Ireland' and lost power twice in vain attempts to introduce Home Rule. Unlike Disraeli, he was not interested in expanding the empire. He felt that the government ought to ensure 'equality of opportunity' for all citizens and many of his reforms were based on this principle.

Attitude to monarchy

Another contrast with Disraeli was the fact that Gladstone was not prepared to flatter the Queen. She, in return, did not hold him in high regard. Disraeli was partly

Gladstone in his old age, still devoted to reform

responsible for the monarchy once again becoming popular. However, Gladstone's more formal and reserved manner was ultimately more beneficial in the development of the modern constitutional position (rule by parliament with the monarch as non-executive head of state) in Britain.

The GOM

Gladstone's second Irish Home Rule Bill was defeated in 1893 and he retired the following year. He had won the respect of all, including his staunch political opponents and earned the title 'Grand Old Man' (GOM) of British politics. His biographer, Philip Magnus, speaks of Gladstone's view of politics as a Christian duty and says that 'he incurred martyrdom for himself as well as for the Liberal Party'.

Parliamentary Democracy

Electoral reforms

In the 19th century, Britain was not a true democracy. It was ruled by the upper classes. Parliament was elected by a minority of better-off citizens who had the right to vote. These were people of position and property.

The **First Reform Act** of 1832 had given the vote to the middle class, while the **Second Reform Act** of 1867 had extended this to the working class in towns. The **Secret Ballot Act** of 1872 made voting secret and diminished the power of landlords and employers at election time.

Democracy

As the number of people entitled to vote grew, the political parties were forced to change. The Liberals and Conservatives set up branches in every electoral area. Gladstone toured the country in 1880 and canvassed actively for support. This would have been frowned on in previous times.

Gladstone's **Third Reform Act** of 1884 gave the vote to all male householders. The number of people entitled to vote rose from three million to five million. The principle of one Member of Parliament per constituency was eventually accepted.

The extension of the franchise (vote) had unforeseen results for the Liberals and Conservatives. It led to the rise of the **Labour Party**. The democratic process was not completed until over forty years later. The **Representation of the People Act**, 1918, gave the vote to all men over 21 and to women over 30. In 1928 women got the same voting rights as men.

The Parliament Act 1911

The cause of democracy was further aided by the **Parliament Act** of 1911. There had been a number of conflicts between the House of Lords and the House of Commons.

The Lords had rejected or tampered with various acts including the Home Rule Bill of 1893 and the Education Bill of 1906.

Many democrats were becoming increasingly annoyed that an unelected and unrepresentative body should have the power of veto over laws passed by the representatives of the people. The Liberal victory in the 1906 election brought matters to a head. Various bills introducing social changes were tampered with or blocked. In 1909, the Chancellor of the Exchequer, **Lloyd George**, introduced the 'People's Budget'. He increased taxation in order to pay for schemes like labour exchanges and children's allowances, to cover debts and to build up the navy. Death duties were increased, a super tax was placed on incomes above £3,000 and a duty was put on the sale of land. The Lords felt that the budget was an attack on private property and they called it a socialist plot.

The Lords rejected the 1909 budget by a large majority. The Liberals called the action a 'breach of the constitution'. They dissolved the parliament and called a general election. The Liberals won the election and set out to destroy the power of the House of Lords. Under the threat that the King would flood the House of Lords with Liberal peers, the Lords passed the Parliament Act of 1911. In future the Lords would have no veto on money bills and a two-year delaying-power over other legislation. The maximum life of a parliament was reduced from seven to five years.

The Liberals also passed a bill through parliament which provided for the payment of salaries for M.P.s. A sum of £400 a year was fixed. This was to have a major impact on the composition of Parliament from that time.

The Suffragettes

Early Years

The name '**Suffragettes**' was first used by the British newspaper the *Daily Mail*. The writer was referring to women who sought the right to vote in elections for parliament. During the nineteenth century some improvement in the position of women in British society had taken place. By 1900 women could retain their own property after marriage. A husband could no longer legally detain his wife in their home, and women were also allowed to claim custody of their children after a marriage break-up. Some women could also vote for the new rural, urban and district councils and seek election to these, but they were unable to vote in parliamentary elections.

Emmeline Pankhurst

In 1898 **Emmeline Pankhurst** and her husband, **Richard**, formed the **Women's Franchise League** in Manchester. In 1903, Emmeline Pankhurst, now a widow, set up the **Women's Social and Political Union**. She was helped by her daughters, **Christabel** and **Sylvia**.

Mrs Pankhurst on her release from prison, 1912

After a while, her followers became more popularly known as the Suffragettes. They received support from Keir Hardie, the Labour M.P., (see page 59) but were viewed with suspicion by many Liberals and Conservatives.

The movement develops

At first, the Suffragettes tried to influence public opinion through marches and peaceful persuasion. These methods failed and they began to interrupt political meetings and sessions of parliament. When attempts were made to arrest them and remove them by force they resisted, sometimes chaining themselves to fixed points like railings.

The window-breaking campaign

To protest against the brutal treatment of women following arrest, and to draw more attention to their cause, the Suffragettes embarked on a policy of breaking windows on a large scale. This policy was also intended to show that they were prepared to use violence. Mrs Pankhurst declared:

> The argument of the broken pane is the most valuable in modern politics.

The campaign intensifies

The more militant wing of the movement, led by Christabel Pankhurst, adopted more extreme policies. They clashed with police, attacked private property and works of art. Arson, bombings, the slashing of paintings (most notably the attack by **Mary Richardson** on the *Rokeby Venus* in the National Gallery in March 1914) and pouring acid on golf course greens were some of the methods used.

Over two hundred women were arrested and imprisoned. They went on hunger strike and were subjected to intensely painful methods of force-feeding by the authorities. In 1913 a new law, nicknamed the 'Cat and Mouse Act', was introduced.

Emily Davidson throwing herself under the king's horse at the 1913 Derby

By the terms of this Act the government could release and later re-arrest prisoners who went on hunger strike.

The 1913 Derby

The most dramatic episode in the Suffragettes' campaign took place at the 1913 Derby. **Emily Davidson**, one of the most fanatical members of the W.S.P.U. ran out in front of the king's horse and was killed. Her funeral was the occasion of a massive demonstration on behalf of the Suffragettes.

Victory

The Prime Minister, Mr. Asquith, was beginning to show sympathy for the Suffragettes' cause when war broke out in 1914. Their campaign then came to a halt, as most women joined in the war effort. Women's contribution during the war was enormous. In 1918 their rights could be denied no longer. The vote was given to women over thirty. In 1928 women finally got the same voting rights as men. A major landmark on the road to democracy had been reached.

Other Reforms

Reforms in the Army

Scandals and setbacks during the **Crimean War** (1853-56) and the Indian Mutiny of 1857 had shown up major faults in the British army. The example provided by Prussia's efficient military machine made the case for reform more urgent. The Liberals hoped to make the army more efficient, more economic and more humane.

In 1871 Gladstone, despite great opposition, succeeded in passing the **Army Regulations Bill.** Up to that time, sons of aristocrats could become officers and gain promotion merely by paying for these privileges. This practice was counter to the Liberal principle of equality of opportunity. The new law opened the army to all and ensured that promotion would depend on ability.

Gladstone, despite opposition from Queen Victoria, ensured that the Commander-in-chief would be subject to the Secretary for War. Flogging was abolished. It was no longer permitted to pay 'bounty money' in order to enlist recruits. The **Short Service Act** meant that a man could now serve for six years and remain at home on reserve for the following six. In addition, the old muzzle-loading muskets were replaced by the new breech-loading rifles.

The result of these reforms was that the British army became modern and efficient by the standards of the day. It played its part in British imperial expansion.

Educational Reforms

In 1870 a report ordered by Gladstone's government revealed that all was not well in the education system. In Liverpool over half the children did not attend school at all, and many who did went to private schools of very low grade. An inspector's report on one of these private schools in Leeds ran thus:

> In the front room of a small dwelling house, half filled with dirty household furniture, I found 35 boys, all of whom were entirely unemployed, except eight who were writing in copy books. The master... was a cloth dresser by trade, and 'took to schooling because work was slack'... He regretted that he 'is not a bit of a singer', for if he were, he would 'learn them a few ditties, and the time would pass away quickly'.

Football match: Cup Final 1891, Blackburn Rovers v. Notts County

Forster's Education Act

Forster's Education Act of 1870 made schools available to every child up to the age of thirteen. Financial aid was increased for existing schools. The new state schools were to be run by district boards. Although parents had to pay fees, a reduction or even free schooling was available for poor families. In 1880 elementary schooling became compulsory, and in 1891 fees were abolished.

Reforms in secondary school education

Secondary education lagged behind other European countries and was enjoyed only by the children of the wealthy. Countries like France and Germany were far more advanced in technical instruction suitable for the new industrial age. The **Technical Instruction Act** of 1898 placed technical education under the control of local government authorities.

By 1907 a scholarship scheme was introduced to help children of poor people avail of secondary education. By 1914 Britain had developed an efficient system of education at elementary and secondary levels.

Health and Welfare

In the nineteenth century it was realised that many diseases were caused by bad sewage, polluted water and poor housing. The **Public Health Acts** of 1872 and 1875 divided the country into districts under health authorities. Local authorities were made responsible for water, sewage, street lighting and housing.

Conditions remained bad for the poor. Low wages and unemployment added to the problem. In 1908 infant welfare centres were set up to give advice to mothers and provide free milk for babies. In the same year Lloyd George introduced **pensions** for people over seventy. In 1911 the **National Insurance Act** gave assistance to unemployed workers. Problems with health and poverty remained, but things would never be as bad again.

Trade Unionism and Labour

Trade Unions

During the 1850s skilled workers began to organise themselves into unions. However, these were not recognised in law. The first **Trades Union Congress** took place in 1868. The Liberals and Conservatives were now looking for the union votes. From 1875 on, unions were allowed to bargain collectively, to strike and to picket peacefully.

The trade union movement develops

Until the 1880s only the skilled crafts were organised into unions. Soon miners, semi-skilled workers and eventually unskilled labourers were organised. The number of strikes increased, the most notable ones being the London match-girls' strike, the gas workers' strike and the London dockers' strike. The dockers' strike was settled when the workers got the famous 'dockers' tanner', or standard wage rate of six old pence (2½p) an hour.

The Trades Disputes Act

The unions had a setback in 1901 in what was known as the **Taff Vale Case.** The railway workers had to pay damages to the railway company following a strike. The **Trades Disputes Act** of 1906 gave the unions protection when strikes were properly organised and picketing was peaceful. Many unions followed the example of the Syndicalists (see chapter 3) in France and militant unions in the United States, and a number of bitter strikes took place between 1910 and 1914. A powerful 'triple alliance' of miners, railwaymen and transport workers was formed. Only the outbreak of war in 1914 prevented a massive national work stoppage.

Keir Hardie

In the General Election of 1874, two trade union candidates were elected to Parliament. In 1888 the Secretary of the Miners' Union, **Keir Hardie**, formed the **Scottish Labour Party** and in 1892 he became the first Labour M.P. A year later he set up the **Independent Labour Party**. (ILP)

The Fabian Society

The cause of labour was also taken up by the **Fabian Society**, a group of middle-class intellectuals. The society was named after a Roman general who avoided defeat by refusing to fight any decisive battles against Hannibal. The most famous members of

Keir Hardie election poster

this organisation were the Irish writer **George Bernard Shaw**, the politician and future prime minister **Ramsay MacDonald**, the novelist **H.G. Wells** and socialists **Sidney** and **Beatrice Webb**. The Webbs suggested ways in which the state could directly fight poverty. The Fabian Society rejected Marx's idea of the class struggle, believing instead that social reforms and equality would come through gradual changes. They produced pamphlets and hoped to influence society and parliament.

The establishment of the Labour Party

In 1900 the ILP and the Fabians joined with a group of socialists to form the **Labour Representation Committee.** Their aim was to get members into parliament. In 1906 they adopted the name **The Labour Party.** That same year they had twenty-nine members elected to parliament. This number increased to forty-two in 1910. The Party continued to grow and with Liberal support formed the first Labour Government in Britain in 1924.

Foreign Policy

Splendid isolation

For most of the nineteenth century Britain followed a foreign policy which can be summed up in the words of one of her Prime Ministers, **Lord Salisbury**, as 'splendid isolation'. She intervened in European affairs only when she thought that her interests were threatened. Her position was based on her industrial might, her empire and her naval supremacy.

The Empire

We have seen that Disraeli was committed to colonial expansion. In 1875 he bought shares in the **Suez Canal** in order to protect the route to India and the Empire in the East. In 1877 he persuaded parliament to grant Queen Victoria the new title of 'Empress of India'. At the Congress of Berlin (see chapter 2) he acquired **Cyprus** and blocked Russian attempts to gain influence in the Balkans. This was done to protect Britain's route to her Empire.

Unlike Disraeli, Gladstone had little interest in colonisation. However, the ordinary Briton had a great pride in the Empire. Gladstone was blamed for the death of **General Gordon** at **Khartoum** after a ten-month siege. The public felt that Gladstone should have sent a relief force earlier and the government lost prestige.

Britain's Empire continued to grow during the final years of the nineteenth century. (see chapter 6) This led to hostility and to disputes with other countries. During the **Boer War**, Britain realised that it had few friends. The poor performance of the army in that war made many politicians feel that Britain needed an ally. This feeling intensified in the light of the rapid growth of the German navy (see chapter 9). The days of splendid isolation were over.

Alliance with Japan

Britain signed an alliance with Japan in 1902. She was worried about growing Russian influence in northern China. Just as she had supported the Turks in the nineteenth century, she now backed Japan in her dispute with Russia.

The *Entente Cordiale*

In 1904 Britain and France signed the ***Entente Cordiale***. By the terms of this agreement, Britain was allowed a free hand in Egypt, while French interests in Morocco were recognised. Three years later Russia joined Britain and France in the **Triple *Entente*** (see chapter 10) designed to resolve their remaining colonial differences. In 1914 this became a military alliance. The 'splendid isolation' of 1870 had been replaced through what has been described as a 'diplomatic revolution'.

Chapter 6

European Imperialism

> Take up the White Man's Burden,
> Send forth the best ye breed,
> Go, bid your Sons to exile,
> To serve your captives' need.
>
> *The White Man's Burden*
> Rudyard Kipling

'New Imperialism'

Imperialism was a belief in the virtues and benefits of having an empire. It was a mixture of idealism, patriotism, pride and greed. European expansion overseas had been going on since the days of Vasco da Gama in the fifteenth century. From about 1870 onwards, Europe began to develop a new and stronger interest in acquiring colonies in Africa and Asia. The term 'New Imperialism' is applied by historians to the process of colonisation by the stronger and more developed nations, of the weaker and less developed areas of the world. Most of this colonial activity took place in the second half of the nineteenth century and in the early years of the twentieth century.

Reasons for Imperialism

Economic ambition

The newly-industrialised countries of Europe lacked essential raw materials. As a result, they established colonies in the countries where these raw materials were to be found. The colonies provided them with a wide variety of goods, such as gold, minerals, bananas, palm oil, rubber, cocoa and tea. They also provided cheap labour and a ready market for manufactured goods.

Lenin saw imperialism as 'the territorial division of the whole world among the greatest capitalist powers'. This may be an extreme view, but it is true that Western businesses now had an opportunity to invest their spare capital and surplus profits in new projects overseas.

Strategic action

In the days of the steamship, it was necessary to establish coaling stations and seaports to enable ships to take on fuel at regular intervals. They were also essential to the operation of a powerful navy. The Great Powers, aware of growing tensions and rivalries, tried to gain strategic naval bases at home and abroad.

Missionary and humanitarian zeal

Explorers and missionaries drew attention to previously unexplored areas. Missionaries like **Dr Livingstone** were also explorers and their charts and records proved useful and inspiring to later travellers. The sentiment of 'the white man's burden' combined humanitarianism with a missionary spirit. Thousands of missionaries went to Africa and Asia to spread Christianity.

These missionaries and humanitarians tried to stamp out many 'evils' such as slavery, barbaric punishments, sacrifices and pagan rites. They believed that they knew what was 'right' for the 'heathens'.

Nationalism

Colonies were seen as a mark of prestige and glory. The more colonies a country had, the more powerful it was considered. Envy and rivalry led to a race or a scramble for land in the colonies. Public opinion, especially in Britain, regarded the acquisition of colonies with great pride. This was seen in the work of writers like **Rudyard Kipling** and **A.C. Benson**. The latter's famous song best captures the sense of pride and enthusiasm:

Land of Hope and Glory, Mother of the Free,
How shall we extol thee, who are born of thee?
Wider still and wider shall thy bounds be set
God who made thee mighty, make thee mightier yet.

Social Darwinism

During the nineteenth century, the population of Europe grew rapidly. While some emigrated to countries with temperate climates, others went to the developing colonies in Africa and Asia. Many found rewarding careers abroad, in the army, the navy and in administration.

'Social Darwinism' was a term borrowed from the theories of **Charles Darwin** who saw the evolution of man in terms of the 'survival of the fittest'. While many saw this as an opportunity for helping what they believed to be 'inferior' people, others took advantage of what they regarded as their 'God-given racial and social superiority'. They treated the colonised people very badly and exploited them terribly. The greed of individuals and of nations was one of the major reasons for colonisation.

The Scramble for Africa

Early explorers

During the nineteenth century, Europeans referred to Africa as 'the dark continent'. It was dark because the people were black, the interior was unknown and the Europeans saw it as a sinister and backward place.

Britain's interest in Africa grew after 1850 because of the feats of her explorers. **Mungo Park** had already explored the Niger river. In 1856, disguised as an Arab, **Richard Burton** visited the holy shrines of Mecca and Medina. Later, with **J.H. Speke**, he penetrated central Africa and discovered Lake Tanganyika. Speke went on to discover the source of the White Nile.

THE COLONISATION OF AFRICA: 1914

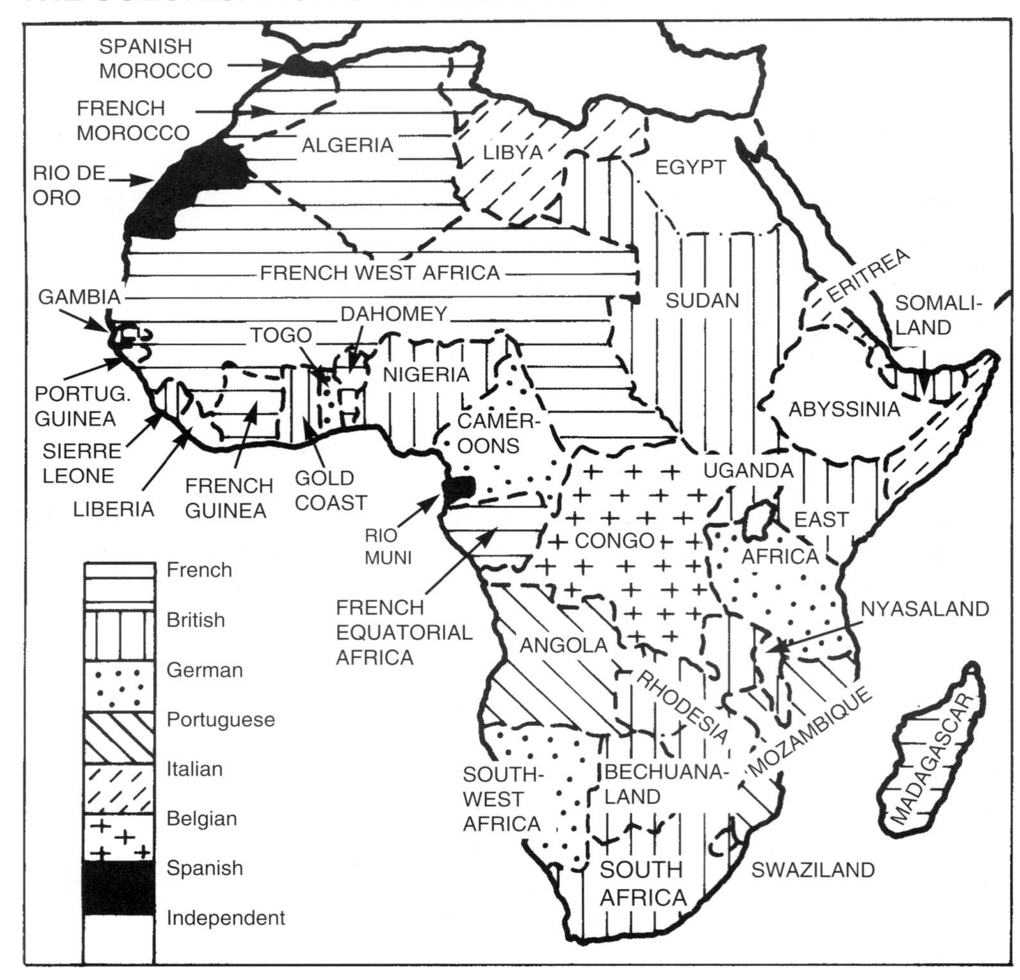

David Livingstone

The most famous of these pioneers was **David Livingstone**. He was a Scottish doctor who went to Africa as a medical missionary. He discovered the Victoria Falls. His writings helped to bring about the abolition of the slave trade. He said that Africa should be opened up to 'Christianity, commerce and civilisation'.

During Livingstone's second series of journeys, he discovered Lake Nyasa. In 1866, he was sent by the Royal Geographical Society to search for the source of the Nile. In 1869, after enduring severe illness, and deserted by his porters, he disappeared. The *New York Herald* sent **H.M. Stanley** to search for him. After a year's search, Stanley found the missing doctor and greeted him with the famous words, 'Doctor Livingstone, I presume'.

Livingstone refused to return home with Stanley. He continued his explorations and died in 1871. His African servants embalmed his body and carried it the 1,500 miles to the coast. He was brought back to England and was buried in Westminster Abbey. Stanley went on to explore the Congo river.

The Berlin Conference

Disputes in South Africa and in the Congo basin prompted Bismarck to call a conference in Berlin in 1884. He hoped that the Great Powers could settle the disputes and agree on future guidelines. The principal decisions of the Berlin Conference were:

- The claims of Leopold II, King of Belgium, to the Congo were recognised.
- When a government effectively occupied African territory and notified other governments, it could annex or take over that area.
- Slavery and the slave trade were illegal.
- The Congo and Niger rivers were declared open to navigation by all.

The Berlin Conference gave a semi-legal basis to the colonisation of Africa. The process whereby Africa fell under European control became known as 'the scramble for Africa'. The phrase was first used by *The Times* newspaper in 1884.

Britain

Suez Canal

The Suez Canal was opened in 1869, funded by French loans. The canal soon became Britain's main route to India. In 1875 the British Prime Minister, **Disraeli**, bought £4 million worth of shares in the canal from the ruler of Egypt on behalf of the British government. In 1882 Britain suppressed a revolt by the Egyptians and established control over the country.

The purchase of Suez Canal shares . . . Britain gets the lion's share.

Sudan

Sudan lay south of Egypt and Britain considered it to be within her sphere of influence. In 1883, a rebellion, led by the ***Madhi***, who claimed that he was a Muslim prophet, broke out in the Sudan. The British Prime Minister, **Gladstone**, sent **General Gordon** to the Sudan to deal with the problem, but Gordon was killed at Khartoum in 1885. Gladstone was blamed for being too slow in responding to Gordon's call for reinforcements. The Madhi died soon after, but his successor, the Khalifa, retained control of the Sudan. Fourteen years later, another British general, **Kitchener**, avenged Gordon's death by defeating the Khalifa's forces in the Sudan.

The Fashoda Incident

The British were determined to control the important Nile waterway and its fertile valley. However, France was also interested in establishing a presence in the region and hoped to challenge British dominance. Kitchener proceeded up the Nile through the Sudan. He found his way blocked by a small French force under **Captain Marchand** at **Fashoda**. Kitchener had a much superior force of 2,000 men and had five gunboats, but the French had reached the fort of Fashoda first.

The generals referred the dispute to their governments. The French recognised

The death of General Gordon at Khartoum

Marchand's weak position and ordered him to withdraw. Britain was allotted the area around the Nile as its sphere of influence, while the French developed French West Africa. The Fashoda Incident could easily have led to war between two European countries.

. . . . 1898. Note the French flag flying from the fort at left.

South Africa

> I contend that we [Britons] are the first race in the world, and that the more of the world we inhabit, the better it is for the human race.
>
> Cecil Rhodes

The **Boers** (Dutch) settled in South Africa in the seventeenth century. In 1815 the British took over Cape Colony and the Boers moved inland. They then established their own states – the **Orange Free State** and the **Transvaal**. In 1886 gold was discovered in the Transvaal and many British settlers were attracted to the area. The Boers called these immigrants ***Uitlanders*** (outsiders). They taxed them very heavily and did not allow them voting rights.

Cecil Rhodes

In 1890 **Cecil Rhodes** became Prime Minister of Cape Colony. He hoped to extend British influence northwards and spoke of building a railway from the Cape to Cairo. He saw **Paul Kruger**, President of the Transvaal, as the main obstacle to his dream. In the Transvaal itself, the gold mining industry grew, and Kruger's government continued to discriminate against the *Uitlanders*.

In 1895 Rhodes encouraged one of his friends, **Dr Jameson**, to lead a raid into the Transvaal. They were hopeful that this would persuade the *Uitlanders* to revolt

Cecil Rhodes like a colossus astride Africa

The Boers manning their trenches outside Mafeking, 1899

against the Boers. The raid was a total failure and Rhodes was forced to resign as Prime Minister of Cape Colony. It was after the Jameson raid that Kaiser William II of Germany sent a telegram to Kruger congratulating him on restoring order 'without appealing for the help of friendly powers'. (See chapter 9)

The Boer War

The Boer War broke out in November 1899. The British had rejected an ultimatum from Kruger to keep out of Transvaal affairs. The Boers had a civilian army of about 70,000 and were equipped with German arms. They were familiar with the terrain and hoped for a quick victory.

The Boers invaded the Cape and Natal. They inflicted defeats on the British at Stormberg and Colenso and besieged Ladysmith, Kimberly and Mafeking. The British appointed **Lord Roberts**, assisted by **General Kitchener**, to take command of the army and sent in over 300,000 reinforcements. They regained control of the besieged towns and also took control of the Boer capital of Pretoria.

The Boers then began a campaign of guerrilla warfare against the British. In retaliation, Kitchener burned their farms and houses. He set up concentration camps. When the over-crowded and unhygienic conditions in these camps were reported in Britain, there was an outcry. However, the harsh methods were successful and the Boers finally surrendered in 1902.

The Treaty of Vereeniging

The **Treaty of Vereeniging** brought the Transvaal and the Orange Free State under British rule. A grant of £3 million was given to the Boers to help them rebuild their houses and restock their farms. In 1910 Cape Colony, Natal, the Orange Free State and the Transvaal joined together in the **Union of South Africa** which received dominion status in the British Empire.

Other British interests

As well as having interests in Egypt, the Sudan and South Africa, Britain established protectorates over or colonised a number of other areas. The most notable of these were Somaliland, Zanzibar, Kenya, Uganda, Rhodesia and Nigeria.

Other Colonial Powers

France

> If trade followed the flag, the flag sometimes followed the cassock.
>
> Alfred Cobban, *A History of Modern France*, vol. 3

During the second half of the nineteenth century, France became the leading Catholic missionary country. Thousands of French priests and nuns spread the faith in Africa and Asia. This often led French government and military authorities to colonise these areas.

In 1830, France occupied part of Algeria in order to eliminate pirates who operated in the area. Eighteen years later Algeria legally became part of France. After France's defeat in the Franco-Prussian War (see chapter 1) the Third Republic sought other colonies abroad in order to restore some of her lost prestige. Bismarck encouraged French expansion in North Africa in an attempt to divert French attention from Alsace-Lorraine. The main French force used in the colonisation process, the **Foreign Legion**, became a symbol of military prowess.

Jules Ferry was the politician who set the Third Republic on the road to expansion. A protectorate was established over Tunisia in 1881 and Italian competition there was eliminated. Between 1870 and 1914, France established effective control over Madagascar, French Somaliland, French West Africa, the French Congo and Morocco.

France also established rule over the whole of Indochina. By 1914, France's overseas possessions were second only to those of Britain. She avoided a war with Britain over the Fashoda Incident but came into conflict with Germany over Morocco, and with Italy over Tunisia.

Germany

During the early 1870s Bismarck concentrated on domestic affairs. He stated that colonies were too costly and were more bother than they were worth. However, he came under great pressure from the **German Colonial Union**, founded in 1882. They campaigned for colonial expansion for patriotic and economic reasons. Germany soon took over South West Africa, Togoland, The Cameroons, German East Africa and some Pacific islands.

William II took a far greater interest in German colonialism than William I had. He saw the possession of colonies as a symbol of Germany's position as a world power. He linked his expansive foreign policy (***Weltpolitik***) with the building of a large navy. But a policy of colonisation was not without its dangers. Germany's disputes with Britain during the Boer War and with France during the Moroccan crisis drove France and Britain together.

Italy

Italy wished to expand into North Africa for reasons of trade and prestige. Following France's seizure of Tunis in 1881, Italy acquired Eritrea, Italian Somaliland and Libya. Most of the Italian possessions were sandy wastelands. Italy suffered the humiliation of defeat by the Abyssinians at the **Battle of Adowa** in 1896 and was forced to acknowledge Abyssinia's independence.

Belgium

The Belgian Congo was extremely rich in natural resources. Belgium had established control of the area by 1882. In 1885, it was recognised as the personal possession of King Leopold II during the Berlin Conference. It was taken over by the Belgian government in 1908.

Asia and the Pacific

India

India was originally controlled by a commercial group, the **East India Company**. Following riots in 1857, it was taken over, in 1858, by the British government which ruled through a Viceroy. In 1876 Disraeli persuaded parliament to declare India an empire. Queen Victoria was crowned 'Empress of India' and India became known as 'the jewel in the crown' of the British Empire. Britain took a controlling interest in the Suez Canal in order to protect her trading interests in India.

China

During the second half of the nineteenth century, countries such as Britain, Russia and France gained bases and trading concessions in China. The United States was afraid that she would be shut out of China by the European states. In 1899 she proposed the '**Open Door**' policy. This meant that all foreign powers should have equal opportunities in China and the Far East. This was accepted and saved China from being carved up in the same way as Africa.

Procession through Delhi to acclaim Queen Victoria as Empress of India, 1877

The **Boxers** was one of a number of secret organisations set up by the Chinese to resist exploitation by foreigners. The Boxer Revolt in 1898 involved rioting, demonstrations, destruction of property and the killing of foreigners. It was put down by the foreign powers with great harshness.

In the early years of the twentieth century Japan gained influence in China. The Chinese continued to resent foreigners in their country but it was not until the communists took over in 1949 that foreign domination in China (with the exception of Hong Kong which remained under British control) was ended.

The French built up a large empire in the Far East in what became known as French Indochina. Cambodia and Vietnam were two French colonies in this area. Britain's colonial influence in the Far East included the control of Burma and Malaysia.

The Pacific

The United States had been a colony of Britain and many liberals were adamant that their country should not become involved in building an empire. However, the imperialists won the argument and the United States gained control over the Philippines, Hawaii, Guam and part of Samoa. The Dutch, British, French and Germans all took over various Pacific Islands.

Effects of Imperialism

Benefits of colonialism

The imperialists benefited economically from the colonies. Raw material and overseas markets led to a great increase in trade and to the creation of wealth. In many cases the colonies too were improved by western civilisation. Law and order, public health, education, roads and railways were a visible result of European influence.

In India, British rule banned female infanticide and the custom in some areas of widows committing suicide on their husbands' funeral pyres. In Africa, some sacrificial rites and bitter wars between tribes were halted. Barbaric practices and slavery were brought to an end.

The French had an official policy of 'assimilation'. This meant absorbing the natives and bringing the two cultures together. Attempts were also made to raise the living standards of the native population. In practice, equality was rarely extended beyond a small native elite. In colonies where indirect rule applied, for example British rule in Nigeria, there was a closer observance of native laws and traditions.

The contribution of Christian missionaries and humanitarians brought great benefits to many of the colonies, especially in terms of religious instruction, education and health care. The European ideals of liberty, equality, fraternity, democracy and nationalism followed. Ironically, it was the acceptance of these ideals that led to revolts against the imperialists and freedom for the new states.

Negative aspects of colonialism

In general, the natural resources of the conquered countries were shamefully abused by the imperialists. Greed was often the main motive and the white man treated the native with terrible brutality. Workers were exploited in factories and mines and were forced to work on plantations. The Congo Free State was the most notorious example of a colony which was ruthlessly exploited.

The rich traditions and cultures of the native people were often destroyed by the white colonists. Missionaries were shocked by some native customs and many of them made little attempt to blend Christianity with the cultures they encountered. The boundaries of various countries were decided by drawing lines on maps, without reference to tribal locations. This led to many civil wars in the twentieth century.

A German satirical cartoon depicting British imperialism and exploitation of black people

While the French and Portuguese had an official policy of treating the natives as equals, in practice the Europeans looked on the natives as inferiors. In general the 'white man's burden' fell heavily on the natives.

Conclusion

It was the English writer Rudyard Kipling who urged his country to 'send forth the best ye breed' to extend Britain's colonies overseas. Yet even he sensed that Europe's dominance over Africa would not last:

> Far called, the navies melt away
> On dune and headland sinks the fire;
> And all our pomp of yesteryear
> Is one with Nineveh and Tyre.
>
> *Recessional*

Decolonisation did not begin in earnest until after World War II. Gradually most of the colonised countries in Africa and Asia regained their independence.

Chapter 7

The Eastern Question

The Question defined

The Ottoman Empire

Woe unto us, woe unto us: the imperial throne is destroyed!
The churches are devastated: the monasteries are wrecked!

The Christians of Constantinople (Byzantium) showed their anguish in these lines from a song about the fall of their city to the Turks in 1453. The Turkish Empire was ruled by the Islamic Ottomans. They soon conquered much of central and south-east Europe. However, in the course of the nineteenth century a number of European states (among them Greece, Serbia and Romania) which had become part of the Ottoman Empire, were able to gain a great degree of independence from the Turks.

'Sick man of Europe'

The problems that occurred in south-eastern Europe as a result of this growing weakness of the Turkish Empire are known collectively as the Eastern Question. These problems can be grouped into three main areas:

- The attitudes of the other empires to Turkey's weaknesses
- Attempts to prevent the break-up of the Ottoman Empire
- The emergence of nationalism in the Balkans, that is the area of south-east Europe still controlled by Turkey.

During the nineteenth century, Turkey became known as the 'sick man of Europe'. The situation was further complicated by the presence of a large number of Christian subjects in the Ottoman Empire.

The Interests of the Great Powers

Britain

As we have seen in chapter 5, Britain strongly opposed Russian expansion in the Balkans after the opening of the Suez Canal in 1869. Britain's main concern was the protection of her route to India. The fall of the Ottoman Empire, with control of

Punch cartoon showing the Turkish rejection of Russian demands

Constantinople passing to Russia, would give Russian ships entry to the Mediterranean and endanger the British shipping route. It was for this reason that Britain decided to support Turkey in the face of growing Russian interest in the Ottoman Empire.

Russia

The Russians were Slavs and members of the Orthodox Church, just like the majority of Turkish subjects in the Balkans. The capture of Constantinople, the centre of Orthodox Christianity, had long been an aim of the Tsars of Russia. They advocated a policy of **'Panslavism'**, that is the claim to act as protector of Slav peoples and the Orthodox faith in the Ottoman and Habsburg empires.

Other powers suspected Russia of trying to extend her influence while using Panslavism as an excuse. She certainly hoped for the break-up of the Turkish Empire and encouraged Balkan nationalism. One of the main reasons for this was the fact that Russia needed an exit for her shipping from the Black Sea to the Mediterranean. The death of the 'sick man of Europe' would help her achieve this.

Austria-Hungary

The Austro-Hungarian Empire was made up of many national groups. She feared Balkan nationalism as it might lead to the disintegration of her own empire. At first the Austrians tried to prop up the Turks. However, when Austria and Russia drifted apart, the Austrians tried to replace Turkish influence in the Balkans with their own in order to stop Russian interference in the area.

Other countries' attitudes

The German Chancellor, Bismarck, tried to prevent trouble between Russia and

Austria-Hungary, both rivals in the Balkans. He was concerned that one of these empires might enter an alliance with France, thereby weakening Germany's position. Under Bismarck, Germany itself had no territorial ambitions in the Balkans, an area described by Bismarck as 'not worth the bones of a single Pomeranian grenadier'. However, under Kaiser William II, Germany's economic interests in Turkey increased significantly. (see chapter 9)

The French were too concerned with domestic affairs and thoughts of revenge on Germany to pursue interests in the Balkans after 1870, although they did not wish to see the extension of Russian influence there. On the other hand, Italy regarded the break-up of the Ottoman Empire as an opportunity for gaining colonies in North Africa, an area of Turkish influence, and viewed the proceedings in the Balkans with interest.

The Near Eastern Crisis, 1875-1878

Revolt in the Balkans

In 1875, a revolt took place in **Bosnia** and **Herzegovina** which were Balkan provinces of the Turkish Empire. The Turks had increased taxes, despite bad harvests and economic problems. Demands by Austria, Germany and Russia for reforms within the Empire had been ignored.

The revolt spreads

By 1876 the revolt had spread to the other Balkan states of **Serbia**, **Bulgaria** and **Montenegro**. The new head of the Ottoman Empire, Sultan **Abdul Hamid II**, attempted to solve the problems by sending a group of irregular Turkish troops, the

Abdul Hamid II, Sultan of Turkey, driving through Constantinople

Bashi-Bazouks, to Bulgaria. Instead of easing the situation, they became involved in a campaign of terror and committed terrible atrocities.

Reaction in Britain

In Britain, the Prime Minister, Disraeli, was following the old policy of preserving the Turkish Empire. He dismissed stories of the Bulgarian massacres as 'coffee house gossip'. However, his political opponent, Gladstone, was better informed about the situation. His pamphlet *The Bulgarian Horrors and the Question of the East*, sold nearly a quarter of a million copies in a month. In this pamphlet he referred to the Turks as 'the one great anti-human species'. Disraeli rejected the criticism that his policy had encouraged the Turks to act with brutality, stating that in his view Gladstone himself was 'worse than any Bulgarian horror'.

Russian intervention

While there was disagreement in Britain about what should be done about the situation in the Balkans, public opinion in Russia was strongly in favour of intervention against Turkey. Panslav agitation and news of the Bulgarian massacres convinced the Tsar that he would have to act against the Sultan. Russian volunteers advanced into Bulgaria and were finally engaged by the Turks at **Plevna.** A five-month siege was proving a drain on the resources of both sides. The Tsar and the Sultan decided that it was time for peace.

The Treaty of San Stefano

The main terms of the **Treaty of San Stefano** in 1878 were:

- Romania, Serbia and Montenegro were recognised as independent states.
- Bulgaria, a new state, independent of Turkey, was set up in the Balkans. It had access to the Mediterranean.
- Turkey had to pay a war indemnity to Russia.

However, the other great powers felt that Bulgaria would be completely dominated by Russia. The Tsar, under the threat of war, agreed to submit the Treaty of San Stefano to an international congress.

The Congress of Berlin

Bismarck did not wish to see Germany dragged into a war between Austria-Hungary and Russia. He decided to act as 'honest broker' between Britain, Austria-Hungary and Russia at the Congress of Berlin in June 1878. The other powers forced Russia to change the terms of the Treaty of San Stefano and to drop their plans for expansion in the Balkans. The main terms of the Congress of Berlin were:

- The 'Big Bulgaria' agreed to at San Stefano was split up into three parts: a small

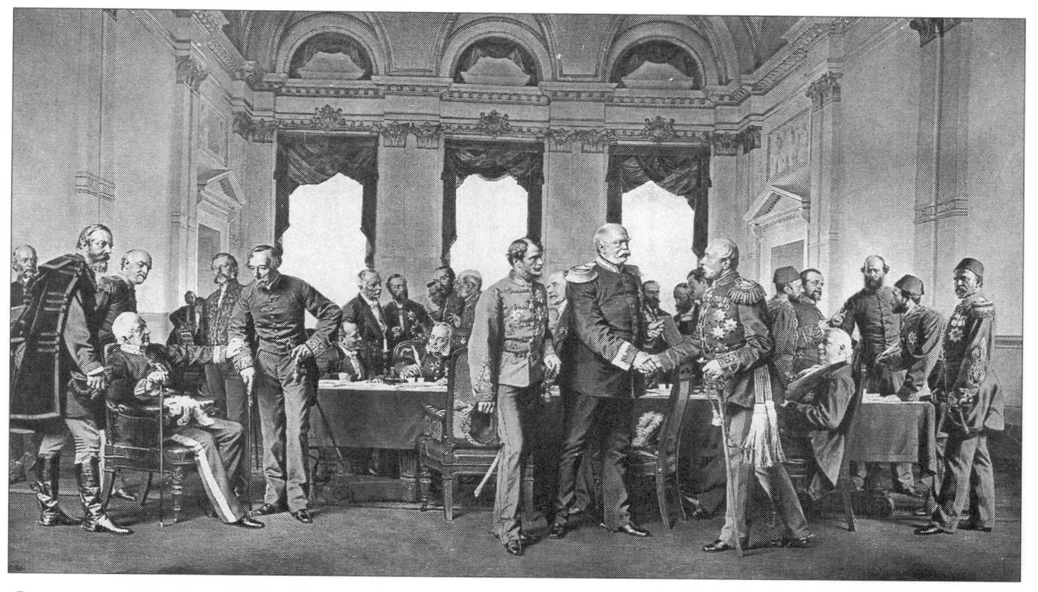

Congress of Berlin, 1878 – Bismarck, Disraeli and other European statesmen agree on rules for the colonisation of Africa

independent Bulgaria; Macedonia was returned to Turkey; Eastern Rumelia, with a Christian governor, also came under Turkish influence.

- Austria was allowed to occupy and administer Bosnia and Herzegovina. However, she could not annex (take over) these provinces.
- Romania, Serbia and Montenegro had their independence confirmed.
- Britain gained Cyprus, Russia got South Bessarabia and France was allowed occupy Tunis.

At the Congress of Berlin, Balkan nationalism was sacrificed to the interests of the major powers. Serbs within Bosnia and Herzegovina were very annoyed when their provinces were handed over to Austrian administration. It could be said that the seeds of Serbian discontent, which burst forth at Sarajevo in 1914, were sown at Berlin in 1878. In addition, the Russians saw the Congress as a diplomatic defeat.

From Berlin to Sarajevo

1885: the Bulgarian crisis

The new state of Bulgaria soon began to resent Russian interference in her affairs. In 1885 Eastern Rumelia proclaimed union with Bulgaria. This was contrary to the terms of the Congress of Berlin. The Russians were annoyed. Britain now saw

THE BALKANS AFTER THE CONGRESS OF BERLIN 1878

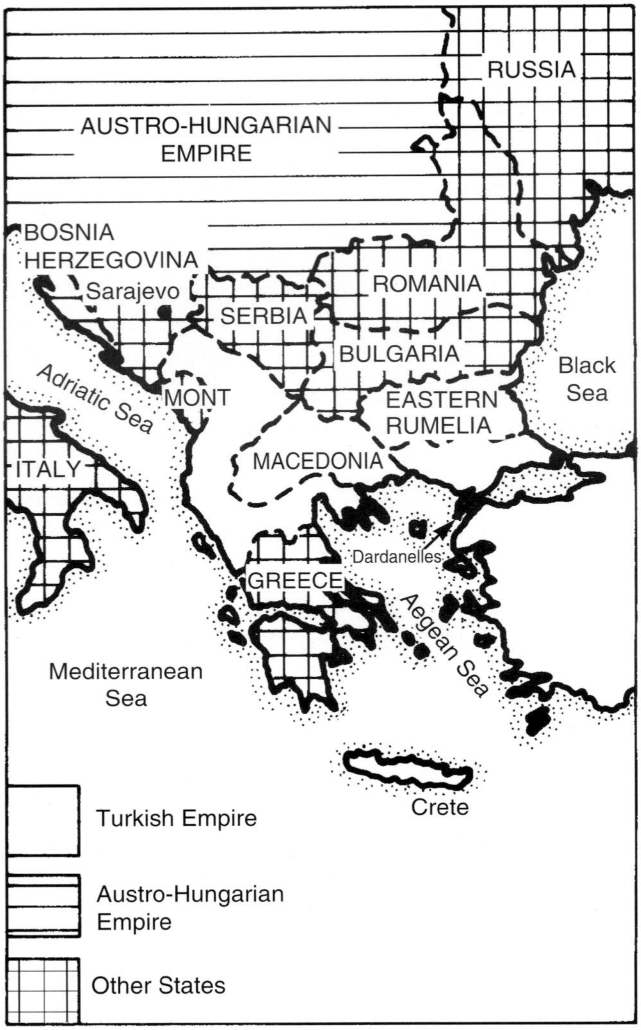

Bulgarian nationalism as a barrier to Russian ambitions in the Balkans and supported the union. The Russians had no option but to accept the situation.

In 1886 the Bulgarian parliament offered the vacant throne to **Prince Ferdinand**, a German Catholic. The Russians angrily opposed the appointment, which was acceptable to Austria-Hungary. It looked as if another war was going to break out in the Balkans, this time between Russia and Austria-Hungary. Bismarck then published the terms of the Dual Alliance of 1879. On reading them, Russia realized that if she attacked, Bismarck would come to the aid of his Austrian neighbours. However, if

Austria-Hungary was the aggressor, she would fight Russia alone. Neither country was prepared to go to war over Bulgaria, and tensions relaxed for a time.

1894-96: the Armenian massacres

Armenia, a province at the south-east corner of the Black Sea, was about 500 miles from Constantinople. Its two million Orthodox Christians were ruled by the Muslim Turks who persecuted the Armenians for their faith. When an uprising occurred in 1894, the Turks carried out a series of massacres which continued until 1896.

International reaction

Kaiser William II of Germany was trying to cultivate good relations with the Sultan. He hoped that his country could forge economic links with Turkey. He was not prepared to speak out against the massacres. British newspapers spoke of 'Abdul the Damned and Unreformed', but the Sultan ignored all British protests. Turkey was saved from united action by disagreements among the great powers.

1896: revolt in Crete

The island of Crete was mainly inhabited by Greeks, but was subject to Turkish rule. The Greeks revolted in May 1896. In 1897 Greece declared war on Turkey and sent troops to the island. Although they were defeated by the Turks, the great powers intervened and Crete was given self-government in 1908. It eventually became part of Greece in 1913.

The Young Turks

The **Young Turks** were a group of ardent nationalists who wished to modernise Turkey. Led by **Enver Bey**, and with the support of many young army officers, they advocated the introduction of a liberal constitution and a parliament elected by the people. They forced the Sultan to introduce these reforms in 1908. However, the Sultan attempted an unsuccessful counter-coup in 1909 and the Young Turks replaced him with his brother.

Austria's reaction

The Austrians took advantage of the trouble in Turkey to take over full control of Bosnia and Herzegovina in 1908. This further frustrated the Serb nationalists. Russia protested, but it was significant that Germany supported Austria-Hungary against the Serbs.

Demise of the Young Turks

As for the Young Turks, the movement proved a false dawn. The promises they made were all broken, and Turkey reverted to her old ways. So hard did life become for the various national groups in the Balkans, that they were driven into a new and threatening unity.

THE BALKANS IN 1914

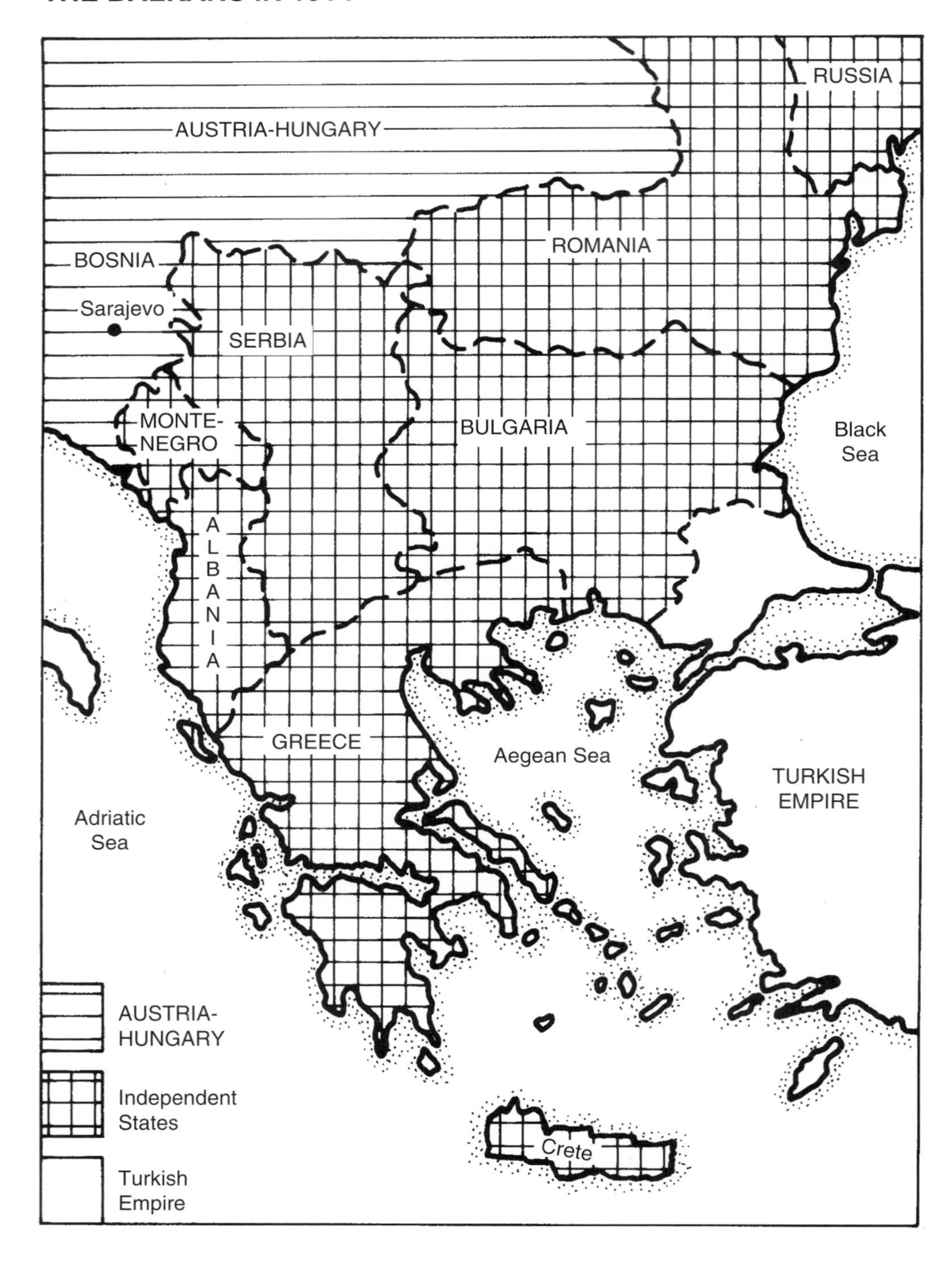

1912: the First Balkan War

Renewed Turkish persecution led to the formation of the **Balkan League** in 1912 by Serbia, Bulgaria, Montenegro and Greece. The League aimed to drive the Turks out of Europe and declared war on Turkey. The united Balkan states had remarkable success and by 1913, the 'sick man' was virtually gone from Europe. Little more than Constantinople remained in Turkish hands when the great powers intervened in the First Balkan War.

The Treaty of London

The **Treaty of London** was signed in 1913. Austria demanded that Serbia and Montenegro should evacuate the Adriatic coastal areas they held. This led to the creation of the new independent state of Albania. Serbia was now a landlocked country. Constantinople and the Dardanelles coast were left to Turkey, but the remainder of her European possessions were divided between Serbia, Greece and Bulgaria.

1913: the Second Balkan War

Shortly after, war broke out between the members of the Balkan League. Bulgaria was unhappy with the terms of the Treaty of London. In June 1913, she attacked Serbia. The Greeks and Romanians were afraid of Bulgaria and helped the Serbs. The Turks intervened in order to recover Adrianople. The odds against the Bulgarians were too great and they agreed to end the fighting. The **Treaty of Bucharest**, signed at the end of the Second Balkan War, meant the loss of large areas of Bulgaria to Serbia, Greece, Romania and Turkey.

Tensions remain high

The Balkans remained very unstable in 1914. Serbia had doubled in size and become proud of her position as the leading Slav state in the Balkans. She continued to encourage her fellow Slavs to seek independence. The House of Habsburg and the Austro-Hungarian government were now convinced that Serbian nationalism would have to be suppressed or her own multi-racial empire would be destroyed.

The end of the Eastern Question

As we shall see in Chapter 10, the Eastern Question and Balkan nationalism, among other factors, sparked off World War I. In the war, Turkey and Bulgaria took the side of the Triple Alliance, while Serbia, Romania and Greece supported the *Entente.* It wasn't until five years after the war had ended, that the event which finally settled the Eastern Question took place. By the terms of the **Treaty of Lausanne**, agreed between Turkey and the Allies in 1923, Turkey gave up all claims to the former European territories of the old Ottoman Empire.

Chapter 8

Austria-Hungary and Italy (1870–1914)

The groups within the empire

The Austrian Empire was ruled from Vienna by the ancient House of Habsburg. It spread across south-central Europe from the Swiss and German borders in the west to the boundaries of Russia, Romania and Bulgaria in the east.

The most remarkable feature of the Empire was the fact that it consisted of eleven racial groups. People of German origin made up about a quarter of the population of the Empire and were dominant in Austria itself. The next largest group was the Magyars of Hungary. The majority of the others were of Slav origin.

Percentage of population of the Empire

Austrians (Germanic)	24%
Magyars	20%
Czechs	13%
Poles	10%
Ruthenians	8%
Romanians	6%
Croats	5%
Serbs	4%
Slovaks	4%
Slovenes	3%
Others	3%

Nationalism

Many of the racial groups were happy to be part of a great empire. However, as nationalism grew stronger in Europe in the nineteenth century, many groups began to demand a greater or lesser measure of independence.

Meanwhile, Austria was also facing attack from outside. Following her defeat at the hands of Prussia in the Seven Weeks War of 1866 (see chapter 1), she was no longer regarded as a major force in German or Italian politics. Nationalists within the Empire hoped that this defeat would bring about the collapse of Austria's multinational structure and end her role as a great European state.

THE AUSTRIAN EMPIRE: NATIONALITIES

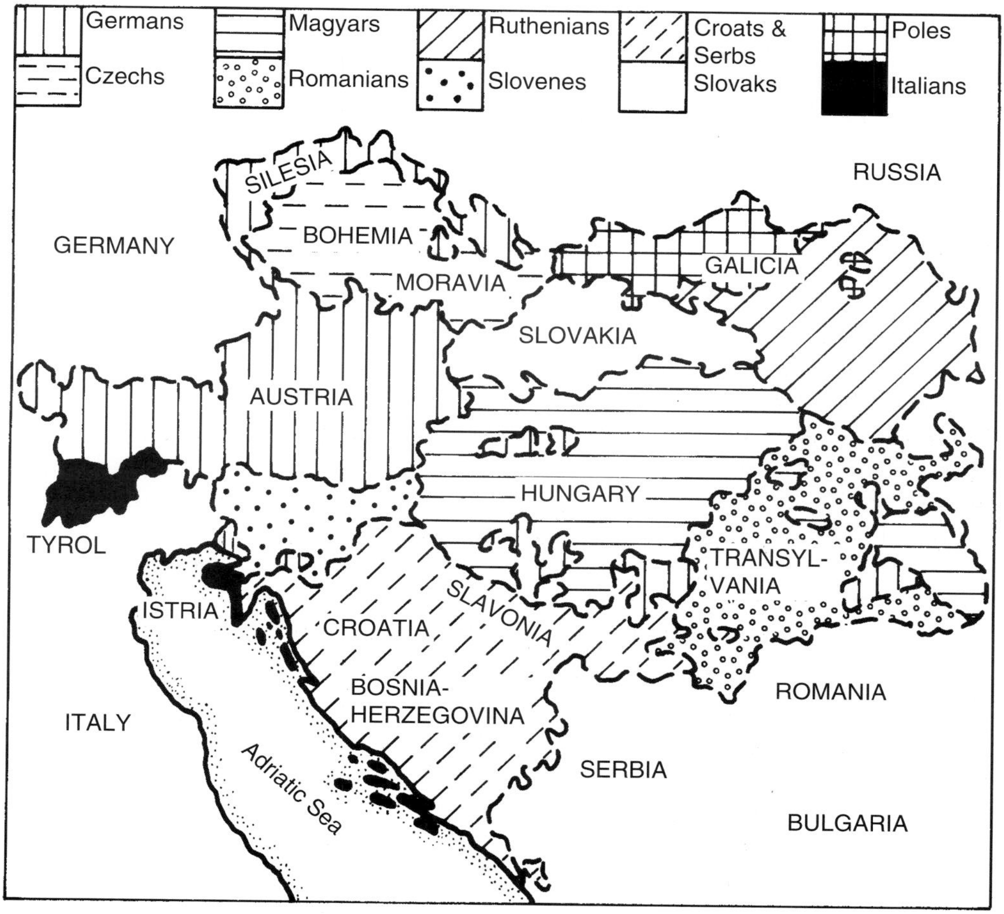

The Magyars

The most vocal and strongest non-ethnic German group in the Empire, the Magyars, called for Home Rule for Hungary after the Seven Weeks War. A revolt against Habsburg rule spread through the country. The Emperor, **Franz Joseph**, decided to grant concessions to the Magyars. Their moderate leader, **Ferenc Deák**, persuaded the headstrong nationalists that Hungary would be better off within the Empire:

> We must not over-rate our strength, and must confess that on our own we are not a great state.

1867: The Dual Monarchy

The agreement between Austria and Hungary was known as the ***Ausgleich*** or compromise. This transformed the Austrian Empire into the Dual Monarchy of Austria-Hungary.

The Dual Monarchy meant that Austria and Hungary had separate parliaments in Vienna and Budapest. They shared a common ruler, a common foreign policy and joint armed and naval forces. Other matters of common interest to the two states, such as customs, the monetary system and railway communications were to be governed by agreements renewed between the two partners at intervals of ten years. The parliaments in **Vienna** and **Budapest** looked after domestic issues like education, road maintenance and local taxes.

Emperor-King

At the centre of this new constitutional arrangement stood the Emperor-King, Franz Joseph. He had become Habsburg Emperor in 1848 at the age of eighteen. A convinced autocrat, he tackled his 'God-given' duties with enthusiasm.

The Emperor led a Spartan life-style and spent long hours at work each day. He was devoted to duty and it is said that he learned to speak nineteen languages in order to communicate with his many subjects. He earned the respect of his people, but many of them had no loyalty to, or respect for, the empire.

Although he was a conscientious and hard worker, the Emperor lacked the imagination and flexibility to continue the process of constitutional development after 1870.

Emperor Franz Joseph, ruler of Austria for 68 years

Personal tragedy

In his personal life, Franz Joseph was a tragic figure. He came to the throne in 1848, the 'year of revolutions'. His brother, **Maximilian**, became Emperor of Mexico, but was executed by Mexican revolutionaries in 1867. Franz Joseph's only son and heir, **Crown Prince Rudolf**, had many disagreements, personal and political, with his father. Rudolf and his mistress, **Maria Vetsera**, committed suicide at the **Mayerling** hunting lodge in 1889. Nine years later, the Emperor's wife, **Elizabeth**, was stabbed to death by an Italian anarchist.

Following Rudolf's death, the heir to the throne was Franz Joseph's nephew, Archduke **Franz Ferdinand**. He was assassinated in Sarajevo in 1914. (see chapter 10) Franz Joseph could truthfully say that he had been 'spared nothing.' He was eighty-four years old when World War I broke out. Most historians blame the Empire's politicians rather than the Emperor for Austria-Hungary's hard-line approach in 1914. In November 1916, the Emperor died peacefully in his palace in Vienna. Two years later his country surrendered and the Empire collapsed (see chapter 11).

Austria 1870-1914

Liberal rule (1867-79)

After 1867 the Liberal Party, which represented German middle class interests, gained control in the Austrian parliament. The Catholic Church soon found itself under attack. It lost control over education and was forced to accept the legalisation of civil marriage. Other reforms included the provision of compulsory and free elementary education. Laws were also introduced to improve the status of the Jews.

However, these liberal measures did not include equality for other nationalities. Czech attempts to gain equality in a 'Triple Monarchy' of Austrians, Hungarians and Czechs failed. The Czechs withdrew from parliament. The Liberals did not want any more Slavs in the Empire and were opposed to Austria-Hungary's growing influence in Bosnia and Herzegovina. The Emperor was annoyed by the Liberals 'interfering' in foreign affairs and he opposed them in the 1879 election.

Count Taaffe (1879-1893)

The Emperor now turned to the support of his conservative friend, **Count Edward Taaffe**. The Count was an aristocrat of Irish origins and he governed the Empire for the next fourteen years. He tried to remain above party politics and to rule by conciliation. Long before it became fashionable in England, he used the term 'muddling through' to describe his method of operation.

Taaffe operated an **Iron Ring** of German Catholics, Poles and Czechs in order to form a government. He made many concessions to the various nationalities, but never worked out a comprehensive programme to deal with the minority question. In Bohemia, Galicia and Slovenia, the Slavic languages as well as German could now be used in education and administration.

Count Edward Taaffe, Austrian Premier 1879-93

But Taaffe's greatest problem was with the Czechs. He encouraged them to give up their boycott and return to parliament. He allowed the Czech language to be recognised as the official language of Bohemia. However, his concessions only made the Czechs bolder. Extremists demanded the complete exclusion of German influence from what they referred to as the 'historic kingdom' of Bohemia.

Taaffe's concessions not only failed to conciliate the Czechs, but they infuriated the Germans who saw their influence steadily eroded in parts of the Empire where they were in a minority. Serious financial problems led to a major increase in the national debt. Government attempts at social reform failed to satisfy the socialist and trade union movements. Taaffe attempted to introduce universal male suffrage in 1893. He lost the support of the conservatives, the middle classes, the Church and the Emperor. Taaffe was forced to resign.

Badeni (1893-1897)

Taaffe was replaced in 1893 by **Count Badeni**. Badeni tried hard to cope with rising German nationalism and the nationalities question among the minority races. In 1897 he introduced a **Language Ordinance** in order to gain the support of the Czechs.

Those seeking positions in the Civil Service in Bohemia had to be fluent in both the German and Czech languages.

This caused great resentment, because while many Czechs understood German, few Germans understood Czech. Disturbances occurred in parliament and there was serious rioting on the streets of Vienna. The stability of the Empire was at stake. Badeni was dismissed. Now rioting took place in Prague. In order to neutralise the situation, the Emperor used his powers to rule by decree.

The minority problem remained a controversial one. The Czechs and Slovaks came together under pressure from the dominant Austrians. But demands for independence did not cease. The Czech philosopher, **Thomas Masaryk**, proposed the creation of a new state of Czechs and Slovaks. When the Empire broke up in 1918, Masaryk became president of an independent Czechoslovakia.

Hungary

Economic situation

Meanwhile many Hungarians, particularly the Hungarian nobles, had been placated by the ***Ausgleich*** of 1867. They gained a large degree of control in domestic matters and control over their national minorities. They also had an important voice in the affairs of the Empire.

Hungary's economy depended mainly on agriculture. She supplied Austria with most of her grain. The Hungarians continued to insist on tariffs to protect grain prices. A gulf arose between the vast majority, who were landless labourers, and the powerful landlord and noble classes. Many of these labourers left the countryside and drifted to the cities in search of work. They had difficulty in adapting to city life and, along with the minorities, constituted another restless group.

The minorities

As in Austria, the nationalities question was one of the great problems facing the government. In 1889 the population of Hungary consisted of Magyars (47 per cent), Germans (14 per cent), Slovaks (14 per cent), Romanians (18 per cent), with Ruthenians, Serbs and Croatians forming 7 per cent. The Austrians attempted to solve minority grievances through concessions. The Hungarians on the other hand favoured repression and a policy known as **'Magyarisation'**. In the long run repression of minorities was no more successful than conciliation in quelling the spirit of nationalism.

In 1868 Deák had promised that all national minorities would be treated fairly. However, this did not happen. The Magyars dominated the parliament in Budapest and filled about 90 per cent of government posts. Magyar was the official language and every effort was made to prevent other languages from being used in official business.

A movement then developed to encourage unity among the Slavs in the south. The racial groups involved were the Croats, the Slovenes and the Serbs. The number of Serbs within the Empire increased when Austria-Hungary annexed Bosnia-Herzegovina in 1908. They looked to the kingdom of Serbia to unite all Serbs, half of whom were living in Austria-Hungary. The failure to solve the nationalities question was a key element in the lead up to World War I and the break-up of the Empire.

Foreign Policy

The main aspects of Austro-Hungarian foreign policy are examined in other chapters:

— The Seven Weeks War and Austria's exclusion from German affairs (chapter 1);
— Austria-Hungary's role in Bismarck's system of alliances, the *Dreikaiserbund*, the Dual Alliance and the Triple Alliance (chapter 2);
— Austria-Hungary's involvement in the Eastern Question (chapter 7);
— Relations between Austria-Hungary and the German Empire under William II (chapter 9);
— The part played by Austria-Hungary in the outbreak of World War I (chapter 10).

Court Ball in Vienna during the reign of Emperor Franz Joseph

The End of the Empire

Conflict with Russia

Austria-Hungary, mainly land-locked and without a great fleet, did not seek an overseas empire. She looked instead for expansion in the Balkans. This policy was known as *Drang nach Osten* (Drive to the East). This led to conflict with Russia (see chapter 7), the break-up of the *Dreikaiserbund* and the signing of the Dual Alliance with Germany.

Russia saw herself as protector of her fellow Slavs. The Serbs also looked to Russia in their dispute with Austria-Hungary. Many Austrians, including Archduke Franz Ferdinand favoured a 'Triple Monarchy' of Austrians, Hungarians and Slavs. His assassination brought the military alliance with Germany into play. The concept of the equality of Austrians, Hungarians and Czechs within the Empire did not survive the war.

A strong economy

For all its problems and divisions, it is easy to overestimate the weaknesses of the Empire in 1914. There was great economic strength, with Hungary's agricultural produce complementing Austria's industrial output. In spite of party disputes and the claims of nationalities, 'no party on either side seriously desired the break-up of Austria-Hungary' (Historian A.J.P. Taylor). It took four years of fierce fighting to bring the Dual Monarchy to its knees.

Aftermath

Austria-Hungary was beginning to crumble by October 1918. World War I ended on

The Glory of Imperial Vienna, 1890

11 November 1918. On the following day an **Austrian Republic** was proclaimed. The Versailles Treaty (chapter 12) did not allow Austria join with Germany.

The independent states of **Yugoslavia** and **Czechoslovakia** were created. The south Slav races which made up Yugoslavia – the Croats, Serbs and Ruthenians – had little in common and were often hostile to each other. The Czechs and the Slovaks were not natural allies. The new states were artificial creations which might in the future be endangered by racial tensions. Hungary became independent and lost a lot of territory. The Austro-Hungarian Empire was no more.

Italy

Domestic Affairs

The new state of Italy was proclaimed in 1861. Italian nationalists, led by the rebel, **Giuseppe Garibaldi**, and the politician, **Count Camillo Cavour**, created a united Italy from a collection of smaller states. The Pope would not give up the Papal states and refused to recognise the new state of Italy. In Rome, the Pope was protected by **Napoleon III** of France. In 1870 France went to war with Prussia (chapter 1) and Napoleon withdrew his troops from Rome. The Italian king, **Victor Emmanuel** II, a firm advocate of the policy of unification, entered Rome with his troops. Italian unification was now complete.

A parliament and King for a united Italy

The following year the first parliament of a united Italy met in Rome. Victor Emmanuel became a constitutional monarch. The vast majority of the citizens of the new state were very poor and did not have the right to vote. Politicians could therefore afford to ignore these people and paid little attention to their needs. Instead, they devoted much of their time to the policy of ***transformismo***. This meant that opponents were 'transformed' into supporters through bribery and political deals.

The rise in population and secret societies

In addition to the growing corruption, Italy also had a growing population. The state was too poor to support all the people, particularly in the impoverished south. Only emigration, mainly to the United States, allowed conditions to remain tolerable. Northern Italy was fertile and was becoming industrialised. This created a gulf with southern Italy where landless labourers lived in poverty and where life was dominated by secret societies and terrorist groups like the **Mafia**.

Despite these problems, progress was made in the fields of education, health and working conditions. The franchise, or right to vote, was extended. Socialists saw no hope of progress through peaceful means and became revolutionary. Attempts to ban socialist societies failed. They grew in strength both inside and outside parliament. Strikes were common, as in France.

Benito Mussolini

When the war broke out in 1914, Italian socialists demanded peace. They insisted that Italy should stay out of the conflict. One of their leaders was **Benito Mussolini**. However, he and a number of others changed their attitude after a few months. They demanded that Italy should go to war against Austria. After the war, Mussolini's new **fascist** party destroyed Italian socialism.

The Church

After the withdrawal of Napoleon III's forces from Rome in 1870, the Pope lost the Papal states and all his possessions except for the Vatican. **Pope Pius IX** withdrew to the Vatican and maintained constant opposition to the state. He instructed Catholics not to take part in Italian politics.

As the Pope refused to negotiate, the government brought in the **Law of Guarantees** in 1871. The Church was recognised as the owner of the Vatican, its role in education was guaranteed and an annual payment was offered to the Pope. Pius IX refused the offer.

In 1878 **Leo XIII** was elected Pope. He adopted a more conciliatory approach, but many problems remained. The government interfered in Church appointments and in the running of Church schools. Civil marriage was introduced. Many Catholics ignored the Pope's instruction and took part in politics. The ban on political involvement was not lifted until **Pope Pius X** withdrew it in 1905. He was worried about the growth of Italian communism and socialism. Relations improved and the **Lateran Treaty** of 1929 between Pope Pius X and Mussolini finally settled the dispute between Church and state.

Foreign Policy

'Italy has a big appetite but small teeth', declared Bismarck. Italy felt that Austria should surrender some border areas which the Italians had claimed for years. The Italians referred to these territories as ***Italia Irredenta*** (Unredeemed Italy). Italy, however, was not strong enough to force Austria to hand over these territories.

The Italians had colonial ambitions in North Africa. They suffered a major setback in 1896 when defeated by the Abyssinians at Adowa. Later they had better fortune in Tripoli and Libya, but never became a major colonial power.

In 1882 Italy had joined Germany and Austria-Hungary in the **Triple Alliance**. (see chapter 2) Her motive was mainly prestige, but it also made her feel more secure, as she was afraid of French ambition. Italy had no real commitment to the Triple Alliance and refused to fight when war broke out in 1914. In 1915 she joined in on the side of France and Britain, who promised that after the war Italy would be given the disputed territories of *Italia Irredenta*.

Chapter 9

Kaiser William II and Germany, 1888–1914

Profile of the Kaiser

Early life

William II was born in Berlin on 27 January 1859. He was grandson of Germany's first Kaiser, William I, eldest son of Princess Victoria and a grandson of Queen Victoria of Britain. He had a deformed left arm and was deaf in his left ear.

William overcame these handicaps quite well, but the effort required seems to have left him little time for formal education. He showed little interest in studying

Kaiser William II, Emperor of Germany

and one of his tutors said he believed that he knew everything without having to learn anything! One of his favourite activities was daydreaming about the military glory of the new Empire. He read magazines, studied war maps and collected photographs of his grandfather's coronation and other glorious events.

The future Emperor was only eleven years old when he took part in the parade through Berlin in 1871 to celebrate victory in the war with France and the creation of the new Empire. He retained a lifelong interest in military uniforms and parades.

William was very impetuous and often lost his temper, especially with his British cousins whom he often visited. His wise old grandmother, Queen Victoria, considered her grandson to be in a 'very unnatural and unhealthy state of mind' and feared that he might 'at any moment become impossible'.

The death of William I

In 1887 William's father, **Crown Prince Frederick**, developed throat cancer. The aged Emperor, William I, was growing weaker, and William was virtually deputy Emperor. He disliked his father, Frederick, and showed his insensitivity in 1887 when he considered the future:

> It is very questionable if a man who cannot speak has any right to become king.

William I died in March 1888 and was succeeded by Frederick. However, the new Emperor's cancer was incurable and he died in June the same year, having been Emperor for only ninety-nine days. Many historians feel that the new Emperor, William II, succeeded to the throne before his time. His lack of political and personal maturity combined with his arrogance and eccentric temperament led his country on the road to ruin.

Domestic Policies

Autocrat

In his first public speech as Kaiser, William II addressed himself not to the German people, but to the army:

> We belong to each other, I and the army. We were born for each other and will always remain loyal to each other, whether it be the will of God to send us calm or storm.

William believed in the divine right of kings and in 1891 he made this clear in an address to the Diet of Brandenburg, one of the states of the Empire:

> You realise that I regard my whole position as having been imposed on me from heaven.

Wilhelm II with his six sons in the streets of Berlin on New Year's Day, 1912

Relations with politicians

William viewed the Reichstag and all elected politicians with contempt. He once boasted that he had never read the constitution. He was particularly annoyed when the Reichstag failed to support his plans. He reminded its members that it was the army that had welded the Empire together.

William's Chancellors

Unlike his grandfather, William II was determined to show that he was the real ruler of Germany. We have seen (chapter 2) how this led to the fall of Bismarck who had virtually ruled the Empire single-handedly for twenty years. During the next twenty-seven years four men served as Imperial Chancellor: **Caprivi** (1890-94), **Hohenlohe** (1894-1900), **Bülow** (1900-09) and **Bethmann-Hollweg** (1909-17).

These men were generally well-meaning, but they did not possess the political skills of Bismarck. They were constantly looking over their shoulders at the Emperor. His wild statements and undiplomatic language made their lives very difficult. Only Caprivi really attempted to stand up to William. Unlike Bismarck, these Chancellors were unable to bully the Reichstag and they were caught in the middle between the Emperor and the parliament. It was usually when they failed to agree with the Emperor that they resigned or were forced to resign.

The growth of German Socialism

Caprivi introduced a number of bills into the Reichstag in line with William's 'new course' in German politics. These were welcomed by socialists and liberals. Working conditions were improved and a fair system of income tax was introduced. The reduction of tariffs on cattle and wheat reduced food prices and pleased the workers.

William soon tired of social reform. He was worried by the continuing growth of socialism at home and abroad. Social reform did not stop the forward march of the SPD, the German Socialist Party. Caprivi and William clashed when the Emperor wanted the introduction of a harsh anti-socialist bill. The Chancellor, convinced that the Reichstag would never agree to the measure, resigned. William vowed to continue the fight:

> Onward to battle for religion, for morality and order, against the forces of revolution.

Caprivi's successors as Chancellor tried to curb the rise of socialism. However, most of the bills which they brought forward on behalf of the Emperor were rejected by the Reichstag. Higher tariffs were introduced. This won the support of conservatives but alienated the socialists and most of the liberals. Despite the best efforts of Bülow, Bethmann-Hollweg and the Kaiser, the SPD continued to grow. They won 110 seats in the 1912 election and became the largest party in the Reichstag.

Economic Policies

Between 1890 and 1914 German economic progress continued at an impressive rate. The population continued to grow and a skilled workforce emerged. The country had an abundant supply of raw materials, especially of coal and iron. Banks promoted growth through credit to industry. State policies of protection and subsidies proved invaluable.

Coal production, especially in the Ruhr, trebled between 1890 and 1914. The production of steel was even more remarkable, with growth expanding ninefold from over 2 million tons to almost 19 million tons.

	Coal in millions of tons			**Steel in millions of tons**			**Railways in thousands of Kms**		
	Germany	*Britain*	*France*	*Germany*	*Britain*	*France*	*Germany*	*Britain*	*France*
1870	38	118	13	0.3	0.6	0.1	19	24	17
1890	89	184	26	2.2	3.6	0.7	43	33	36
1900	149	228	33	6.7	5.0	1.6	51	35	37
1913	279	292	41	18.9	7.8	4.7	61	38	50

Krupps steelworks, Essen

Much of Germany's growing steel production was used by the shipbuilding industry. The growth in the chemical, electricity and machine tool industries was also impressive. Names such as **Krupps**, **Siemens** and **AEG** in the engineering industry, **Daimler**, **Diesel** and **Benz** in the motor industry and **Zeppelin** in aviation became known throughout the world. By 1914 Germany's share of world trade equalled that of Britain and was twice that of France.

Foreign Policy

Germany and Russia

William promised a 'new course' in foreign affairs as well as in domestic issues. Bismarck had always sought to avoid a two-front war. Even after the collapse of the *Dreikaiserbund,* he had signed the Reinsurance Treaty with Russia. Caprivi and William took the advice that this treaty was not really compatible with the Triple Alliance of Germany, Austria-Hungary and Italy. They allowed the Reinsurance Treaty to lapse. All Bismarck's work was undone. He pointed out the folly of the 'new course' from his place of retirement.

William continued the policy of forbidding German banks to loan money to Russia. This was a kind of blackmail through which he hoped to influence Russia's foreign policy. Russia needed the capital for industrial development, especially for the Trans-Siberian Railway (see chapter 4). French bankers took the place of the Germans and a Franco-Russian friendship was born.

In 1891 Tsar Alexander III welcomed the French fleet to Kronstadt. The results of William's policy were seen in 1894 when the **Franco-Russian Alliance** was signed. Bismarck's old policy of isolating France was now totally in ruins. This was the beginning of the system of alliances which developed into the two armed camps which would come into conflict in 1914.

Weltpolitik

Weltpolitik was the policy of establishing Germany as a world power. During the 1890s the Kaiser began demanding his country's 'rightful place in the sun'. He saw the policy as a means of uniting all groups in the Empire. He called on workers and industrialists, on socialists and capitalists to rally behind the policy of colonial expansion.

The Pan-German League and the Colonial Union

New sources of raw materials, larger markets and bigger profits were put forward as reasons why Germany should pursue a more active colonial policy. William looked on colonies as a matter of prestige. The **Pan-German League** and the **German Colonial Union** were established.

Relations between Germany and Britain

During the 1890s the policy of *Weltpolitik*, along with the Kaiser's rash statements, led to some problems. In 1895 William was angered by the British raid, led by **Dr Jameson**, on the Boers of the Transvaal (see chapter 6). The Boers were of Dutch and German stock. William considered sending troops and establishing a German protectorate in the area. When the Jameson raid was repulsed, William sent a telegram to the Boer President, **Paul Kruger:**

> I send you my sincere congratulations that without calling on the help of friendly powers you and your people have succeeded by your own efforts in restoring peace in face of armed bands.

In Britain the telegram was seen as interference in Britain's affairs.

Germany continued acquiring naval bases and colonies in Africa, Asia and in the Pacific. This led to further problems with Russia and Britain. In 1897 Germany took the naval and coaling base in **Kiao-Chow**, China. In 1899 the Kaiser supported German construction of the **Berlin-Baghdad** railway. Britain was not pleased at the sight of Germany displacing her as an ally of the Turks.

William II's attitude to the navy

The policy of *Weltpolitik* led logically to naval as well as colonial expansion. William was aware that his country's fleet was no match for Britain's Royal Navy. During the Boer War he had reflected that a force of well-trained German troops could have pushed either the Boers or the British into the sea and given him the colonial Empire he craved. The trouble was that there had been no means of getting the soldiers to South Africa. William now believed that he had the man to build up a new German navy. His name was **Alfred von Tirpitz**.

Alfred von Tirpitz and the Navy League

After a successful career at sea, von Tirpitz became Secretary to the German Navy at the end of the nineteenth century. He felt that what Germany needed was warships. Von Tirpitz's strategy involved what was called the 'risk theory'. This meant that if Germany built up her fleet, it could be used as a diplomatic weapon. A superior fleet, such as that held by Britain, would come to terms with Germany rather than risk battle. Von Tirpitz was also concerned about the risk to Germany if Britain was provoked and advocated that she should avoid war while building the fleet.

William felt that he would gain Britain's respect if he had a strong navy. He loved naval titles and saw himself as 'Admiral of the Atlantic'. He supported von Tirpitz's **Navy League**, which had a membership of a million. This led to the propaganda campaign to convince the country and the Reichstag of the need for naval expansion. After much difficulty, the Reichstag was persuaded to vote for the monies required.

Alfred von Tirpitz, Admiral of the German Fleet

The start of the naval race

The first Naval Law of 1898 provided for a vast programme of shipbuilding. This included nineteen battleships together with a number of cruisers and torpedo boats. Germany became involved in a 'naval race' with Britain between 1898 and 1914. **Sir John Fisher**, British Lord of the Admiralty, decided that Britain needed a superior weapon. This came in 1906, when ***HMS Dreadnought*** came into service. She had more speed, greater firepower and an altogether better design than any previous warship. Once *Dreadnought* had been launched, everything else became obsolete. Germany decided that she too had to have dreadnoughts. By 1911 the armaments policy was taking up 90 per cent of the national budget.

International Crises, 1905-1911

1905: First Moroccan crisis

In 1904 Britain and France came to an agreement over their affairs in North Africa. Britain would turn a blind eye to France's determination to take over control of Morocco, and France would not interfere in British dealings in Egypt. Neither side had consulted William. He decided that, somehow, the German presence must be felt, if only to disrupt the growing understanding which seemed to be developing between Britain and France.

In 1905 the French gained more concessions from the Sultan of Morocco. The chancellor, Bülow, persuaded William to stop off at **Tangier** during a Mediterranean cruise in order to assure the Sultan of German goodwill and support for Moroccan independence. The Kaiser also told the French Consul in Tangier that Germany would defend her own interests in Morocco. He called for an international conference to discuss the future of Morocco. In so doing William hoped to drive a wedge between the British and the French.

The conference took place in January 1906 at **Algeriças** in Spain. France was supported by Britain, Russia, the USA and Italy. Germany could only rely on Austria-Hungary. The Algeriças Conference recognised French predominance in Morocco. Naval and military cooperation between the French and British against the Germans was discussed. Meanwhile, the Triple Alliance between Germany, Italy and Austria-Hungary was weakened by Italy's failure to support Germany. The Germans were drawn closer to their remaining ally, Austria-Hungary.

The crisis passed, but it left a sour taste. King **Edward VII** of Britain's words are an indication of British annoyance over the first Moroccan crisis. It was, he believed:

> the most mischievous and uncalled for event which the German Emperor has ever been engaged in since he came to the Throne.

Daily Telegraph interview

Britain would be very wary of the Kaiser in future years. The hostility was increased by an interview given by William in 1908. While it seems he was making a bid to win British sympathy, part of the article printed in the *Daily Telegraph* caused grave offence in Britain:

> You English are mad, mad as March hares... What has come over you that you are so completely given over to suspicion unworthy of a great nation?

1908: Bosnian crisis

In 1908, during what became known as the **Bosnian crisis**, Germany showed that she was prepared to back Austria-Hungary in a dispute with Russia. Bosnia was taken over by Austria-Hungary. Serbia objected and appealed to Russia. The Russians would not agree to the takeover of Bosnia unless her own warships were allowed through the Straits of the Dardanelles. Russia backed down when faced with Germany's support of Austria-Hungary.

When another dispute between Serbia and Austria-Hungary arose in 1914, the Russians were determined that they would make a stand rather than face another humiliation. Austria-Hungary felt that Germany would again show her support.

The German Gunboat "Panther" which arrived at Agadir on July 1st 1911 (p 103)

1911: Second Moroccan crisis

In 1911 the Moroccan problem flared up again. The French sent troops to **Fez,** capital city of Morocco. Germany pointed out that this violated the Algeriças Agreement of 1906 and demanded compensation. In July 1911 the German gunboat ***Panther*** arrived off the port of Agadir to 'protect' Germans living there. Some days later the German government announced that in return for recognising French control of Morocco it wanted the French Congo, a vast colony in Central Africa. **Lloyd George**, the British Chancellor of the Exchequer, issued a warning to the Germans and ordered the British fleet to prepare for action.

The Kaiser backed down and negotiated with France. He recognised French claims to Morocco in return for a large tract of waste land in the French Congo. William was even more unpopular at home after the *Panther* incident than he had been in 1906. 'Have we become a generation of women?' asked the influential newspaper, *Der Post*. The Kaiser was determined that he would never again be accused of failing to stand firm.

Questions

ORDINARY LEVEL – D

Answer the following questions briefly. One or two sentences will be enough.

1. Explain one reason why Bismarck provoked a war with France, 1870.
2. What was the Dreikaiserbund, 1872?
3. Describe one way in which Bismarck attempted to combat socialism in Germany.
4. How did the Panama Scandal affect the Third French Republic?
5. Why did Emile Zola write an open letter 'J'Accuse' (I accuse) to the President of the French Republic, 1898?
6. Explain briefly why Sunday, 22 January 1905, is known as 'Bloody Sunday' in Russian history.
7. Describe one important contribution made by Sergius Witte (1849-1915) to the Russian economy.
8. In connection with Russia in the early 20th century explain any two of the following: Bolshevik, Menshevik, Soviet, Iskra, Duma.
9. What was the aim of the Suffragette Movement?
10. Why did Disraeli purchase a major shareholding in the Suez Canal, 1875?
11. Give reasons why the Great Powers of Europe became involved in the scramble for colonies in the late 19th and early 20th centuries.
12. What was the Fashoda incident?
13. Explain briefly what is meant by the Eastern Question?
14. Explain one reason why the Congress of Berlin, 1878, weakened Russian friendship with Germany.
15. What, in your opinion, was the most important problem facing Austria-Hungary in the period 1867-1914?
16. Explain one major problem confronting Italy after unification?
17. Give one reason why Great Britain fell behind Germany as an industrial power in the period 1870-1914.
18. Explain one problem facing the Turkish Empire in 1870.
19. How did the Boer War affect Anglo-German relations?
20. Why did Great Britain build the Dreadnought which was launched in 1906?

Questions

ORDINARY LEVEL – E

Write a short paragraph on each of the following:

1. Why France lost the Franco-Prussian War, 1870-71
2. The Kulturkampf
3. The Boulanger Affair
4. Nicholas II, Tsar of Russia
5. The Boer War, 1889-1902
6. The Scramble for Africa
7. The problems affecting the Austro-Hungarian Empire during the period 1870-1914
8. The arms race between Great Britain and Germany up to World War I
9. Reform legislation in Victorian Britain
10. Kaiser William II

ORDINARY LEVEL – F

Write an essay on each of the following:

1. The Franco-Prussian War under each of the following headings:
   (i) Circumstances and events leading up to the war
   (ii) The military events and reasons for the defeat of the French army
   (iii) The peace settlement
2. France, 1870-1914 under each of the following headings:
   (i) Why France lost the Franco-Prussian War
   (ii) Relations between the Catholic Church and the State
   (iii) Political scandals
   (iv) Cultural developments
3. The Eastern Question
4. Europeans and Africa, 1870-1914, under each of the following headings:
   (i) Why Europeans were drawn to Africa
   (ii) The European countries involved and the colonies acquired
   (iii) The Berlin Conference, 1884-1885
   (iv) How the coming of Europeans affected the native peoples of Africa

Questions

HIGHER LEVEL

Write an essay on each of the following:

1 Account for the ease of the Prussian victory in the Franco-Prussian War, 1870-71. (80)

2 Discuss the domestic policies pursued by Bismarck as Chancellor of the German Empire, 1871-1890. (80)

3 'Bismarck's diplomatic juggling was triumphantly successful.'
Discuss. (80)

4 'Although France was a deeply-divided country between 1870 and 1914, the period saw considerable achievement at home and expansion overseas.'
Discuss. (80)

5 Compare and contrast the 1905 and 1917 revolutions in Russia under such headings as the following: background, revolutionary groups, role of the Tsar, international involvement, outcomes etc. (80)

6 Treat of the part played by Lenin in Russian history. (80)

7 Treat of social and political reform in Britain during the period 1870-1914. (80)

8 Consider the extent to which European colonisation in Africa was responsible for international tension in the period 1870-1914. (80)

9 Account for the rapid rate of European expansion overseas in the period 1870-1914. (60)
Assess the impact of that expansion on Europe. (20)

10 'The greatest political problem facing the Habsburg Empire during the period 1870-1914 was the nationalities question.'
Discuss. (80)

11 Show how the recurring crises in the Balkans became a major cause of World War I.

12 'Kaiser William II of Germany, 1888-1918, by his personality and actions, contributed to the outbreak of World War I.'
Discuss. (80)

SECTION TWO

1914-1924: WAR AND PEACE

Chapter 10

Origins of World War I

Introduction

Germany blamed

World War I ranks second only to World War II as the most bloody and most costly war in modern history. Two pistol shots in the streets of **Sarajevo** on 28 June 1914 signalled the start of the war, a conflict which did not end until an armistice was signed on 11 November 1918.

Article 231 of the **Versailles Treaty** of 1919 blamed Germany for causing the war and demanded her admission of war-guilt. The nation was gravely offended by the suggestion that it alone caused the war. Writers from Germany and from other countries soon pointed the finger of guilt at others.

International tensions

The horrors of war convinced people that such a catastrophe must never occur again. President Wilson of America based his peace plan in 1918 on the reasoning that if only the causes of the conflict could be discovered, remedies could be applied and such a widespread war would not happen in future.

The published literature on the causes of the First World War is immense. We have already seen various tensions arise between states in our study of France, Germany, Britain, Russia and Austria-Hungary from 1870 to 1914. However, like President Wilson, we will understand the conflict better if we can isolate the origins and see how individually and collectively they led to war.

Summary of causes

Most writers have fixed the blame for the war on the system of alliances, on militarism, colonial and economic rivalry, nationalism, domestic problems and the public attitude to war. These were highly sensitive and emotional issues; all that was needed was a spark to light the fire. That spark was provided by the pistol shots in Sarajevo in June 1914, which ignited the 'powder-keg' of nationalism in the Balkans.

System of Alliances

The Dual Alliance

The division of Europe's major powers into two armed camps was the result of alliances and understandings made in the years following the Franco-Prussian war.

SYSTEM OF ALLIANCES, WORLD WAR I

Germany's attitude was summed up by **von Moltke**, the German commander, after the war:

> What we gained by war in six months, we must protect by arms for fifty years, unless it is to be torn from us again.

Bismarck was not afraid of France if she stood alone, but he did fear the development of a coalition of powers that would encircle Germany, forcing her to fight a two-front war. As a result, he did all he could to keep France isolated. He allied himself with Austria-Hungary and Russia. However, these two powers clashed over their interests in the Balkans.

Bismarck then had to choose between Austria-Hungary and Russia. He selected Austria-Hungary, primarily because he could exert a more direct influence over events there. In 1879 a defensive alliance was signed between Germany and Austria-Hungary. It was known as the **Dual Alliance**, but its terms were not published until 1888. By the terms of the agreement, each partner promised to help the other if attacked by Russia or another power supported by Russia. This alliance was regularly renewed and was still in force in 1914. It is beyond dispute that the Dual Alliance emboldened Austria-Hungary, whose attitude often became dangerously reckless.

The Triple Alliance

In 1882 Italy was in dispute with France over French actions in Tunisia. In May of that year she joined Germany and Austria-Hungary in what became known as the **Triple Alliance**. By the terms of this agreement, Germany and Austria-Hungary agreed to support Italy if she were attacked by France. In return, Italy promised to assist Germany in the event of a French attack.

The Franco-Russian Alliance

The drawing together of France and Russia began with a French loan to the Russians in 1888. Further loans were made over the years right up to 1914. Both countries felt threatened by the Dual Alliance. In 1891 France and Russia pledged to consult with each other if there was a threat to peace.

A formal Franco-Russian alliance in 1893 declared that each would come to the other's help if attacked by Germany. If any member of the Triple Alliance mobilised, France and Russia would mobilise immediately. The members of the Triple Alliance did not believe that the agreement between France and Russia would endure.

The *Entente Cordiale*

The Boer War had given Britain a severe shock. She realised that her army was weak by European mainland standards. She would have to give up her policy of 'splendid isolation' and find an ally. Germany's naval expansion and anti-British attitudes made it natural for the British to turn to the French. **King Edward VII** loved France. His visits to Paris led to the signing of the ***Entente Cordiale*** in 1904. The two countries drew closer together between 1904 and 1914. Their generals attended conferences together and drew up plans of campaign. However, a full military alliance was made only after World War I began.

The Triple *Entente*

It seemed unlikely that Britain would ever tie herself to Russia, France's partner in the 1893 alliance. British suspicions of Russian moves in the Balkans made matters difficult. But both countries feared Germany's growing influence in Turkey and the threat to their interests in Persia. Encouraged by France, Britain and Russia resolved their colonial difficulties. This enabled the three countries to come to an understanding known as the **Triple *Entente***, in August 1907.

Secrecy

One of the aspects of European political policy criticised by President Wilson after World War I was the existence of secret agreements. The Dual Alliance of 1879 was

"ALL'S WELL!"
British Lion and Russian Bear (together). 'What a pity we didn't know each other before!' December 1894

not published until 1888. And although Italy signed the Triple Alliance in 1882, her continuing dispute with Austria over the South Tyrol and Dalmatia led to her secret decision in 1900, not to join in an aggressive war against France. Nor did Italy join the conflict with her allies Germany and Austria-Hungary in 1914.

Britain was also inclined to conduct her diplomacy in great secrecy. She was almost exclusively concerned with her own interests. Her Foreign Secretary, **Sir Edward Grey**, never made it clear that Britain intended supporting France, or intervening if Belgium was invaded. As a result, the Germans thought that they could exclude the British from their military calculations.

The secrecy that surrounded diplomacy and agreements led to suspicion and fear. It was this fear and suspicion that resulted in Lloyd George's claim that countries 'stumbled and staggered' into World War I.

Militarism

The Arms Race

After 1900 army general staffs were involved more and more in political decisions. They were able to persuade their governments to spend ever-increasing amounts on armaments. The expenditure on armaments made by the great powers rose steadily. Between 1900 and 1910, Germany increased its army budget by 20 per cent, Russia by 65 per cent, Italy by 50 per cent, France by 30 per cent, Austria-Hungary by 25 per cent and Britain by 30 per cent. After 1910 these war preparations became even more feverish, with military expenditure by Germany and Austria-Hungary doubling between 1910 and 1914.

From the late nineteenth century, European armies grew enormously, not so much in terms of the standing armies, but in terms of reserves of trained men. These reserves were established by the introduction of compulsory military training for all able-bodied men. They were conscripted into the army for a period of regular service and then joined reserve units. By developing this system of reserves, the European powers would be able to call on an additional 10 million men in the event of war.

When the drums began to roll in 1914, the European powers could mobilise large well-trained and well-equipped armies. When reservists were included, France and Germany were each able to put about 3.5 million men into the field and Russia about 4 million.

Growth in Armaments
(Cost in dollars per head of population)

	Britain	Germany	France	Russia	Austria-Hungary
1890	4.03	2.95	4.82	1.32	1.56
1910	7.56	4.17	6.70	1.91	1.77
1914	8.53	8.52	7.33	2.58	3.48

Source: Quincey Wright, *A Study of War*

The Naval race

In 1889 the British government began to modernise its navy. Britain believed in the importance of a strong navy to an island nation with a far-flung empire. She set down the principle that the Royal Navy should possess more battleships and cruisers than the next two largest navies in the world together. This became known as the 'two-power standard'.

German activity

What concerned Britain most was the decision taken by Germany in 1896 to build a high-seas fleet. The architect of the new German navy was **Admiral von Tirpitz**. He aimed to build a fleet that would be about two thirds the size of the Royal Navy. His strategy was to establish a German navy that would be large enough to deter the British in wartime, because any attempt to destroy it would be too costly for Britain. William II stated that 'our future lies on the water'. The German naval laws of 1898 and 1900 provided finance to double the number of battleships and expand the rest of the fleet considerably.

The arrival of the *Dreadnought*

To meet the German naval challenge, **Admiral Sir John Fisher** began to reorganise and expand the British fleet. A new class of battleship, *HMS Dreadnought* was launched in 1906. She was the first 'all big gun' ship and made all previous battleships obsolete. Then Germany launched her first *Dreadnought*-class ship in 1907. The naval race intensified in 1908 thanks to a German naval law which provided for four new *Dreadnoughts* to be built each year between 1908 and 1911.

The reaction in Britain

British public opinion demanded that the government step up its own *Dreadnought* construction programme. 'We want eight and we won't wait' was the slogan of the day. In 1912 Germany demanded British neutrality in Europe as the price for a truce on naval expansion and negotiations broke down. The challenge to British naval supremacy from Germany was one of the main reasons why Britain allied herself with France and Russia. The British viewpoint is best summed up in the words of Winston Churchill who became First Lord of the Admiralty in 1911:

> The British navy is to us a necessity and, from some points of view, the German navy is to them more in the nature of a luxury. Our naval power involves British existence. It is existence to us; it is expansion to them.

HMS Dreadnought, 1907

War plans

An important factor in the outbreak of war in 1914 was that military plans and joint plans of campaign became part of the alliances. So in many ways the politicians were to a large degree at the mercy of the generals.

The situation in Germany

In Germany the army and navy were free from the authority of the government and were responsible to William II alone. The Kaiser had long feared the awful prospect of having to wage war simultaneously against France in the west and Russia in the east. As early as 1905, **Count Alfred von Schlieffen** had devised a plan to cope with such a situation. His intention was to quickly destroy one enemy (France) and then speedily transfer the German armies to deal with the other (Russia).

Plan XVII

After the *Entente Cordiale* of 1904, British and French military authorities consulted frequently on possible joint action against Germany. **Marshal Joffre**, the French chief-of-staff, drew up **Plan XVII**. This involved a French advance into Germany supported by a British expeditionary force on the left-wing. It also relied on a Russian attack in the east.

Russia and Austria-Hungary had also drawn up their war plans. It should be noted that all these plans involved rapid mobilisation of huge armies. Once the plans were put into action, they could not be stopped without the risk of almost certain chaos. In 1914, nobody, politician or general, would take that risk.

Nationalism and Rivalry

Introduction

The most far-reaching development in the nineteenth century was the growth of nationalism. This was the belief that people linked by such factors as language, religion, heritage or territory should form self-governing states. In 1914 there were various national groups which had not gained full unity and independence. The attempts by such groups in the Balkans to set up nation states were a major cause of the war.

The growth of nationalism

The growth of nationalism led to rivalry between states. This was a source of commercial, colonial and territorial conflict. A number of international incidents took place in the early years of the twentieth century; they caused great tension and could have led to an international war. The situation was not helped by William II of Germany who saw Germany and Britain as the great rivals in terms of European and world power.

Commercial rivalry

Economic Development – Germany and the Great Powers

Population in millions					
	Germany	Britain	France	U.S.A.	Russia
1870	41	32	37	40	87
1890	49	38	38	63	111
1900	56	42	39	76	130
1913	67	46	40	95	174

Coal in millions of tons					
	Germany	Britain	France	U.S.A.	Russia
1870	38	118	13	37	1
1890	89	184	26	150	6
1900	149	228	33	244	16
1913	279	292	41	517	29

Steel in millions of tons				
	Germany	Britain	France	U.S.A.
1870	0.3	0.6	0.1	0.1
1890	2.2	3.6	0.7	4.3
1900	6.7	5.0	1.6	10.4
1913	18.9	7.8	4.7	31.8

Railways in thousands of kilometres				
	Germany	Britain	France	U.S.A.
1870	19	24	17	90
1890	43	33	36	250
1900	51	35	37	309
1913	61	38	50	384

An examination of these charts shows the rapid growth of the German economy between 1870 and 1913. Although Britain and France also developed, they were envious of the Germans' great successes. In 1916 Lenin wrote that World War I was the result of the clash between the capitalist interests of rival states. Not all historians agree with this analysis. However, industries like **Krupps** in Germany, **Schneider-Creusot** in France, **Skoda** in Austria-Hungary and **Vickers** in Britain profited enormously from the build-up of arms prior to 1914. They could not be said to have directly caused the war, but they formed powerful pressure groups which resisted any attempt to limit the manufacture of weapons.

Colonial rivalry

Some historians have argued that as long as there were colonial conflicts between the powers, attention was diverted from Europe. There is no doubt that the Kaiser was envious of the British Empire and determined to build a new German Empire. This ambition, along with the commercial rivalry that existed between these two powers, led to the naval race. In his interview with the British newspaper, the *Daily Telegraph*, in 1908, William set forth his views:

> Germany is a young and growing empire. She has a worldwide commerce which is rapidly expanding, and to which the legitimate ambition of patriotic Germans refuses to assign any bounds. Germany must have a powerful fleet to protect that commerce and her manifold interests in even the most distant seas.

Churchill, First Lord of the Admiralty

Churchill's view of Britain's position in 1911 also showed the possibility of conflict in the future:

> The whole fortunes of our race and empire, the whole treasure accumulated during so many centuries of sacrifice and achievement would perish and be swept utterly away if our naval supremacy were to be impaired.

Britain and Germany

Britain resented William's support for the Boers during the Boer War. It seemed that in the early years of the twentieth century, the Kaiser was prepared to test the strength of Anglo-French friendship by meddling in North Africa. This led to some colonial disputes.

1906: the Algeçiras Conference

It was William II's visit to Tangier in 1905 which caused the **First Moroccan Crisis** (see chapter 9). This was a direct challenge to French interests in Morocco. At the Algeçiras Conference in 1906 only its ally, Austria-Hungary, supported Germany. The outcome of the conference resulted in the strengthening of the understanding between Britain and France, and the drawing up of the **Schlieffen Plan**, the plan adopted at the start of World War I, by the Germans.

1911: the *Entente* strengthened

In 1911 France occupied large parts of Morocco. Germany sent the gunboat *Panther* to Agadir (see chapter 9). Britain supported France and Germany backed down. The ***Entente*** was further strengthened, but Europe moved nearer to war. The Germans were determined that they would not be humiliated again.

Balkan nationalism

The war that broke out between Austria and Serbia in 1914 was a direct result of Balkan nationalism. The provinces of **Bosnia** and **Herzegovina** had been taken over by Austria-Hungary in 1908. The **Serbs** wanted these provinces to become part of a united Serb state called **Yugoslavia**. They called on Russia for support.

Russian reaction

Russia saw herself as protector of all the Slav races, including the Serbs. Germany warned Russia that it would not restrain its ally, Austria-Hungary, from attacking **Serbia**. Russia was not prepared for war and urged Serbia to give in to Austro-Hungarian demands.

Austria-Hungary's attitude to Serbia

Austria-Hungary regarded Serb nationalism as a threat to the continuing existence of its empire. The collapse of the Turkish Empire left a power vacuum in the Balkans.

Both Russia and Austria-Hungary were willing to fill the vacuum. This created a state of permanent tension between the two empires. The Serbs and other Slav races in the Balkans hoped to use this conflict for their own purposes. Austria-Hungary was determined to crush Serbia to prevent this happening. The opportunity to attack the Serbs arose in 1914. The Russians had stood back during the Bosnian crisis in 1908, but by 1914 they were prepared to take action.

Domestic Issues

Political aspects

Some modern historians argue that domestic issues were just as important in determining the actions of the various states in 1914 as the foreign policy issues which have been traditionally blamed for the war. Socialism had caused turmoil in countries like Britain, France and Germany, resulting in social upheaval and strikes. But as the reaction to the outbreak of the war showed, patriotism was stronger than class differences.

Domestic disputes disappear

In Britain the threat of civil war in Ireland and the suffragette agitation were temporarily solved by war. In France the crises over income tax and the length of military service became less relevant once the war broke out. In Russia, the war seemed to rally the nation behind the Tsar. The government thought that a 'short victorious war' would divert the people's attention from their domestic problems and boost the prestige of the autocracy.

Attitudes to the war

The various crises and small wars that occurred in the late nineteenth and early twentieth century led to a high level of jingoism and chauvinism. This in turn led to a desire for war, or at least a feeling that war was inevitable.

Most people thought that the war would not last long, that 'it'll all be over by Christmas.' Some thought that it would be a great adventure. Others were idealistic and thought of war in terms of blood sacrifice. A great wave of patriotism and a spirit of national pride sent millions of young men to fight and die 'for King and Country' in 1914.

The Events of 1914: Immediate cause

28 June 1914: assassination in Sarajevo

On the morning of Sunday 28 June 1914, the heir to the Austro-Hungarian throne, **Archduke Franz Ferdinand** and his wife, **Sophie**, were assassinated in **Sarajevo**. The

Assassination of Archduke Franz Ferdinand and his wife, Sophie, Sarajevo, 28 June 1914

assassin was **Gavrilo Princip**, a young Bosnian student and a member of the group **Black Hand**, a secret Serbian organisation working for Slav independence.

Archduke Franz Ferdinand

Archduke Franz Ferdinand was the nephew of **Emperor Franz Joseph of Austria-Hungary**. As heir to the throne, he was seen as the enemy of Serbia and of south-Slav nationalists. Serbia's national day was 28 June. The visit on that date was seen as a message to Slav nationalists that Bosnia and Herzegovina would remain part of the empire. The 'Black Hand' saw it as an opportunity for making a gesture of revenge and defiance.

The assassin, the nineteen-year-old Princip, was quickly arrested and dragged to the police station for questioning. Soon after, his accomplices were also captured. The assassins claimed that they acted alone and without the knowledge of the Serbian government.

Reaction in Vienna

The assassination in Sarajevo gave Austria-Hungary the opportunity to move against Serbia and to resolve its Balkan problems once and for all. The government decided to hold Serbia responsible for the Archduke's murder. The military leaders in Vienna pressed for immediate action. The politicians wished to secure German support before any action was taken. This would provide protection in case Russia came to the aid of Serbia. The Emperor, Franz Joseph, wrote to the Kaiser:

> Serbia must be eliminated as a political factor in the Balkans...Friendly settlement is no longer to be thought of.

On 6 July the Kaiser assured Franz Joseph of German support if Russia intervened. This gave Austria-Hungary a 'blank cheque' and allowed her to take a bold role in the critical early stages of the crisis.

Ultimatum to Serbia

On 23 July the government in Vienna presented the Serbs with an ultimatum. It contained ten demands and was a strongly-worded document. Nevertheless, the Serbian government accepted all points except the one which required them 'to open a judicial enquiry against those implicated in the murder, and to allow delegates of Austria-Hungary to take part in this'. The Serbs maintained that this would mean a violation of their independence and their constitution. They were willing to refer the matter to the Hague Court. The Germans advised Austria-Hungary to accept the Serbian reply and to negotiate a settlement. Even the Kaiser was impressed and declared that because of the positive Serbian reaction 'every cause for war falls to the ground'.

28 July: Austria-Hungary declares war on Serbia

On 26 July the Tsar ordered partial mobilisation of the Russian forces on the Austro-Hungarian border. As Serbian acceptance of the Austro-Hungarian ultimatum was not total, the Austrians took the action they had planned from the beginning. On 28 July they broke off diplomatic relations with Serbia and declared war. On 29 July the Serbian capital, Belgrade, was bombarded by artillery.

Mobilisation of Russian troops

The position of Russia became crucial. It now seems that French politicians promised Russia full support in the event of war. Added to this was the encouragement given to the Russian military chiefs by the French general staff, especially **General Joffre**. It appears that at this stage there was an amount of diplomatic blundering in Paris and St Petersburg. On the evening of 30 July the Tsar, after much hesitation and under pressure from the military, ordered full mobilisation.

Germany's demands

Germany now demanded that Russia cease her mobilisation within twenty-four hours. She also demanded that France announce her neutrality and that she hand her most important fortresses over to Germany as a sign of goodwill. The Russians failed to reply and France refused to comply.

1 August: Germany declares war on Russia

A hesitant Kaiser and his chancellor, **Bethmann-Hollweg**, had to contend with militant politicians and army chiefs like **General von Moltke**. The generals felt that any further delays would reduce the chances of a quick victory over the *Entente* powers. On 1 August Germany declared war on Russia. On the same day both France and Germany ordered general mobilisation.

3 August: Germany declares war on France

On 2 August the German army occupied Luxembourg. Germany next demanded that the Belgian government permit the free passage of the German army across Belgium into northern France. Belgium was a small country whose neutrality had been guaranteed by Germany, France and Britain in a treaty signed in 1839. Belgium rejected the German demand on 3 August. On the same day Germany declared war on France and crossed the Belgian frontier.

Vague reaction in Britain

The British cabinet was divided over the question of a possible involvement in a European war. As a result, **Sir Edward Grey**, the British Foreign Secretary, sent quite vague messages to France and Germany. The Germans were informed that British neutrality could not be taken for granted, but they were not told that Britain would go to war to aid her allies. At the same time, Grey told France that British participation on the side of France could not be assumed.

Reasons for British reaction

The British government was not willing to back Russia in the Balkans or to automatically aid France. Britain was only prepared to act when her own interests were clearly threatened. British interests lay in the security of the English Channel. Thus the fate of Belgium and of the northern coast of France became a matter of British concern, purely because it threatened British security. The issue which decided British involvement in the conflict – Belgian neutrality – was important to Britain not because German occupation of Belgium would deprive that country of its independence, but because German occupation of Belgium placed British security in danger.

4 August: Britain and Germany on the brink of war

The Germans scathingly referred to the Treaty of London, which guaranteed Belgian neutrality, as a 'scrap of paper'. The Kaiser was confident that the British would not

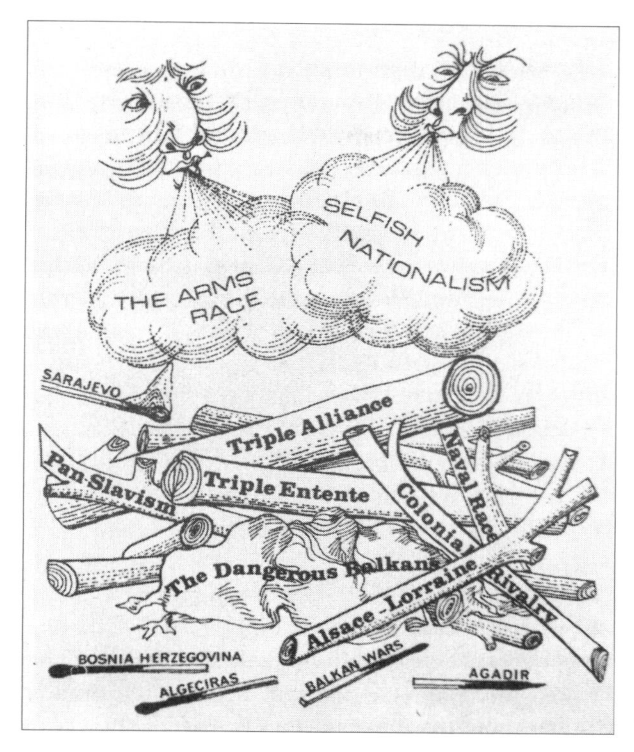

Causes of World War I: the bonfire

honour it. Sir Edward Grey demanded the withdrawal of German troops from Belgian soil by midnight on 4 August. The hour passed and Britain was at war with Germany.

Responsibility

Who was responsible for the outbreak of war?

And so we return to the view that Germany was responsible for the war. An examination of the issues involved has shown that this was a simplification and therefore unfair. Austria-Hungary, Russia, Britain and France were all concerned with their own selfish interests in 1914.

Germany's role in the affair

Germany must bear responsibility for a very hostile attitude towards Britain. The Kaiser seemed envious of Britain's empire and was inclined to make rash, undiplomatic comments. Germany's challenge to Britain's supremacy at sea was always dangerous. The Kaiser's support in 1914 gave the Austrians the confidence to go to war. In addition, Germany's decision to invade Belgium and attack France are clear evidence of German aggression.

The part played by Austria and Russia

The Austrians sent a set of demands to Serbia which that country could not accept. The government sought revenge for the assassination of Franz Ferdinand in Sarajevo and hoped that a victorious war in the Balkans would revive the fortunes of the Habsburg monarchy. The Russians also hoped for a short victorious war to help the Romanov dynasty. They promised unlimited support to Serbia and had their own agenda in the Balkans. They were first to mobilise and this spurred other powers to do likewise.

French political and military ambitions

The appointment of **Raymond Poincaré** as French President in 1913 marked a fresh revival of French political and military ambitions. France was ready once again for *'La Révanche'*, the war of revenge to recover the lost provinces of Alsace and Lorraine. The visit of the President and Prime Minister to Russia in July 1914 did nothing to moderate Russian support for the Serbs. Indeed, many historians argue that the visit 'fanned the flames' and was partly responsible for provoking the Germans and Austrians.

British self-interest

As we have seen, Britain was mainly concerned with her own selfish interests. The ***Entente Cordiale*** was not a formal agreement and it did not demand automatic British involvement in a war on the French or Russian side. The British cabinet was divided and Grey did not make the British position clear until her interests were threatened. It was then too late to stop the spread of the conflict.

Conclusion

So all the major powers must take a portion of the blame, some, perhaps, more than others. As well as the attitude of the politicians, the war also had its origins in the

German soldiers in Berlin marching off to war

system of alliances, the militarism, nationalism and old rivalries, commercial and colonial.

The popular attitude to the war was clearly visible on the streets of London, Paris, Berlin and Vienna during July and August 1914. Crowds cheered and jostled as they queued at the recruiting centres. It all seemed a big adventure. However, Sir Edward Grey was not fooled. When, on 4 August 1914, Britain and Germany were about to enter the war, he spoke the well-known words:

> The lamps are going out all over Europe, and we shall not see them lit again in our lifetime.

Chapter 11

World War I

Introduction: Total War

World War I was entered into casually and with light hearts by some, but with dreadful fears by others. The Kaiser said his army would be home 'before the leaves fall'. However, a few, like **Lord Kitchener**, British Secretary of State for War, estimated that the war would last four years. There had been longer wars in the past, but never such a total war as this.

World War I was fought on four continents. Millions of people, civilians, as well as soldiers, sailors and airmen, lost their lives in the conflict. New weapons of war destroyed homes, factories and churches. For the first time in history, aerial and submarine warfare were used. The countries involved committed all their resources, human, industrial and agricultural, to the achievement of victory. Thirty million

British War Posters

people are estimated to have lost their lives, directly or indirectly, as a result. This was total war and became known as the Great War.

Although the war was being fought simultaneously on land, at sea and in the air, it will be understood more easily if we examine each aspect separately. The struggle was dominated by trench warfare and this feature merits detailed consideration.

World War I has also been called the war of artists, poets and writers. Examination of their output will convey something of the hopes and the horrors of this terrible war.

Nineteen-Fourteen: Land War

Schlieffen Plan fails

The Schlieffen Plan (see chapter 10) worked on the theory that the ponderous Russian army would take at least six weeks to mobilise fully. Because of this it was thought that Germany would need only one eighth of its army in the east, while the main army would achieve a quick victory against France. The Kaiser summed up the Schlieffen Plan neatly, if somewhat over-ambitiously, in his declaration: 'Lunch in Paris, dinner in St Petersburg'.

In August 1914, five German armies hurtled through Belgium and into northern France. They met unexpected resistance from the Belgians who held out in such fortresses as **Liège** and **Namur**. The French army, as Schlieffen had correctly assumed, attacked the German frontier but was easily contained. Schlieffen had planned that the main army would quickly sweep through northern France, proceed south and east past Paris and attack the French army in the capital from the rear. The plan failed for a number of reasons.

THE SCHLIEFFEN PLAN

THE SCHLIEFFEN PLAN IN PRACTICE

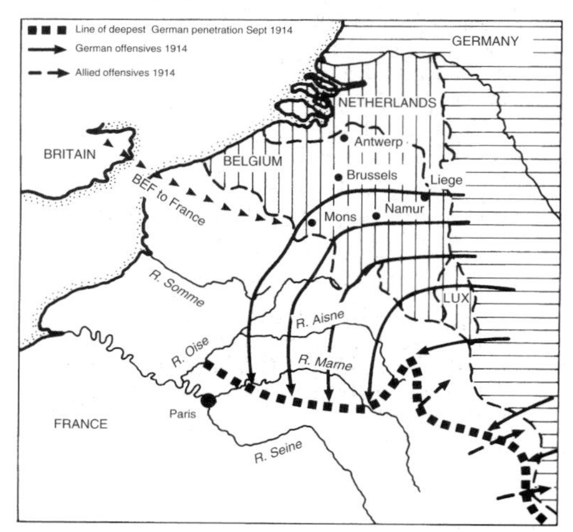

The Schlieffen Plan was brilliant because it was so daring. 'Keep the right-wing strong' said Schlieffen, and 'let the last man on the right brush the Channel with his sleeve'. **General Helmuth von Moltke**, who succeeded Schlieffen, was not very daring. He had altered the plan by taking troops from the right-wing and strengthening the forces on the left-wing who were defending the Franco-German border. He also switched two corps from France to meet the Russian forces who had mobilised more quickly than expected and were invading East Prussia.

The Western Front

In the meantime, a **British Expeditionary Force**, under **Sir John French**, had crossed the Channel. The Schlieffen Plan had presumed that Britain would be neutral. The British Expeditionary Force linked up with the French. They were able to delay but not halt the German advance. By 2 September one German army was close enough to Paris to see the Eiffel Tower. However, they had met with unexpected resistance, were nearing exhaustion and their supply lines were breaking down.

The Battle of the Marne

With his army exhausted and reduced in strength, von Moltke decided that it would be very difficult to pass to the west of Paris. He decided to depart once more from Schlieffen's original plan and to pursue the retreating French to the east of the city. This was a costly blunder. **General Galieni**, the military governor of Paris, assembled reservists and sent them to the front in taxis, a force which became famous as the 'taxis of the Marne'.

Trench positions

The French Commander-in-Chief, **General Joffre**, now launched a counter-attack. The French troops, assisted by a small British army turned about and forced the exhausted German army to retreat from the River Marne to the **River Aisne**. There the Germans held their ground. They dug a defensive line from which they could not be moved. They took up the positions which they were to hold, with few changes, for the remainder of the war. Both sides 'dug in' and the trench lines on the Western Front stretched from Switzerland to the English Channel.

Importance of the Battle of the Marne

Six weeks after the war had commenced the Germans had failed in their objective of achieving a quick victory. Their strategy and tactics had failed. The **Battle of the Marne** was decisive because it forced the Germans into a two-front war. After the battle, von Moltke is supposed to have told the Kaiser 'Germany has lost the war'. He was relieved of his command and replaced by **General von Falkenhayn**.

Race to the sea

The Allies and the Germans now turned their attention to the Channel ports. In a 'race to the sea,' the two sides tried to outflank one another. Both sides suffered enormous casualties during the **First Battle of Ypres**. In this battle, the Belgian town was reduced to rubble. The Germans continued to hold on to most of the Belgian coast, but the Allies retained the vital ports of **Dunkirk**, **Calais** and **Boulogne**. Both sides dug trenches along a line stretching from Belgium to Switzerland (see page 133). It took almost four years before a decisive breakthrough was made by either side.

The Eastern Front

In the decade before the war, the French had lent the Russians money for the purpose of building railways which led to the German frontier. The Russians were, accordingly, able to mobilise quickly and were able to mount a major offensive within two-and-a-half weeks.

Von Hindenburg and von Ludendorff

In mid-August the Russian armies invaded East Prussia and threatened the capital, Konigsberg. The Germans fell back and called **General Paul von Hindenburg** out of retirement. He appointed **General Eric von Ludendorff** as his Chief-of-Staff. These two men were soon to dominate German military campaigns and from then on became virtual masters of Germany while she was at war.

August 1914: Battle of Tannenberg

Hindenburg saw that the Russian army had split into two and was separated by the Masurian Lakes. He left a small covering force to hold the northern army, and concentrated his attack on the Russian forces to the south of the Lakes. In late August, at the **Battle of Tannenberg**, the Russians were badly defeated in a decisive battle. More than two-thirds of the Russian troops were killed or captured, together with vast quantities of guns and munitions.

September 1914: Battle of the Masurian Lakes

Hindenburg now turned his attention to the Russian forces on the other side of the Lakes. The resulting battle, known as the **Battle of the Masurian Lakes**, was a German victory, but not the rout that Tannenberg had been. The Russians retreated and were never able to set foot on Prussian soil again during the war. Their campaign, which had cost so much in terms of lives and weapons, probably saved France from defeat.

Nineteen-fifteen: Land War

The Western Front

During 1915 the French and British made a number of attempts at breaking through the German lines. They gained a few miles at **Neuve-Chapelle,** but casualties were very heavy. The Germans made another attempt to break through at the **Second**

WORLD WAR I: THE EASTERN & BALKAN FRONTS

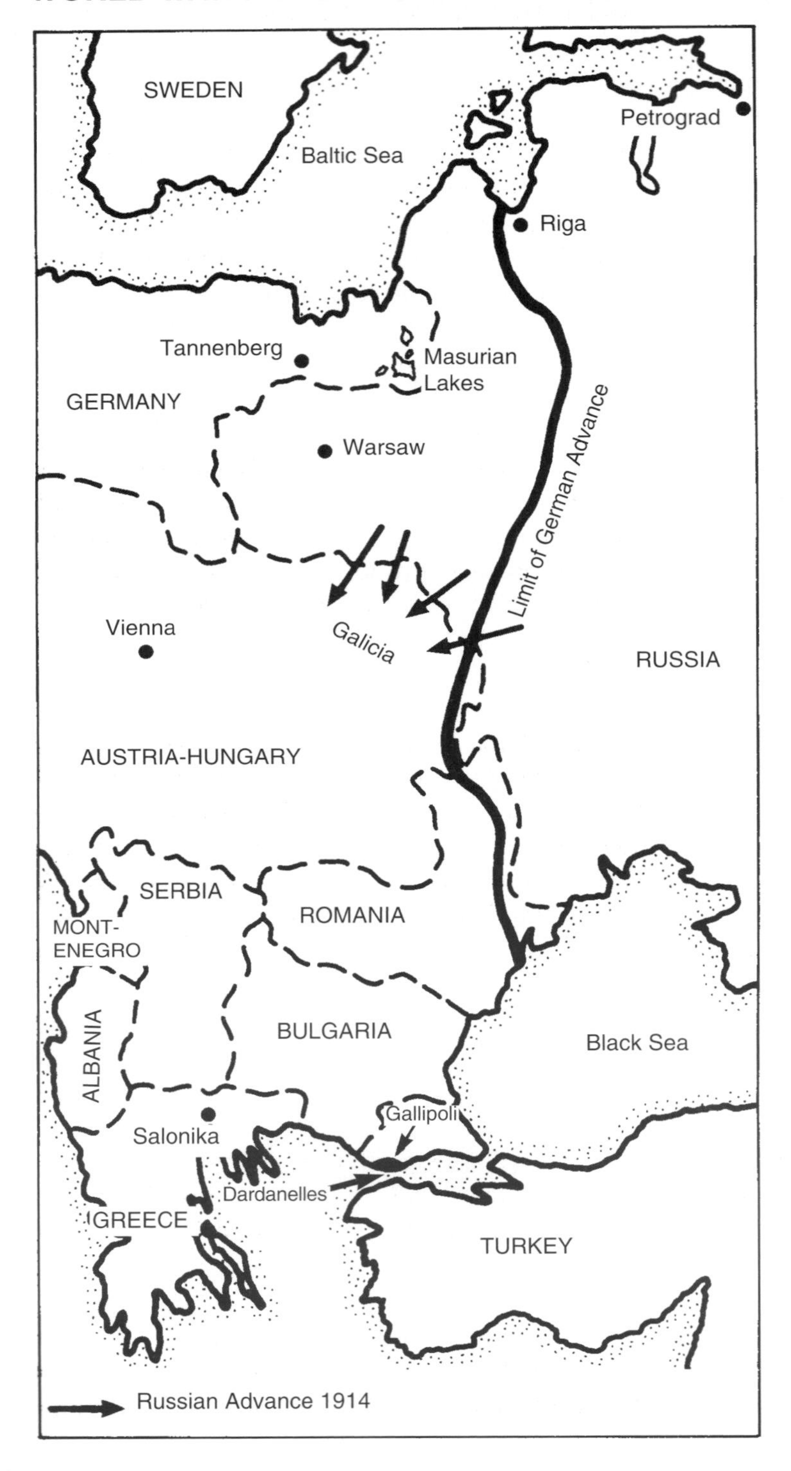

Battle of Ypres. This battle saw the Germans using poisonous gas for the first time. The Allies saw that they were making no progress and considered the option of opening a new front.

Gallipoli

The Tsar of Russia sent a request to the British government in 1915, asking them to organise an attack on Turkey, one of Germany's allies. This would relieve the pressure on the Russians by opening up a supply route to Russia through the **Dardanelles. Winston Churchill**, First Lord of the Admiralty, convinced the British cabinet of the benefits that would come from an attack in the Dardanelles.

The British navy gave insufficient support to the plan. The French generals, fearful of a German breakthrough at home, refused to divert sufficient troops to the area. The British naval attack on **Gallipoli** failed and only alerted the Turks to the danger. A landing of British, French and Anzac (Australia and New Zealand Army Corps) troops took place in April 1915. Within eight months 200,000 of them had been killed by the well-positioned Turks. A second landing at **Suvla Bay** also failed. During December 1915 a successful evacuation of Allied troops was completed. The cost of the failure was considerable. The Black Sea route to Russia remained firmly closed to Allied shipping. Once again thoughts turned to breaking the stalemate on the Western Front.

The Eastern Front

The Russians had some successes against the Austrians in 1914. However, after Tannenberg and the Masurian Lakes, the Germans sent troops to help their allies. During 1915 the Russians lost Galicia and Poland. Two Austro-German armies advanced into Russia in an enormous pincer movement. One million Russian troops were killed. Tsar Nicholas II decided to take control of the army. This was a terrible blunder. For the remainder of the war the Russian army drifted without proper leadership or any real strategy.

Nineteen-Sixteen: Land War

The Western Front

General Falkenhayn thought that the best way to win the war was to concentrate on the destruction of the main allied military force, the French army. He decided to attack the famous fortress of **Verdun,** intending to bleed France white and force her out of the war.

Battle of Verdun

On February 21 1916 the Germans opened the offensive with an artillery barrage of hundreds of thousands of shells during the first twelve hours. French officers knew that Verdun was of little military value, but the politicians insisted that French honour was at stake. Wave after wave of French infantrymen fell in attack and counter-attack over the wilderness of mud. The ground was poisoned by gas shells and pitted with innumerable craters.

The French commander, **General Petain**, was ordered not to give in at any cost. *Ils ne passeront pas* (they shall not pass) became the French motto. Each day fresh troops made their way along the *Voie Sacrée* ('Sacred Way'), the only open road leading to Verdun.

By early July, when the fighting died out, the Germans had captured 130 square miles of territory but had not taken the fortress. In all, the French lost more than 300,000 men, but the Germans lost almost as many. The defence of Verdun inspired the French nation to resist Germany, but it shattered the morale of the French army which was now on the verge of mutiny.

The Battle of the Somme

The command of the British army had passed to **Sir Douglas Haig**, Scottish-born and a member of the famous family of whiskey distillers. He was to become one of the most controversial generals of the war. He appeared to work on the theory that given enough men and enough goods, a breakthrough would be inevitable.

From 1 July until the middle of November 1916, the allies fought a series of battles known as the **Somme,** where a major attempt was made to breach the German lines. Preliminary raids by British troops led the Germans to strengthen their defences. When

THE WESTERN FRONT 1914-1918

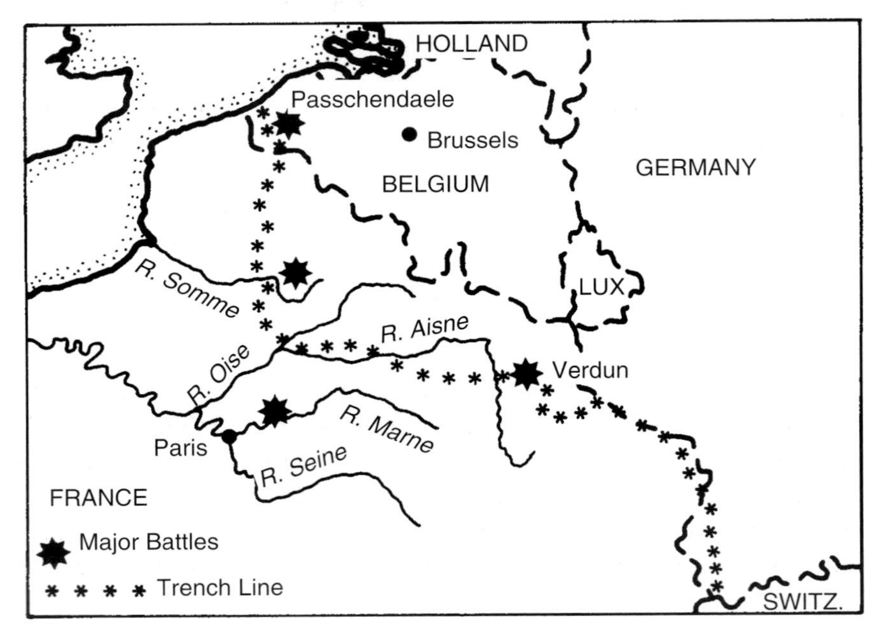

the British opened the offensive on July 1, they had to attack crests of hills heavily protected with barbed wire. Behind the hills, in dug-outs up to forty feet deep, the German troops waited, secure from artillery bombardment. The ground became so churned up and cratered that orderly advance by the soldiers was almost impossible.

Continuing British attacks

By the end of the first day no territory had been gained. The British had lost 20,000 dead and 40,000 wounded. Attacks continued throughout the summer and autumn. The British introduced tanks, but they were too few and often sank in the mud. Haig and his generals would not give up. By mid-November the British front-line had advanced no more than five miles. The British and Germans had lost over 400,000 men each and the French over 200,000. Over a million men had died and Haig's strategy had failed.

'Lions led by donkeys'

Verdun and the Somme showed the stupidity of generals and the incredible waste of manpower. The amazing courage of the ordinary soldier was also revealed. Historians have spoken of 'lions led by donkeys'. However, soldiers were losing their idealism and faith in their leaders.

The Eastern Front

From June to September 1916, a huge Russian army, led by **General Brusilov**, drove west against the Austrians. The attack was only halted when the Austrians were joined by German reinforcements. Russian losses were very heavy, but the Brusilov offensive had helped the allies to survive in the west during 1916.

Romania enters the war

Romania also joined the allies in 1916. She was knocked out of the war by the Germans and suffered terribly. The Germans used Romanian supplies of wheat and oil for the remainder of the war.

Trench Warfare

Trench structure

World War I was dominated by trench warfare, fought between two armies dug in along a trench line which stretched about 750 kilometres from Belgium through northern France to Switzerland. Some sections of trenches were quite deep and had elaborate living quarters. However, trenches were mostly a little over two metres deep and a little less than two metres wide. The sides were supported by sandbags. A raised section or parapet of sandbags at the front gave an overall depth of about three metres to a trench.

A TRENCH

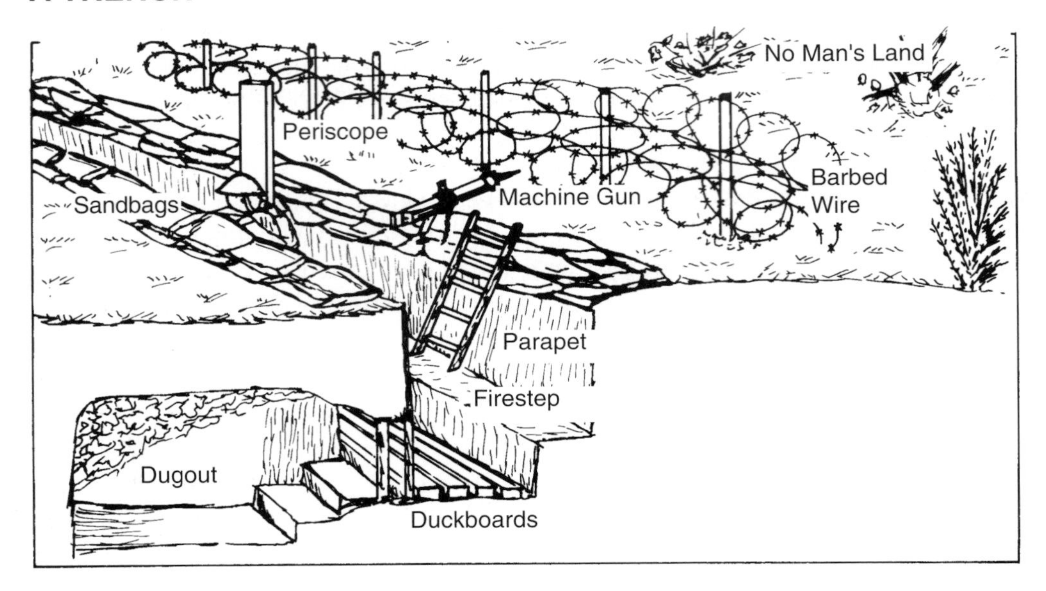

The lines had three parallel trenches, a front line trench, a centre line trench and a reserve trench all linked by communications trenches. The reserve lines were usually out of enemy range. A 'firestep' was built to allow soldiers shoot over the top of the trench. At the back a 'dugout' provided a small room-like shelter. The bottom of the trench was usually lined with wooden planks called 'duckboards'. The ground between rival trenches was called 'no-man's land', an area that quickly developed huge shell craters. The front of the trenches was protected by barbed wire.

New weapons of war

Trench warfare came about because of the rapid advances in the weapons of war. Heavy guns like the new 75mm **Howitzer** had a range of over thirteen kilometres. It delivered shells that caused huge craters or exploded into flying shrapnel. The **Vickers** machine-gun fired 600 rounds per minute. The British **Lee-Enfield** rifles and the German **Mausers** had a range of more than a kilometre and could fire up to twenty-five rounds a minute.

In 1916 tanks were used for the first time. At first they were slow-moving and often stuck in the mud. They became more effective later in the war. When fighting at close range, soldiers used bayonets fixed to the end of their rifles. From 1915 on, grenades were used. To this range could be added mortars and flame-throwers. The Germans used chlorine and mustard gas for the first time in 1915. It caused terrible panic and suffering. Later in the war, gas masks provided some protection against this horrible weapon.

Life and death

Troops usually spent eight days in the front lines followed by four days in the reserve trenches. But in some cases they could be kept in the front line for over a month without a break. The food was monotonous – a diet of bread, biscuits and a type of corned beef called 'bully beef'. Lice, mites, flies and rats continually irritated the soldiers. They developed 'trench foot' and 'trench fever' from standing in wet and muddy trenches. Many suffered from 'shell-shock', and developed stammers and uncontrollable nervous reactions, and even went mad. These men were often harshly treated and told to 'pull themselves together' before being sent back into the line. Some hoped for a 'blighty one', a wound severe enough to cause them to be sent home.

The object of the exercise was to break through the enemy lines. An attack began with the bombardment from the heavy guns in an attempt to kill front-line troops and break down the barbed wire. Then a whistle-blowing officer would lead the soldiers out of the trenches and 'over the top'. They would charge towards the enemy with fixed bayonets. Those who were not mown down by enemy fire would find shelter in a crater and later, under the cover of darkness, crawl back to their lines.

Millions of men lost their lives in trench warfare. The failure of the generals to achieve a lasting and significant breakthrough meant that the war lasted for over four years.

Trench warfare: British troops go 'over the top'.

War at Sea

Skirmishes in 1914-1915

The Royal Navy had to protect Britain from attack and imposed a blockade on Germany by crippling her overseas trade. Being an island nation, Britain was in a vulnerable position and German strategy was to impose a counter-blockade on Britain and starve her into submission.

The German Pacific Squadron

When war broke out, German warships on the high seas began to prey on British merchant ships. The most powerful fleet was the **German Pacific Squadron** under **Admiral von Spee**. In November 1914, von Spee destroyed a British squadron at the **Battle of Coronel**, off the Chilean coast. A month later, at the **Battle of the Falkland Islands**, the British gained their revenge. The German fleet was destroyed and von Spee went down with his ship.

BLOCKADE & U-BOAT WARFARE 1914-1918

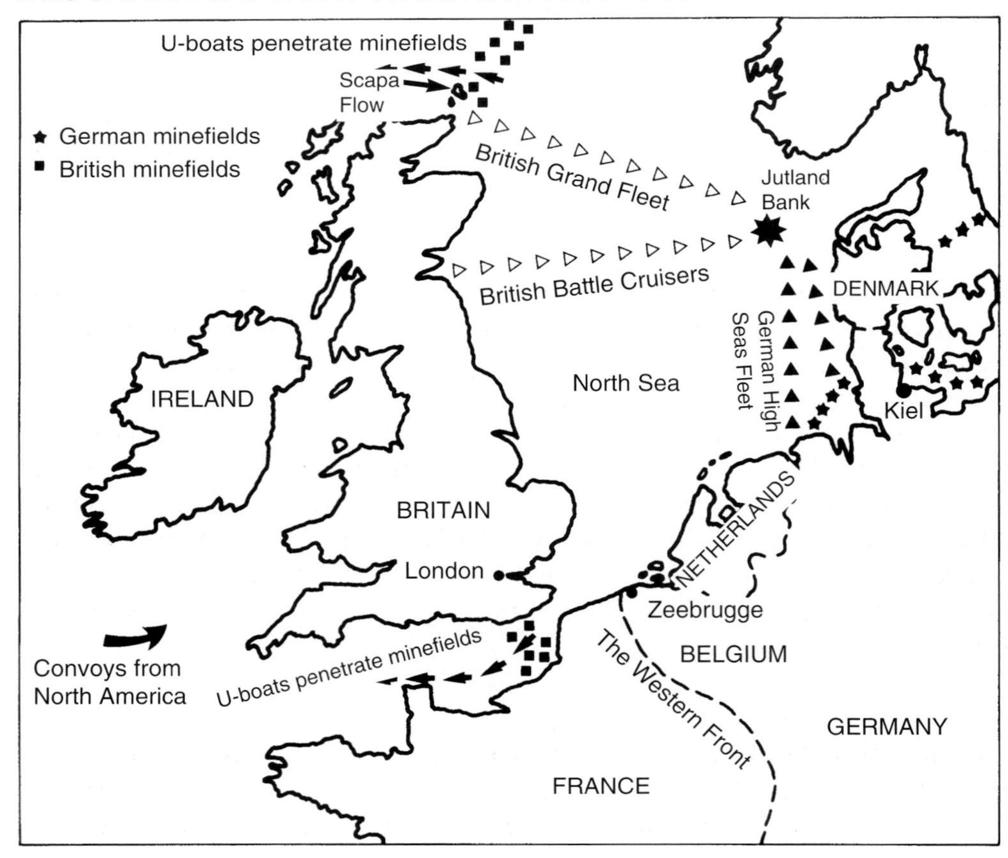

British cartoon: The Kaiser and Admiral von Tirpitz show their approval, as German mother nurses her 'baby', the U-boat.

March 1915: start of German U-boat campaign

In March 1915 Germany began her submarine campaign against Britain. She declared the sea approaches to Britain to be a war-zone in which all shipping was liable to be sunk without warning.

The sinking of the Lusitania

Admiral von Tirpitz felt that a U-boat campaign could break the British blockade. Heavy losses were suffered by the allies. In retaliation Britain ordered her merchant ships to ram submarines. The British also hid guns in merchant ships. In May 1915, the British passenger ship, the *Lusitania,* was on her way from New York to Liverpool. The German embassy had published a warning that the ship was liable to be attacked. On 7 May, she was torpedoed off Kinsale, on the south coast of Ireland. Over one thousand passengers lost their lives, including 128 American citizens. In the United States there was an outcry, as anti-German feeling swept the country. President Wilson threatened war unless Germany called off its unrestricted submarine campaign. Further sinkings occurred before the Germans reluctantly suspended the campaign.

May 1916: the Battle of Jutland

The **British Grand Fleet** remained at its base at Scapa Flow in the Orkneys, while the **German High Seas Fleet**, protected by minefields, waited at Kiel. In May 1916, the one naval battle of the war in European waters took place at **Jutland,** off the coast of

Admiral Scheer, Commander-in-Chief of the German High Seas Fleet, Battle of Jutland, 1916

Denmark. **Admiral Reinhard Scheer** devised a plan to lure out part of the British fleet and destroy it. The Germans used a small force under **Admiral Hipper** to bait the British, while the main fleet, under Scheer, remained out of sight. The British were also careful and a squadron of cruisers under **Admiral Beatty** sailed ahead of the British Grand Fleet. Hipper's gunnery proved superior and Beatty retreated towards his Grand Fleet, which was under the command of **Admiral Jellicoe**.

The German and British fleets became involved in conflict off the coast of Jutland. The battle was indecisive, although Britain lost fourteen ships and 6,087 men while the Germans lost eleven ships and 2,551 men. Admiral Jellicoe was very cautious and was not prepared to lose the British fleet, and perhaps the war, in one afternoon. Despite the losses, Britain claimed victory because she continued to control the seas, while the German fleet was firmly bottled up in port for the remainder of the war.

Unrestricted submarine campaign renewed

In Germany, the British blockade was causing acute shortages and suffering. The Germans calculated that if they could double the losses of British shipping they would reduce Britain to a state of collapse before any American assistance would arrive. They expected that sooner or later America would take the allied side.

On 31 January 1917, Germany announced that, from February, she would resume unrestricted submarine warfare in a zone around Britain and in the Mediterranean. **President Wilson** broke off diplomatic relations with Germany. On 6 April 1917, the United States declared war on Germany.

The German gamble almost paid off. In the month of April, German submarines sank a million tons of allied and neutral shipping. Finally, Prime Mister **Lloyd George** ordered the British navy to adopt a convoy system.

A convoy consisted of a group of merchant ships sailing together with a naval escort. New methods of hunting and destroying submarines were also used. Results were dramatic. Losses in British shipping were reduced from 25 per cent to one per cent. By the end of 1917, the United States and Britain were building ships faster than Germany could sink them and were sinking submarines faster that Germany could replace them. Germany's failure to break the British blockade was a major factor in her ultimate defeat.

War in the Air

German superiority

At the outbreak of war, aircraft were used only for reconnaissance purposes. Pilots could spy on enemy trenches, spot targets and report troop movements.

By 1915, command of the air was firmly held by the Germans. They had developed the **Fokker monoplanes**. These had mounted machine guns and could fire between their own revolving propeller blades. The Allies matched the Germans by 1916 with the **British Scout** and the **French Nieuport**. The **Fokker triplane** restored Germany's lead by the end of the year.

Air battles

Aerial warfare became general along the length of the Western Front. Tactics were quickly learned. Aerial battles became known as 'dogfights'. Both sides had their star pilots or 'aces', like **René Fonck** of France and **Billy Bishop** of Canada. Most famous of all was the German **Manfred von Richthofen**, called the '**Red Baron**'. He is reputed to have shot down over eighty Allied aircraft before he was killed in 1918.

Bombing campaigns

During the war, the Germans tried to bomb British towns and cities. At the beginning they used **Zeppelins**. There were over fifty raids on Britain by these huge hydrogen-filled airships, but they were responsible for less than 2,000 casualties. They were vulnerable to attack by aircraft and became blazing infernos when shot down.

Later the Germans used bomber aircraft and these proved far more destructive. The most famous German bomber aircraft, the **Gotta**, caused 800 casualties in London in 1918. British bombers were less effective, although they did carry out raids on German cities.

As the war developed the spotter plane developed into an impressive war plane. However, aerial warfare was not as decisive in World War I as it was in World War II. In the 1914-18 war, the main method of warfare centred on the trenches.

Nineteen-Seventeen: Land War

American reaction to the war

American industry had benefited from the war, especially through the sale of arms. German submarine activity on neutral shipping had been suspended when America protested strongly after the sinking of the ship the *Lusitania* in 1915. **Woodrow Wilson** was re-elected in 1916, and he promised to keep America out of the European conflict. Then, in January 1917, Germany announced that she would resume unrestricted submarine warfare. She hoped to reduce British shipping to the point of collapse before the United States could enter the war. American investors in Europe stood to lose billions of dollars and were pressing the government to take the allied side.

6 April 1917: America enters the war

After Germany's announcement in January 1917, Wilson broke off diplomatic relations with Germany. Germany then committed the act that finally brought America into the war.

Zimmerman, Germany's Foreign Minister, sent a telegram to the German Ambassador in Mexico. It directed him to promise that, in return for a Mexican Alliance, Germany would help Mexico to recover all the territory she had lost to the United States. The British deciphered the Zimmerman telegram and informed the United States of its contents. American opinion was outraged. On 6 April 1917, the United States declared war on Germany.

The Western Front

In February 1917 the Germans eased back a little to the Hindenburg Line which was easier to defend. The British became involved in another engagement in the **Ypres** area. In November, with the aid of the Tank Corps, the British were victorious at **Cambrai,** South of Ypres. However, ten days later the small amount of territory won at Cambrai was lost.

Aftermath at Ypres, France, World War I, 1917

The new French commander, **General Nivelle**, believed he had the secret for success. For over two weeks he persisted in a campaign of frontal attacks. His troops were mown down by German machine guns. This resulted in serious mutinies in the French army. Some of these soldiers were court-martialled and officially shot. Others were executed without sentence. Petain, the hero of Verdun, replaced Nivelle. He decided to bide his time until 'Yanks and tanks' were ready to support his troops.

Haig tried to take some pressure off the French. He ordered the blowing up of the **Messines Ridge** and its German observation posts. The **Battle of Passchendaele** lasted from August to November 1917. Haig lost 300,000 men, but gained only 11 kilometres of mud.

Russia opts out of the war

Conditions on the Eastern Front had become desperate. The Russian soldiers ran out of arms and ammunition. Many of them simply walked off the battle lines and started for home. Tsar Nicholas II was forced to abdicate. Kerensky (see chapter 4) was persuaded by the French and British to keep Russia in the war.

The Germans allowed Lenin to go home in a sealed train. They correctly believed (see chapter 14) that he would take Russia out of the war. Before Christmas 1917, a truce was signed between Russia and Germany.

The Italian Front

Italy had joined the allied side in the war in 1915 because she was promised a number of territories. In October 1917 the Germans and Austrians inflicted a crushing defeat on the Italians at **Caporetto**. Italy lost 200,000 in battle, while 400,000 deserted. The British and French decided to send help to the Italians. The combined armies forced the Austrians to retreat and soon a position of stalemate was also reached on the Italian front.

The campaigns against Turkey

The allies were also engaged in a number of campaigns on the fringes of the Ottoman Empire. They occupied the Turkish province of Mesopotamia in order to secure oil. They sent forces to protect the strategic Suez Canal. They invaded Turkish Syria, which included Palestine, and captured Damascus.

Meanwhile, in Africa, the allies occupied all the German colonies.

Nineteen-Eighteen: End at last

The Ludendorff offensive

General Ludendorff decided to launch one final massive attack on enemy lines before too many Americans reached France. He was now able to use troops transferred from the east when Russia had opted out of the war. The Germans advanced on a fifty-mile front between the British and French lines. They crossed the Somme and the Marne rivers, created a huge wedge between the allied lines and were within forty miles of Paris.

Marshal Foch of France was now accepted as Commander-in-Chief of the allied armies. During July and August 1918, French and American forces launched counter-

Allied tanks and infantry go into battle, 1918

attacks. At the **Second Battle of the Marne**, the Germans were forced to withdraw behind this river. Nearly three quarters of a million fresh and well-armed American troops were beginning to turn the tide. On August 8, the allies began to force the Germans back. Ludendorff called this the 'black day of the German army'. By the middle of August, Ludendorff knew that the Germans would have to negotiate and would not achieve a victory.

The Central Powers surrender

In September 1918 the British and French armies on the Turkish front combined with the Serbs and forced Bulgaria to surrender. On the Italian front the allies shattered the Austrian army at **Vittorio Veneto**. On 3 November, Austria signed an armistice. Hungary signed a separate truce on 7 November.

9 November: abdication of the Kaiser

Ludendorff and Hindenburg had already realised that Germany's position was hopeless and they sought an armistice on fair terms. The allies insisted on unconditional surrender and the Kaiser's abdication. The German navy mutinied at Kiel. Food was in very short supply and disturbances spread. On 9 November, the Kaiser abdicated and fled to neutral Holland.

11 November: the end of the war

A Republic was proclaimed in Germany. Representatives of the new Republic signed an armistice in the presence of Foch in a railway carriage at Compiègne in northern France. The fighting stopped at 11am on 11 November 1918. World War I was over.

Aftermath

About 8½ million soldiers died in World War 1. It is estimated that over 5 million civilians lost their lives, through famine, disease, or attack. Europe was in no condition to fight an outbreak of Spanish influenza, which occurred towards the end of the war; it resulted in the deaths of a further 6 million people. The US War Department estimated in 1924 that over 20 million people were wounded or maimed in the Great War.

World War 1 destroyed the Romanov, Hohenzollern, Habsburg and Ottoman Empires. It dislocated the economies of most countries involved. After 1918, Europe could no longer claim to be the world's centre of power and influence.

There was a levelling effect on social attitudes and manners. Women had solved labour shortages by working in agriculture and industry. After the war, they were no longer prepared to remain in the home, or to be treated as second-class citizens.

Writers and Artists

The Artists' War

It has been said that World War I was the war of writers and artists. Certainly many writers, poets and artists died in combat in this war.

The poets

The English poet, **Rupert Brooke**, volunteered for duty at the outbreak of war. His most famous poem, *1914*, expresses the innocence and idealism of the volunteers. Brooke died on active service in 1915, aged twenty-eight.

1914

I Peace

Now, God be thanked who has matched us with His hour,
And caught our youth, and wakened us from sleeping,
With hand made sure, clear eye, and sharpened power,
To turn, as swimmers into cleanness leaping,
Glad from a world grown old and cold and weary.

V The Soldier

If I should die, think only this of me:
That there's some corner of a foreign field
That is forever England. There shall be
In that rich earth a richer dust concealed;
A dust whom England bore, shaped, made aware,
Gave, once, her flowers to love, her ways to roam,
A body of England's, breathing English air,
Washed by the rivers, blest by suns of home.

Rupert Brooke

Wilfred Owen was one of the most talented of Britain's war poets. His poems show his revulsion from the cruelty and horror of war. Owen was killed in November 1918, in the last week of the war. He was twenty-five years old. His poem *Dulce et Decorum est* (It is Sweet/Right and Proper) is a powerful indictment of war:

Dulce et Decorum est

If in some smothering dreams, you too could pace
Behind the wagon that we flung him in,
And watch the white eyes writhing in his face,
His hanging face, like a devil's sick of sin;
If you could hear, at every jolt, the blood

Come gargling from the froth-corrupted lungs,
Bitter as the cud
Of vile, incurable sores on innocent tongues –
My friend you would not tell with such high zest
Of children ardent for some desperate glory . . . .

Wilfred Owen

Another British poet who served in the war was **Siegfried Sassoon**, who was badly wounded and received medals for bravery. His poetry expressed his cynicism about the war. He wrote some particularly bitter poems about the conduct of war and its effect on men. He survived the war and became a pacifist.

Suicide in Trenches

I knew a simple soldier boy
Who grinned at life in empty joy,
Slept soundly through the lonesome dark,
And whistled early with the lark.

In winter trenches, cowed and glum
With crumps and lice and lack of rum,
He put a bullet through his brain.
No one spoke of him again.

You smug-faced crowds with kindling eye
Who cheer when soldier lads march by,
Sneak home and pray you'll never know
The hell where youth and laughter go.

Siegfried Sassoon

After the war, thousands lived on, crippled, blinded or perhaps mad. Their plight was captured by Wilfred Owen:

Disabled

Millions of people died before the Armistice came.
Thousands lived on, blind or crippled or mad or hungry.
Now, he will spend a few sick years in Institutes,
And do what things the rules consider wise,
And take whatever pity they may dole.
Tonight he noticed how the women's eyes
Passed from him to the strong men that were whole.
How cold and late it is! Why don't they come
And put him into bed? Why don't they come?

Wilfred Owen

A dead soldier in a trench, unburied for months

Others were lucky to survive unscathed. Most of them returned to civilian lives, but few could forget the horrors endured. The memory is best captured in the last verse of a poem by the Australian, **Vance Palmer**:

The Farmer Remembers the Somme

I have returned to these:
The farm, and the kindly Bush, and the young calves lowing;
But all that my mind sees
Is a quaking bog in a mist – stark, snapped trees,
And the dark Somme flowing.

Vance Palmer

About 200,000 Irishmen fought in World War 1, of whom 60,000 lost their lives. Among them were the poets Francis Ledwidge from Co. Meath and Tom Kettle from Co. Dublin. Their poems express their Christian idealism, the cruel conditions endured by the ordinary soldiers and their longing for home. Both lost their lives in the war.

Of course, some of the poems and marching songs written by ordinary soldiers in the trenches were crude and not intended for publication. *We're here because we're here*, sung to the tune of *Auld Lang Syne*, shows the frustration of the troops and their belief that the war had become meaningless:

We're here because

We're here because we're here because
We're here because we're here
We're here because we're here because
We're here because we're here...

The Menin Road: painting by Paul Nash

Artists

The most famous artist of the war was **Paul Nash**. His paintings show the brutality of war and the destruction on the Western Front. His best known works are *The Menin Road* and *We are making a New World*.

Another painting, *Trenches near Notre Dame de Lorette* by the French artist, **Francois Fleming**, conveys some of the misery of life in the trenches.

Prose-Writers

Life in the trenches was also described in the moving work entitled *All Quiet on the Western Front*, by the German novelist, **Erich Remarque**. Remarque depicts the suffering and hopelessness experienced by the troops and this book is thought to be one of the best anti-war books ever written. In this excerpt, a German infantryman has stabbed a French soldier. Because of heavy bombardment, he has to share a crater with the wounded Frenchman.

> The silence spreads. I talk and must talk. So I speak to him and say to him: 'Comrade, I did not want to kill you. If you jumped in here again, I would not do it, if you would be sensible too...'
>
> His tunic is half open...Irresolutely I take the wallet in my hand. Some pictures and letters drop out...There are portraits of a woman and a little girl, small amateur photographs taken against an ivy-clad wall. Along with them are letters. I take them out and try to read them...But each word I translate pierces me like a shot in the chest – like a stab in the chest.

> By afternoon I am calmer...'Comrade,' I say to the dead man, but I say it calmly, 'today you, tomorrow me. But if I come out of it...I promise you, comrade, it shall never happen again.'

In Flanders Fields by Leon Wolff and *Death of a Hero* by Richard Aldington also give vivid insights into the agonies of warfare. One of the characters in *Death of a Hero,* Winterbourne, surveys the scene in 'No Man's Land' after a battle:

> He came on a skeleton violently dismembered by a shell explosion; the skull was split open and the teeth lay scattered on the bare chalk; the force of the explosion had driven coins and a metal pencil right into the hip-bones and femurs. In a concreted pill-box three German skeletons lay across their machine-gun with its silent nozzle still pointing at the loop-hole. They had been attacked from the rear with phosphorus grenades, which burn their way into the flesh, and for which there is no possible remedy. A shrunken leather strap still held a battered wristwatch on a fleshless wrist-bone. Alone in the white curling mist, drifting slowly past like wraiths of the slain, with the far-off thunder of drum-fire beating the air, Winterbourne stood in frozen silence and contemplated the last achievements of civilised men.

Many of those involved in the peace settlements and in the establishment of the League of Nations also hoped that such massive and futile destruction would not happen again. Yet the very terms of the peace settlement were to sow the seeds of discontent which would eventually lead to an even more widespread and horrific war in 1939.

Chapter 12

The Paris Peace Settlements

Problems for the peacemakers

President Wilson's hopes

President Woodrow Wilson had led America into World War I in 1917, as an Associate of the Allies. He was deeply disturbed by the death and destruction which resulted from the war. Wilson believed that future wars could be avoided if a just peace was concluded between the victorious and defeated powers. He hoped that if a settlement was based on a number of principles or 'points', drawn up for his own speech to Congress on 8 January 1918, the peacemakers could 'make the world safe for democracy'.

Background to the peace talks

Wilson's **Fourteen Point Plan**, as his principles became known, were accepted by the Allies in November 1918 as a basis for peace. After the Armistice the Allies made arrangements for a conference to discuss a lasting peace settlement.

18 January 1919: Paris Peace Conference begins

The Peace Conference opened in Paris on 18 January 1919. This was only nine weeks after the First World War had ended. Europe was in chaos, with emotions running high.

The Conference was made up of Allied and Associated powers. There was no place for the defeated nations. The negotiations were dominated by the leaders of the **'Big Three'**, **Wilson** of the U.S.A., **Clemenceau** of France and **Lloyd George** of Britain. Prime Minister **Orlando** of Italy sought an expansion of the inner circle to four, but found himself excluded by the other three leaders.

Woodrow Wilson's difficulties

President Woodrow Wilson of America was stern and idealistic, and had little understanding of European politics. To further complicate matters, the American Congress was controlled by Republicans and the President, a Democrat, did not fully represent the people of his country.

Wilson's belief in self-determination

Wilson's plan was based on the principle of self-determination, that is the right of each ethnic or racial group to form a nation state and to choose their own

'The Big Four' – Vittorio Orlando (Italy), David Lloyd George (Britain), Georges Clemenceau (France), Woodrow Wilson (U.S.A.)

government. By the terms of his plan, the defeated countries should be treated fairly, militarism should be ended and nations encouraged to disarm. World peace would, in future, be preserved by the formation of a **League of Nations** (see chapter 13).

Clemenceau's policy

The policy of France and of Clemenceau followed logically:

> A peace of magnanimity or of fair and equal treatment, based on such 'ideology' as the Fourteen Points of President Wilson, could only have the effect of shortening the interval of Germany's recovery and hastening the day when she will once again hurl at France her greater numbers and her superior resources and technical skill.

This was a summary of Clemenceau's attitude, made by J.M. Keynes, one of the British officials who attended the Paris Peace Conference, in his book *The Economic Consequences of the Peace*.

Clemenceau had seen Germany invade his country twice in less than fifty years. After the 1914-18 war the devastation in northern France was terrible and over a million French citizens had been killed. It is understandable that many French people should seek revenge and complain that Clemenceau was too mild.

But Clemenceau was also known as 'The Tiger'. He quickly made it clear that France's main aim was to cripple Germany so that she could never again invade his country. He proposed that Germany should pay compensation for the damage done to French territory. He also insisted that the border between France and Germany be made secure.

Verdun: a part of the town which was completely wrecked by bombing

Lloyd George's standpoint

Britain had not been invaded, but over 1,000 civilians had lost their lives in Zeppelin or aeroplane raids. About a million British soldiers had been killed. In addition, because of the war the country was heavily in debt. Anti-German feelings ran high and cries of 'Hang the Kaiser' dominated the general election after the war.

Lloyd George who had replaced Asquith as Prime Minister in 1916, won the 1918 election with the slogan 'Make Germany pay'. The British Prime Minister did not seek revenge at a personal level, but he had to pay heed to public opinion at home.

Orlando and the Italian view

Vittorio Orlando, the Italian Prime Minister, represented his country in Paris. He was anxious to claim all territory promised by the Allies to Italy when she entered the war on their side in 1915. The areas she claimed in the Alps and on the Adriatic were known as *Italia Irredenta* – Unredeemed Italy.

Some of these claims could not be justified on grounds of race or the principle of self-determination. Wilson and Lloyd George had little sympathy with Italy. Orlando did not have much interest in other aspects under discussion. With his demands not receiving the response he desired, Orlando felt that his country had been betrayed and for a time he withdrew his delegation from the discussions.

Crowds gather outside the Chateau de Versailles on 28 June 1919, the day the treaty was signed.

The Treaty of Versailles

28 June 1919: Germany signs the treaty

The main treaty was signed by representatives of the new Republic of Germany in the Hall of Mirrors in the palace of Versailles, where the German Empire had been proclaimed nearly fifty years before. The Germans were shocked by the harshness of the terms now before them. The Allies refused to lift the naval blockade on Germany and she was threatened with a renewal of the war if she did not agree. The German delegates felt that they had no real choice and signed the treaty on 28 June 1919.

German delegated forced to agree to the tenets of the treaty:
a.) Allies refused to lift naval blockade
b.) Threat of renewed war.

Territorial settlement

- Alsace and Lorraine were given back to France.
- The areas of Euphen and Malmedy were handed over to Belgium. (Belgium neutrality remained for British security).
- The independent state of Poland was re-created. She received the province of Posen. Germany had to surrender territory in West Prussia in order to create the Polish Corridor, which gave Poland access to the Baltic Sea. Culture
- Northern Schleswig was ceded to Denmark.
- The port of Danzig (Gdansk) became an international free city, under the auspices of the League of Nations.
- The Saar coalfields were to be administered by the League of Nations for fifteen years, but economic control was given to France. At the end of the fifteen-year period the people of the Saar region were to vote on whether or not they wanted to return to Germany.

Saar Coalfields = heart of German economy & accounted for its rapid economic growth from 1870-1913. Though France & Britain developed, they were envious of Germany's success.

GERMAN TERRITORIAL LOSSES 1919

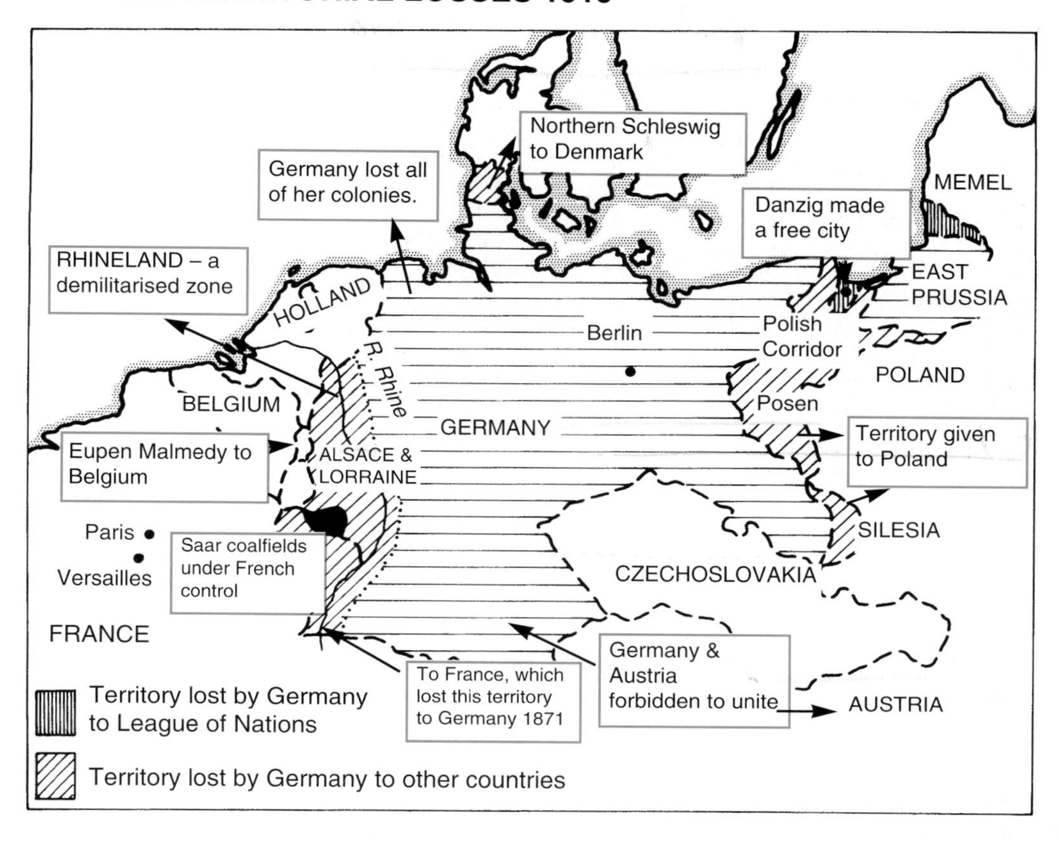

- Germany lost all her overseas possessions. Her colonies became mandates of the League of Nations. This meant that they were to be administered by members of the League and prepared for self-government.

The German armed forces

Germany's army was to be reduced to 100,000 men. No conscription was allowed. She could not have military aircraft, tanks or heavy guns. The navy was allowed only a limited number of ships, could not exceed 15,000 sailors and was to have no submarines.

The left bank of the Rhine and a fifty kilometre strip on the right bank were to act as a buffer zone separating Germany from her neighbours. No German armed forces were allowed in these areas.

Reparations

Germany had to pay not only for the damage caused in France and Belgium, but also for shipping losses. Service pensions and allowances for the Allied soldiers and their

'Swords into ploughshares'
A German tank is broken up for scrap.

families were also demanded. In 1921, the Reparations Commission decided on a sum of £6,600 million, much of it in the form of goods, to be paid in regular instalments.

War guilt

In addition to demanding reparations from Germany, the peacemakers also blamed her for starting the war. Article 231, known as the 'war guilt' clause, stated that Germany was solely responsible for causing the war:

> The Allied and Associated Governments affirm and Germany accepts the responsibility of Germany and her allies for causing all the loss and damage to which the Allied and Associated Governments and their nationals have been subjected as a consequence of the war imposed upon them by the aggression of Germany and her allies.

Other aspects of the settlement

The **Kiel Canal** was opened up to international shipping and the river Rhine was internationalised. ***Anschluss***, that is the union of Austria and Germany, was forbidden. The Kaiser and 'other persons accused of having committed acts in violation of the laws and customs of war' were to stand trial for war crimes. A League of Nations was to be set up in order to promote peace and harmony.

The Settlement in Eastern Europe

Separate treaties

One reason why Paris was chosen as a venue for peace talks was because it had a number of fine palaces which were used during negotiations. These gave their names to the separate treaties made between the Allies and the defeated countries.

The Treaty of St Germain

The Allies signed the **Treaty of St. Germain** with Austria on 10 September 1919. The Austro-Hungarian Empire was broken up. Austria lost territory to the new states of Czechoslovakia, Poland and Yugoslavia. Part of the Austrian Tyrol was given to Italy. Hungary became an independent state. The Austrians had to pay reparations to the Allies and limit the size of their army.

The Treaty of Trianon

The Allies signed the **Treaty of Trianon** with Hungary on 4 June 1920. As in the other agreements, reparations were agreed and the army was limited in size. The old

TREATIES OF ST. GERMAIN (AUSTRIA) & TRIANON (HUNGARY)

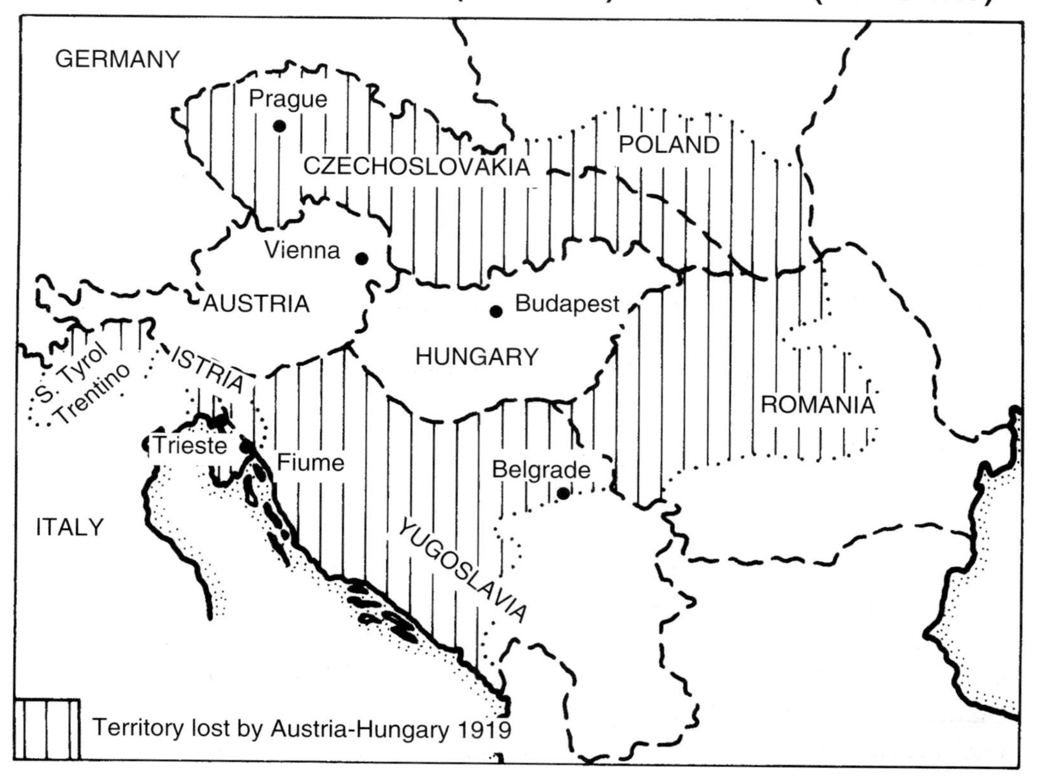

kingdom of Hungary was reduced to about a quarter of its size. This was because of the territory lost to Romania, Yugoslavia and Czechoslovakia.

The Treaty of Neuilly

The Allies signed the **Treaty of Neuilly** with Bulgaria on 27 November 1919. Like all of the defeated powers, Bulgaria had to pay reparations and reduce her armed forces. Although she lost western Thrace to Greece, she retained most of her territory.

The Treaty of Sevres

The Allies signed the **Treaty of Sevres** with Turkey on 10 August 1920. By the terms of this treaty, Turkey lost all her former territory in Europe to Greece, Bulgaria and Italy. The non-Turkish territories of the Ottoman Empire in the Middle East were to be administered as mandates by the Allies. The Straits between the Aegean and the Black Sea (the Dardanelles) came under international control.

The Treaty of Lausanne

Turkish nationalists under **Mustapha Kemal** overthrew the Sultan and successfully resisted the Treaty of Sevres. They then negotiated the **Treaty of Lausanne** with the Allies in 1923. This dropped the requirement of Turkey to pay reparations. However, Turkey accepted the loss of all non-Turkish parts of the old Ottoman Empire, although she retained Constantinople.

Mustapha Kemal (Kemal Ataturk), centre

Analysis of the Agreements

Limiting factors

After the war, Europe was in chaos. The Austro-Hungarian, Russian and Turkish Empires had not survived. Civil War raged in Russia and Communist uprisings had taken place in Munich and Budapest. There was unrest in Berlin and Vienna. Strikes threatened stability in Italy, France and Britain. A quick settlement was needed.

The peacemakers were limited by wartime agreements such as the **Treaty of London**, signed between Italy and the Allies. In Eastern Europe new states were being set up after the collapse of the Habsburg and Ottoman Empires. In such circumstances the 'Big Three' followed rather than influenced events.

The Treaty of Versailles – a non-negotiated settlement

The main criticism of the Treaty of Versailles was that it was not a negotiated one. In later years German propaganda concentrated on the fact that the treaty was 'dictated' and therefore not binding. Germany had been given little opportunity to dispute its clauses or to state her case.

Blame placed on Germany

The Allied demand that Germany should acknowledge that she alone was guilty of having caused the war was unfair and does not stand up to inspection (see chapter 10). The German Foreign Minister, Rantzau, said that 'such a confession in my mouth would be a lie'. However, Germany was forced to accept the clause.

Reparations linked with 'war guilt'

In the Treaty of Versailles the reparations were directly linked with Germany's war guilt. The British economist, J.M. Keynes, has shown that the amount of reparations was beyond Germany's ability to pay. She too had been drastically affected by the war. The reparations were the chief cause of the catastrophic inflation of the German mark in 1923.

Failure of the Fourteen Point Plan

The basis on which the peace was supposed to be made was President Wilson's Fourteen Point Plan. However, the principles on which the plan was based failed to survive during the negotiations. Points such as open diplomacy and freedom of the seas were abandoned during the bargaining process. Self-determination led to many problems and the League of Nations which emerged was a weak organisation.

The outcome of putting self-determination into action

The applications of the principle of self-determination created many national minorities within states. The conference was not fully aware of the complex pattern of racial groupings in parts of Eastern Europe. The peacemakers were less inclined to apply the principle to Germans than to other racial groups. This meant that there were many Germans living in places like the Sudetenland in Czechoslovakia and in the Polish Corridor. In later years Hitler used his propaganda machine to highlight this problem.

A 'flawed settlement'

Another major criticism of the peace settlement was the choice of Paris as the venue for the conferences. It was certainly not a neutral city and ensured a strong French influence on the deliberations. The **Treaty of Brest-Litovsk** (see chapter 14) was cancelled, but the Russians were not consulted about the new arrangements. It has often been said that the decision of the 'Big Three' to take all matters into their own hands and to act without proper consultation, led to a flawed settlement.

Hopes for the future

The reparations and territorial adjustments may seem harsh. However, at the time, treaties were usually dictated rather than negotiated. The Germans can hardly have helped their cause by the harsh terms they imposed on the Russians at Brest-Litovsk.

President Wilson hoped that the settlement would bring lasting peace. He linked future peace with collective security through the League of Nations. America's failure to take her place in the League weakened the peace treaties. If the League were to fail, the peace settlement itself would not survive.

Chapter 13

The League of Nations

Origins of the League

Background to the establishment of the League

Merely to win the war was not enough. It must be won in such a way as to ensure the future peace of the world.

President Wilson

The American President told his audience in Paris that peace could be maintained through a **League of Nations**. In 1918, **Jan Smuts**, the Prime Minister of South Africa, published a book on the idea of a league which would encourage cooperation between nations. Lloyd George spoke about the formation of an 'international organisation' before he came to Paris for the peace talks.

When the Peace Conference met in January 1919, President Wilson insisted that the setting up of the League should have priority. Under his chairmanship a

Headquarters of the League of Nations, Geneva

committee drew up a **Covenant of the League** within three months. This was a type of constitution and was accepted by the Peace Conference. It set out the rules and aims of the organisation. This Covenant was written into the peace treaties made with each of the defeated nations.

Aims of the League

The Covenant of the League of Nations defined four main aims for its organisations:

- To prevent future wars by the peaceful settlement of international disputes
- To promote disarmament
- To supervise and administer territories referred to it by the peace conference
- To promote international cooperation and human welfare by means of its organisations for social and economic work.

Procedures of the League

Members who were in dispute with one another were obliged to refer their disagreements to the League. Failure to do so, or failure to take account of the League's decisions, meant that a member was liable to sanctions. The blockade of Germany during the war had ultimately proved effective, so economic sanctions were chosen as the League's main weapon. The Covenant also referred to the possibility of military sanctions. However, these were vague and undefined as the League had no army of its own.

The League of Nations relied on the principle of **collective security.** Article XVI stated that 'should any member of the League resort to war...it shall *ipso facto* be deemed to have committed an act of war against all other members of the League.'

The Structure of the League

The General Assembly

The **Assembly** was the debating chamber for the members. It admitted new members and elected non-permanent members of the Council. All members had one vote each. It was a world forum that could debate issues but could not make laws. It was to meet once a year in **Geneva**, Switzerland. It could deal with any matter affecting the peace of the world and suggest action to the Council. Its decisions had to be unanimous. One of the most important functions of the Assembly was to vote the Budget.

The Council

The **Council** consisted of permanent and temporary members. The four permanent members in 1921 were **Great Britain, France, Italy and Japan**. Temporary members

The First Sitting of the League of Nations in Great Britain in 1920
Left to right: M. Calamanos (Greece), Senhor Da Cunha (Brazil); Mr. Matsui (Japan); M. Leon Bourgeois (France); Mr Balfour (Great Britain); Sir Eric Drummond (Secretary-General); Senatore Ferraris (Italy); M. Paul Hymans (Belgium) and Señor Quiñones de Leon (Spain).

were elected by the Assembly for a period of three years. The number of temporary members varied from four in 1920 to ten in 1936.

The Council had the power to take action if peace was threatened, but decisions had to be unanimous. Real power rested with the Council. It was a compromise between the reality of recognising the most powerful states and the ideal of democracy.

The Secretariat

The **Secretariat** was an international civil service which carried out the administrative duties of the League. It prepared reports and kept records of the Assembly and the Council. This clerical staff of about 500 collected data required by the League and registered treaties between member states. The first secretary of the League of Nations was **Sir Eric Drummond** of Britain.

The Court of Justice

The **Permanent Court of International Justice** was based at the Hague in Holland. It was referred to as the **World Court** and was set up to settle legal disputes between nations. Such disputes were usually about breaches of treaties or interpretations of international law. The power of the Court was limited, as it had neither the power to bring states before it nor the means to enforce its decisions.

The International Labour Organisation

A number of subsidiary organisations were created to carry out the aims of the league. The **International Labour Organisation** (ILO) had as its aim the improvement of conditions for workers all over the world. Representatives of governments, employers and workers considered topics such as wage rates, conditions at work, safety, health hazards, employment of women and children, and unemployment.

Other Commissions

The Mandates Commission supervised the former German colonies and some former provinces of the Turkish Empire. The **Disarmament Commission** was set up to control the manufacture and sale of arms and to bring about a reduction of armaments throughout the world. Other bodies dealt with such matters as health, drugs, standards of nutrition, refugees, transport, communications and intellectual cooperation.

Successes of the League

Human welfare

It is easy to criticise the League of Nations for its failures. However, much valuable work was done by its subsidiary organisations and special commissions.

Many people had been forced to flee their homes during the Great War. The problem was not easily solved, but the League worked hard on behalf of these

Disinfectors at work in a League of Nations refugee camp

refugees. Vast numbers of prisoners were held in confinement and could not be simply turned free with no means of returning to their own country. **Fridtjof Nansen**, the former Norwegian explorer, set up the '**Nansen Relief Fund**' and with the help of the League returned nearly half a million prisoners to their homelands.

A **Slavery Commission** worked diligently to stamp out slave-dealing. The League also tried, with mixed results, to get the cooperation of members in controlling the traffic in drugs. League publications educated the public in the dangers of using these substances. The energy of the first director of the ILO, the Frenchman, **Albert Thomas**, was mainly responsible for progress in labour matters. A great deal of information was collected and published by the ILO and conditions of work generally improved.

Territorial settlements

Between 1920 and 1932 the League proved successful in solving a number of territorial disputes, mainly between the less powerful states.

- Following a plebiscite organised by the League, Germany and Poland accepted a partitioning of disputed areas in **Upper Silesia.**
- It appointed a commissioner for the city of **Danzig.**
- Sweden and Finland submitted a dispute over control of the **Aaland Islands.** The League decided in favour of Finland. Sweden accepted the verdict under protest.
- Lithuania accepted the League's proposal that the Baltic port of **Memel** should have home rule within Lithuania.
- The Turks agreed to a decision of the World Court at the Hague that most of the **Mosul** region should go to the newly-created Iraq.

Failure of the League

The Corfu Incident

The League is all right when sparrows quarrel. It fails when eagles fall out.

Mussolini

Greece and Albania were in dispute over ownership of frontier territory between the two states. In 1923 five Italians working for the League on deciding the border between the Greeks and Albanians were assassinated on Greek territory. Mussolini failed in his demand for compensation from the Greeks, and he ordered the occupation of the island of **Corfu.**

Mussolini refused to submit the dispute to the League and the matter was settled by the **Conference of Ambassadors**. This Conference was a meeting of diplomats or foreign ministers of a number of major powers, and was completely independent of

'The Doormat'. The Japanese were openly flouting the League by their actions in Manchuria.

the League. Greece was ordered to pay fifty million lire in compensation to Italy. Although Mussolini withdrew his forces from Corfu, the Greeks were bitter at the result. The League Covenant had been defied and the Conference of Ambassadors seemed more powerful than the League. It was a sign of things to come.

Japan's actions in Manchuria

In September 1931 Japan, while still a member of the League, invaded **Manchuria**, a province of China. China appealed to the League, but Japan set up the puppet state of **Manchukuo** under Japanese control.

The League Commissioner, **Lord Lytton**, condemned the Japanese invasion in his report to the League. He recommended that Manchuria should have home rule under China and that Japanese economic rights in the province be recognised. Japan refused to accept the Lytton Report and was condemned by the Assembly as an aggressor. Japan left the League in 1933. Half-hearted attempts to apply sanctions did not have any success. The League had failed totally to prevent aggression.

Italy's takeover of Abyssinia

In 1935 Mussolini invaded Abyssinia, a member of the League. The **Emperor of Abyssinia, Haile Selaisse**, appealed for help to the League. Italy was condemned as an aggressor. Mild sanctions, which excluded the sale of coal and oil to Italy, were applied. It seemed that the British and French were more concerned with not offending Mussolini than with the principle of collective security.

The League fails to stop Mussolini

In 1936 **King Victor Emmanuel III** was proclaimed the new Emperor of Abyssinia and Haile Selaisse went into exile. Two months later sanctions were quietly

The Emperor Haile Selaisse addresses the League.

withdrawn. The British and French Foreign Ministers devised a plan which became known as the **Hoare-Laval Pact**. They suggested that Italy should be given two thirds of Abyssinia if the war stopped. The Pact was so outrageous that the Foreign Ministers were forced to resign. It was obvious that Britain and France were prepared to ignore the League when it suited their interests. The Italian victory in Abyssinia was a blow from which the League never recovered.

The League ignored

In 1936 the League of Nations merely condemned Hitler's re-militarisation of the Rhineland. It adopted a policy of neutrality in the Spanish Civil War later that year. Hitler, Mussolini and Stalin completely ignored this policy. The years between 1937 and 1939 also saw the Nazi invasions of Austria, Czechoslovakia and Poland. It is significant that during this time neither France nor Britain invoked the Covenant of the League of Nations against Hitler. They declared war on Germany in 1939 because of their treaty obligations to Poland – not to uphold the principle of collective security.

The end of the League

The League expelled the Soviet Union in 1939 after she had attacked Finland, but by then hardly anybody noticed. Before the end of World War II, a new organisation, the United Nations, had been structured. The League ceased to exist on 8 April 1945. It transferred its assets to the United Nations.

Reasons for the League's Failure

No military sanctions

The main reason for the failure of the League of Nations was its inability to enforce its decisions. Its main weapon of economic sanctions was impractical. The economies of those attempting to apply sanctions were liable to suffer from as much damage as those of offending states. The League had no army and military sanctions could only be applied if a member state agreed to put its own forces at the disposal of the organisation. During its existence, the League never once sought to apply military sanctions.

United States refuses to join

In November 1919 the United States Senate rejected the Treaty of Versailles and decided that America should not become a member of the League. The Senate feared that America was going to be plunged into European politics and regular military action. The retreat by the U.S. into its traditional isolationism was an early body blow to the League.

The gap in the bridge

Germany, Russia and the League

The Covenant of the League of Nations was written into each of the peace treaties. The defeated nations saw the League as an instrument used by the British and French to enforce the terms of the treaties. Germany was not allowed to join until 1926. To Hitler, the League was simply the living symbol and reminder of the hated Treaty of Versailles and Germany's humiliation. He withdrew Germany from the League in

1933. It seems that the British and French were not keen on the Soviet Union, with its Communist government, becoming a member. She did not join the League until 1934.

Other difficulties for the League

Decisions of the Assembly and the Council had to be unanimous. It was very difficult to reach agreement when states could exercise a veto. The League also suffered from a lack of financial resources. Member states suffered from post-war economic problems and were not prepared to give priority to the financial needs of the League.

Britain, France and the League

Britain and France dominated the League in the absence of the other great powers. Britain had to look after a large Empire and was not anxious to commit herself to other responsibilities on behalf of the League. She tended to put her own interests before those of the organisation. France was disappointed when the principle of collective security was not pursued vigorously. Because the League of Nations did not have an international peace-keeping force, France would not agree to disarmament. Enemies of Britain and France accused them of using the mandated territories for their own benefit. Little progress was made between the wars in promoting the mandates to independent states.

Final verdict

> The League of Nations has not been tried and found wanting. It has been found inconvenient and not tried.
>
> Lord David Cecil

The verdict of **Lord David Cecil**, a former British MP, who had given up his seat to devote himself to the League, was entirely accurate. It has often been said that it was not the League itself that failed but the nations which were in it.

Despite the early successes and the humanitarian work of its agencies, the League did not succeed in its main aim of keeping the peace. However, valuable lessons were learned and passed on to the United Nations. The last words of the President of the League, as the final session closed in April 1945, captured the spirit of hope for the future:

> We part as we have met, delegates of governments, servants of a great idea; and as we break up from the last meeting of the League we all know that 'its soul goes marching on'.

Chapter 14

The Growth of Communism: Russia 1914–1924

The Road to Revolution

Vladimir Ilyich Lenin

Vladimir Ilyich Ulyanov was born in 1870 the son of a school inspector in Simbirsk, a small town on the Volga. His brother, Alexander, was hanged in 1887 for plotting to assassinate the Tsar. A month after the execution, Vladimir left the high school at Simbirsk with a gold medal for being the best student 'in ability, development and conduct'. He vowed to make the authorities pay for his brother's death and was expelled from Kazan university for demonstrations and political activity.

Ulyanov went to Petrograd (St Petersburg), was arrested for subversion in 1897 and sent to Siberia for five years. In 1898 he married a young schoolteacher, **Nadezhda Krupskaya**. She acted as his faithful companion and secretary for the remainder of

Lenin addressing a crowd, Red Square, Moscow, 1919

his life. Not long after his marriage, Ulyanov wrote a revolutionary pamphlet and signed it **'Lenin'**, the name he kept for the rest of his life. This name may derive from the Lena River in Siberia.

Leader of the Bolsheviks

After he had served his sentence, Lenin moved on to Germany and Switzerland where he edited the revolutionary newspaper ***Iskra (The Spark)*** which was smuggled into Russia. In 1903, when the Russian Social Democratic Labour Party split into two groups, the Bolsheviks and the Mensheviks (see chapter 4), Lenin was in England. He became the leader of the Bolshevik group, but did not return to Russia until the autumn of 1905 and did not play a leading part in the revolution. He was not prepared to subject his revolutionary Bolsheviks to domination by the liberals. Lenin saw the revolution of 1905 as a training ground – the lessons learned would be applied when the real revolution took place.

World War I

Before the outbreak of war in 1914, Russia was ready for another revolt. This was due to dreadful factory conditions, rural discontent and hunger for land. However, when Russia entered the war there were mass demonstrations of loyalty to **Tsar Nicholas** and divisions were forgotten.

It soon became clear that the country was totally unprepared for war. Major confusion existed over mobilisation. There was a massive expansion in the production of goods for the war front, but other industries declined because of the shortages of raw materials and labour.

Russian defeats

Two Russian armies invaded Germany. Early successes were deceptive. By the end of 1915, Russia had lost over a million men (see chapter 11). In September 1915 Nicholas dissolved the *Duma* (which did not meet again until February 1916) and took over as Commander-in-Chief of the army. He was now often absent from Petrograd and the German-born Tsarina, **Alexandra**, was left in charge. She was suspected of having pro-German sympathies and was greatly influenced by the mystic, **Gregori Rasputin**.

Gregori Rasputin

Rasputin was a peasant from Siberia, who had travelled around Russia as a 'holy man', claiming to make prophecies and heal the sick. Rumours of his powers of healing reached Nicholas and his wife. They grasped at this chance of help for their son, the **Tsarevich Alexei**, the heir to the throne. The boy had been born with

Rasputin surrounded by his court followers

haemophilia, which meant that a bruise or cut might lead to death from bleeding. His parents believed that Alexei could be helped by Rasputin, who was soon established at court as a trusted adviser.

Nicholas and Alexandra were among the few who believed in Rasputin's holiness. Most others considered him evil, dirty, and a drunkard, and he had a scandalous love life. He made great use of his influence over the Tsarina and soon replaced the absent Tsar as the real ruler of Russia. In less than two years, twenty-one ministers were replaced by Rasputin's choices, men who were unfit for their positions.

The death of Rasputin

The Tsarina became more unpopular and was even accused of treason. By 1916 even the aristocrats had lost faith in Alexandra and a group of princes decided that Rasputin must be killed. He was invited to a party and poisoned with wine and cakes. As this seemed to have no effect, he was shot, but he did not die. He was then battered with a steel bar and his body was thrown into the river Neva. The post-mortem stated that he died from drowning.

The February Revolution

Hunger and strikes

The situation in Russia grew desperate. Conscription led to manpower shortages. Industrial output slumped and agricultural productivity also declined. The war was blamed for the lack of food and rising prices. Peasants refused to forward grain to the cities as inflation made their payments almost worthless.

On 22 February 1917 (according to the Russian calendar which was thirteen days behind the west)* a large crowd of marchers came out on the streets of **Petrograd**. Workers at the **Putilov** factory demanded a 50 per cent increase in wages and went on strike. This unrest spread to other factories. People broke into bakeries and stole food. When these disturbances broke out in Petrograd, the Tsar left for the front to inspect his troops. He believed that the food riots and strikes were no more serious than others in the previous years of the war.

The *Duma* had ignored the Tsar's earlier attempts to have it dismissed and pressed the government for emergency supplies of food to quieten the city. The Prime Minister, **Golitsin**, informed Nicholas that disorders were growing and that some of the crowd carried red flags and chanted 'Down with the German woman'. The Tsar replied by telegram:

> I command that the disorders in the capital shall be stopped tomorrow.

Mutiny and abdication

The Tsar's command was not enforced. Some regiments of the army mutinied and refused to disperse the rioters. The President of the *Duma* warned Nicholas about the situation, but his only reaction was to order that the *Duma* should be disbanded immediately. The *Duma* leaders decided to disobey the order. **Alexander Kerensky**, a socialist and member of the **Petrograd Soviet** (see p 172) announced that the *Duma* had set up a 'provisional government' to replace the Tsar and his ministers.

Nicholas then decided to return to Petrograd, but railway workers blocked the line and he was unable to reach the city. He was asked by two members of the *Duma* to abdicate in favour of **Alexei**. Rather than place his delicate thirteen-year-old son on the throne, Nicholas decided to abdicate in favour of his brother, the **Grand Duke Michael**. However, the Grand Duke did not receive the support of the Provisional Government and refused to accept Nicholas' offer. The February Revolution had ended the Romanov dynasty.

---

* Russian Calendar: Until February 1918 the Russian calendar was thirteen days behind the Gregorian calendar used in the West. This sometimes leads to confusion. The October Revolution took place in November according to the Gregorian calendar. In this book dates are as used by the Russians, i.e. based on the old Russian calendar until 14 February 1918 and on the western system after that date.

The Provisional Government

Liberal reforms

The Prime Minister of the first **Provisional Government** was an aristocrat, **Prince Lvov**. The government itself consisted mainly of liberals who sincerely wanted to turn Russians into free men. They reformed the criminal law system, brought in measures to protect minority interests and released political prisoners. The only socialist member was Alexander Kerensky, who became Minister for Justice.

Growing discontent

Despite their record of reform, the Provisional Government made two serious errors:

- they continued the unpopular war against Germany;
- they failed to transfer land from the landlords to the peasants.

Discontent among the people intensified further. Soon after the February Revolution, trade union and socialist leaders combined with the military to set up the Petrograd Soviet. Similar **soviets** or workers' councils began forming in cities and towns throughout Russia. The Petrograd Soviet took charge of running the city and limited the power and effectiveness of the Provisional Government.

Troops marching through Petrograd, 1917. The banner is inscribed 'Liberty, Equality, Fraternity'.

Lenin returns

News of the February Revolution reached Lenin in Switzerland. He telegraphed the Bolsheviks:

> Our tactics: absolute distrust; no support of the new government; Kerensky particularly suspect; to arm proletariat only guarantee; no agreement with other parties.

German agents approached Lenin and other Bolsheviks, offering to arrange passage through Germany, via Sweden and Finland, to Russia. They hoped that he would cause trouble for the Provisional Government and hamper Russia's war effort. In April 1917 Lenin arrived in a sealed train at the Finland Station in Petrograd. In his **'April Theses'** he outlined a plan for a Bolshevik takeover. He called for an ending of the war and the overthrow of the capitalist system. In its place he wanted a socialist government in which land and the banks would be nationalised. He summed up his policy in two neat slogans:

'Peace, Land, Bread',
'All power to the Soviets'.

The July Days

Troops at the front began to desert in large numbers and huge demonstrations for bread and an end to the war took place in Petrograd. Kerensky replaced Lvov as Prime Minister. During a period known as the 'July Days' Kerensky used troops to curb the demonstrations. Trotsky (see chapter 4) and other Bolshevik leaders were arrested. Lenin escaped to Finland, disguised as a fireman on a locomotive.

The Kornilov Affair

Kerensky's triumph over the Bolsheviks did not last long. Many landowners and army officers blamed the socialists for Russia's troubles at home and for her defeat in the war. They looked for someone who would destroy the power of the soviets. They found their man in **General Kornilov**, Commander-in-Chief of the army since July.

Kornilov decided to overthrow the Provisional Government and move troops close to Petrograd. Kerensky set up a committee to defend the city. Some Bolshevik leaders, including Trotsky, were released from prison and appointed to the committee. Workers put up barricades, and soldiers and sailors loyal to the Soviet got ready to resist. A new force appeared. These were the Bolshevik **Red Guards**, about 25,000 workers who had been armed by the Bolshevik leaders.

Kornilov's troops never reached the city. The railway workers sabotaged the lines. Bolshevik agents persuaded Kornilov's troops to mutiny. Within a week his bid for power was over and he was arrested. Trotsky emerged as the hero of the event and was elected chairman of the Petrograd Soviet.

Cruiser 'Aurora' in Petrograd, 1918

The October Revolution

The Bolsheviks seize power

After the failure of the right-wing coup, Lenin decided that it was time to seize power. He secretly returned to Petrograd. Aided by Trotsky, he drew up plans for the overthrow of the Provisional Government. Trotsky armed city workers with rifles he had coaxed from some discontented soldiers. Lenin and Trotsky planned a coup for 25 October, the day before the second **Soviet Congress**. They wanted it to appear to the congress as a soviet seizure of power, rather than a single party takeover.

The October Revolution all happened very quickly. The siren of the cruiser ***Aurora*** gave the signal for the start of operations at 2am on the night of 24 October. The Bolshevik Red Guards seized the railway stations, power houses, telephone exchanges, the state bank and the main bridges. Kerensky had taken over the Tsar's **Winter Palace** as headquarters for his government. On the morning of 25 October he found the city in the hands of the Bolsheviks. Kerensky fled from Petrograd and tried in vain to raise troops at the front. Most of his ministers were arrested at the Winter Palace, which was occupied by the rebels. The coup had been almost bloodless.

A Communist Government

The **National Congress of Soviets** met next day. Lenin told the delegates that the Bolsheviks were now in power. A **Council of People's Commissars** was given power

Red Guards shooting from an armoured car in Petrograd, October 1917

to govern by decree. Lenin announced that negotiations for peace with Germany would begin immediately. A land decree stated that landlords' estates were to be confiscated. Social, military and naval ranks were abolished and all Russian citizens became 'comrades'.

Lenin's first task was to organise the Bolshevik government. He took the position of President or Prime Minister. Trotsky was Commissar for Foreign Affairs, while **Joseph Stalin** became Commissar for Nationalities. A whole series of decrees brought in reforms designed to bring popularity to the party. The land decrees won support from the peasants. Working hours were reduced and workers were allowed control the factories. Russia was to be a classless society based on the economic principle of 'he who does not work, neither shall he eat'.

Elections took place in November, but the Bolsheviks received only about 25 per cent of the votes. The Bolsheviks quickly closed the Assembly and never again allowed opposition parties to exist. Lenin had proved that he was no democrat. In 1918, the Bolsheviks were renamed the **Russian Communist Party** and moved the government to the **Kremlin** in the old capital of Moscow.

The Treaty of Brest-Litovsk

In December 1917 Russia concluded an armistice with Germany. This was done to give the communists a breathing space and an opportunity to consolidate power. Trotsky, as Commissar for Foreign Affairs, was given the task of negotiating with the Germans. The peace-talks took place in the fortress of **Brest-Litovsk**. The Germans demanded control of Poland, the Ukraine, Finland and the Baltic States of Latvia, Estonia and Lithuania. This meant a loss of a third of Russia's arable land and sixty-two million people. In addition, three quarters of her coal and iron resources were within these areas.

Many Bolsheviks felt that the terms were humiliating and wanted to recommence the war. Lenin was determined to make peace at any price. On 3 March 1918 the Russian delegation agreed to the treaty. Trotsky could not bring himself to sign and appointed a deputy to take his place.

The Civil War

Reign of Terror

The Bolsheviks set up a secret police force, the ***Cheka***, to deal with enemies of the revolution. The humiliating peace treaty caused great dissatisfaction and anti-Bolshevik groups began to organise. Torture and executions were the favourite weapons of the ***Cheka.*** Landlords, former Tsarist army officers and Orthodox priests were shot. A local *Cheka* group organised the murder of the Tsar and his family at **Ekaterinburg** in the Urals in July 1918. Their bodies were burnt. The Bolsheviks had been afraid that they would become a focus for anti-Bolshevik groups.

The Tsar and his family in captivity

The Whites versus the Reds

Those who revolted against the Bolsheviks were collectively known as the **Whites.** Ex-army officers, nobles, landlords and liberals joined together in forming armies which tried to overthrow the Communist government. Many of the Social Revolutionaries and Mensheviks now saw the opportunity for revenge on those who had totally excluded them from power. The situation was complicated by the intervention of foreign powers on the side of the Whites. The victorious powers in World War I, Britain, France and the USA, were annoyed by Russian refusal to pay their debts. They also feared the spread of Communist ideas to their own states.

Trotsky was appointed Commissar for War and Commander of the **Red Army.** He moved about during the war in a special train which contained a printing press, supplies of ammunition and living quarters for his staff. As enough trained officers did not exist among the Communists, Trotsky appointed many officers of the old Tsarist army. To ensure loyalty, each officer was placed under a member of the Communist Party. Large numbers of factory workers were recruited into the ranks of the Red Army. They were ready to fight and die for the government.

The war in Siberia

The war began when 30,000 members of the **Czech Legion** seized control of the Trans-Siberian Railway. These Czechs were ex-prisoners and deserters from the Austro-Hungarian army. The Bolsheviks failed in their attempt to disarm the Czechs who formed the nucleus of the White army in Siberia. In 1919 the Czechs joined with **Admiral Kolchak**'s force of mainly ex-Tsarist officers. This army had early success but was finally defeated at Omsk. Kolchak was captured and executed.

The southern offensive

In 1919 **General Anton Denikin**, took control of the White army in south-east Russia. His troops made major advances, captured the cities of Kharkov and Kiev and even threatened the Bolshevik seat of government in Moscow. A major counter-attack by the Red army stopped the advance and forced Denikin's forces to retreat to the Crimea.

The attack from the Baltic

General Yudenich advanced from the Baltic and reached the suburbs of Petrograd. A brave defence of the city by the Red Army saved Petrograd and Yudenich was forced to withdraw.

The Polish front

In 1920 the Poles, helped by the French, attacked the Ukraine. The Polish army took the city of Kiev. The Russians then pushed them back to Warsaw. The Poles reorganised and advanced again. Peace was concluded in 1921 at Riga and Poland acquired a large slice of the Ukraine.

THE RUSSIAN CIVIL WAR

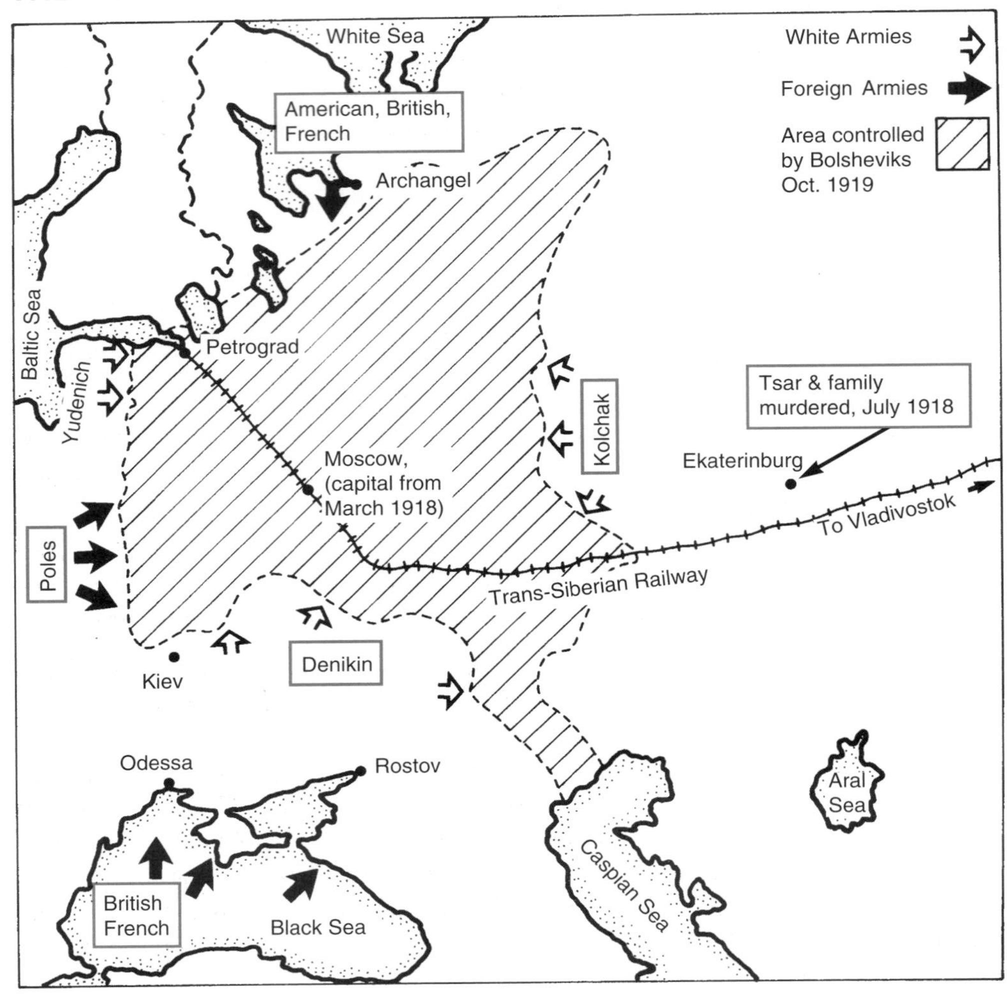

Reasons for the Bolshevik victory

- Great distances separated the various White armies. They did not co-operate properly with one another and were not united in their objectives.
- The allied intervention was only half-hearted. Their troops were war-weary and their governments were more concerned with making peace in Europe after the war.
- The Reds controlled Petrograd and Moscow. Their lines of communication were shorter and they could quickly bring their troops to the danger zones.
- The peasants were opposed to a White victory, which they felt would reinstate the landlords and return the land to them.

- The intervention by foreigners on the side of the Whites resulted in a wave of patriotism and support for the Reds.
- The ability of Trotsky was a decisive factor. He created the army, devised the strategy and was ruthless in his methods. Special *Cheka* units followed troops into battle and machine-gunned any members of the Red Army who fled from the enemy.

Economic Affairs

War Communism

The shortages of the Civil War led the Communists to nationalise all industries in the Red area. A regular supply of food was needed for the Red Army and for industrial workers in towns and cities. To solve these problems, Lenin adopted a policy known as **War Communism**. Teams of soldiers and *Cheka* were sent to the villages to requisition grain. Lenin was not prepared to tolerate opposition from the peasants who resisted this policy:

> It is necessary to organise an extra guard of well-chosen, trustworthy men. They must carry out a ruthless mass terror against the Kulaks (farmers), priests and White Guards. All suspicious persons should be detained in a concentration camp.

Workers were conscripted forcibly into industry, private trade was banned and strict rationing was introduced. War Communism may have been harsh, but it ensured Bolshevik victory in the Civil War.

Between 1914 and 1921 an estimated twenty-eight million people died in Russia, many from fighting but even more from starvation and disease. The system of War Communism gave the peasants no incentive to produce surplus food. Workers conscripted from rural areas proved unsuitable for work in the factories in the cities and large towns. The result was economic collapse. The Communist Party was blamed for the depression, and workers established opposition groups to protest about the conditions.

The end of War Communism

The sailors of the naval base at **Kronstadt**, near Petrograd, had revolted in 1905. They rose again in 1921, this time against the Communist government they had helped put in power. The mutiny was put down ruthlessly by Trotsky, but not without many deaths among the sailors. After this it was clear to Lenin that War Communism had to end if the Communists were to stay in power. He said that they must build 'new gangways to communism'. Lenin's gangway was the **New Economic Policy**.

The New Economic Policy

The **New Economic Policy**, or **NEP**, was a return to a system of private ownership and private trading. The government kept control of banking, power supplies and large industries. Smaller industries were returned to individual owners or to co-operative groups of workers.

The grain requisitioning was ended and peasants were allowed to sell surplus produce for their own benefit. They had to pay part of their crops in tax. Most of the buying and selling was done by traders called '***Nepmen***'. It took some years for NEP to show its full success, but by the end of the 1920s it seemed that the country had recovered from the misery of war.

Although the New Economic Policy was opposed by some pure Marxists in the party, it saved Russia from total ruin. Lenin saw that the country was not ready for communism. He said that the Bolsheviks must 'take one step back in order to take two steps forward at a later stage'.

Lenin: An Assessment

Political Genius

There can be little doubt but that Lenin was a political genius. An excellent orator, he understood the people and could influence his audience by the clear manner in which he set forth his ideas. He was completely dedicated to the revolution and sacrificed his life in order to achieve his goal.

Lenin was convinced that he alone was capable of leading the Communist Party and the revolution. From this conviction came the ability to carry his colleagues with him. Trotsky called him the 'engine-driver of revolution'.

However, it must be said that Lenin was no democrat. He was not concerned with the ideas of the ordinary people. He thought that they needed the guidance of an élite small band of dedicated revolutionaries. He ignored the results of the 1917 elections in which the Bolsheviks got only 25 per cent of the vote. Lenin was ruthless in the pursuit of his goals and fully supported the 'Red terror' of his party during the Civil War.

Above all, Lenin was a realist. This led him to make the correct decisions at vital times during the revolution. His influence was mainly responsible for keeping his party from being swallowed up by the Liberals after the February Revolution of 1917. He knew in October 1917 that the time was right for a Bolshevik coup. When the system of 'War Communism' was leading to famine, he changed course and introduced the New Economic Policy.

Lenin's terror tactics and his creation of an élite are sometimes blamed for the dictatorship that followed. He must shoulder some of the blame, but he can hardly be held responsible for all the ills that befell his beloved country after his death.

Following a stroke in 1922, Lenin was unable to play much part in government. He died in January 1924, at the age of fifty-three. He was greatly mourned in Russia and the old city of Petrograd was renamed Leningrad in his honour.

Questions

ORDINARY LEVEL – D

Answer the following questions briefly. One or two sentences will be enough.

1. Set down the principal members of the two hostile alliances in Europe leading up to World War I.
2. Explain why World War I was generally greeted with enthusiasm.
3. Why was the assassination of Archduke Franz Ferdinand at Sarajevo important?
4. What was the Schlieffen Plan?
5. How did the military situation on the Western Front develop during the early months of World War I?
6. What was the outcome of the Battle of Jutland?
7. Why was the convoy system adopted by Great Britain in 1917?
8. What explanation did the German High Command offer for the sinking of the *Lusitania*, 13 May 1915?
9. Set down and explain one major criticism made of the Versailles Settlement.
10. Set down two of President Wilson's 'Fourteen Points'.
11. Explain one weakness of the League of Nations.
12. What part did Rasputin play in the fall of the Romanov dynasty?
13. Why did the German government help Lenin return to Russia in 1917?
14. Why did Lenin decide to adopt the New Economic Policy in 1921?
15. How did Trotsky help the Bolsheviks win the Civil War in Russia?

Questions

ORDINARY LEVEL – E

Write a short paragraph on each of the following:

1. The events from the assassination of Archduke Franz Ferdinand, 28 June 1914, leading to the invasion of Belgium, 4 August 1914
2. Naval warfare in World War I
3. Life in the trenches on the Western Front during World War I
4. One of the following from World War I: The Dardanelles Campaign; the Battle of Verdun, 1916; the Battle of Jutland, 1916
5. Changes in the political map of Europe as a result of the Versailles Settlement, 1919
6. The failure of the League of Nations
7. The 'Big Three' at Versailles, 1919
8. The Russian Revolution, 1917
9. Lenin's Economic Policies, 1917-1924.

ORDINARY LEVEL – F

Write a short essay on each of the following:

1. How the system of alliances and the growth of nationalism, militarism and colonialism contributed to the outbreak of World War I.
2. World War I under each of the following headings:
   - (i) The opposing sides
   - (ii) French warfare
   - (iii) The war at sea
   - (iv) The ending and cost of the war.
3. Russia, 1917-1924, under each of the following headings:
   - (i) The Bolshevik Revolution, 1917
   - (ii) The Civil War, 1918-1921
   - (iii) Lenin's New Economic Policy.
4. The League of Nations

Questions

HIGHER COURSE

Write an essay on each of the following:

1 Treat of the reasons for the involvement of Great Britain in World War I. (80)

2 Discuss the causes, course and consequences of Russia's involvement in World War I. (80)

3 Discuss World War I under **one** of the following headings:
(i) The quality of military leadership
(ii) The effects of total war on the civilian population
(iii) The war at sea and in the air. (80)

4 'World War I revealed the horror of total war in the industrial age.' Discuss. (80)

5 Treat of the main developments in warfare during World War I, 1914-1918. (80)

6 Discuss the opinion that the Versailles Settlement was a 'conspicuous failure'. (80)

7 (i) Discuss the treatment of Germany in the Versailles Settlement. (40)
(ii) How did the settlement contribute to the rise of Hitler? (40)

8 (i) Treat of the origins and aims of the League of Nations. (40)
(ii) Estimate the extent to which the League of Nations achieved its aims. (40)

9 Analyse the part played by Lenin in the history of Russia, 1917-1924.

10 Why were there revolutions in Russia in 1917? (80)

SECTION THREE

1919-1939: THE RISE OF FASCISM

While fascism began earlier in Italy, it ultimately took on a more extreme form in Germany:

Chapter 15

Fascist Italy

The Growth of Fascism

Diplomatic Failure

> 'For Italy...it was a complete shattering of ideals. We had won the war; we were utterly defeated in the diplomatic battle'
>
> (Mussolini)

Italy contributed little to the Allied victory in World War I. President Wilson of America was not prepared to honour in full the **Treaty of London**, made between Italy and the Allies in 1915 before America became involved in the war. Italy gained **South Tyrol, Trentino, Istria** and **Trieste** at the Paris Peace Conference, but failed to get **North Tyrol, Dalmatia** and the city of **Fiume**, as promised. The Italians were not even given a German colony as a mandate. They were not allowed to expand their African colonies despite the assurances of 1915. President Wilson said that Italy's demands did not conform to the principle of self-determination.

Gabrìele D'Annunzio

In September 1919 the writer and nationalist, **Gabrìele D'Annunzio**, and 2000 of his supporters seized for Italy the city of Fiume. It had been given to Yugoslavia at the Paris Peace Conference. D'Annunzio was a colourful and eccentric figure who ruled the city for fifteen months. During that time he introduced the Roman raised-arm salute, hysterical mass-meetings, mob oratory and parades of his black-shirted followers.

When he was forced by the government to withdraw from Fiume, many Italians felt humiliated. They turned away from their weak government, which had accepted the Paris Peace Settlement, and were ready to support extreme nationalists. Among those impressed by D'Annunzio's direct action and his colourful parades was Benito Mussolini.

Social Unrest

Italy faced crippling debts after the war. As a result of inflation, the prices of most goods rose steeply. The production of armaments stopped and unemployment increased. Strikes and demonstrations were everywhere. In August 1920 many workers in heavy industry were locked out by their employers. The workers decided to occupy the factories. The Prime Minister set up a commission to investigate the unrest. Peasants in some areas seized land from the large landowners. Conditions in Italy resembled those in Russia in 1917. Returning soldiers could find no jobs. They were bitter and many joined right-wing paramilitary groups. Some Italians feared a communist takeover. They began to lose faith in the democratic system and looked elsewhere for a solution to the country's problem. Benito Mussolini was among those who offered a solution.

Benito Mussolini: Profile

Benito Mussolini was born in 1883 in Romagna in central Italy. His father, the village blacksmith, was a socialist, and his mother was a teacher. The young Mussolini was an intelligent pupil. However, he was very unruly and was expelled from two schools. He left school in 1901 with a teaching diploma. After only four months of teaching, he left for Switzerland, the home of exiled socialists such as Lenin.

In Switzerland, Mussolini joined various revolutionary groups. In 1903 he was arrested for advocating violence to achieve workers' demands. In 1904 he returned to Italy and made his name as a skilled journalist and agitator. In 1912 the **Socialist Party** appointed Benito Mussolini editor of *Avanti*, the official socialist newspaper. When the Great War began in 1914, he strongly condemned militarism and said that Italy should not become involved.

Mussolini soon changed his mind and set up his own pro-intervention newspaper, *Il Popolo d'Italia*. He was expelled from the Socialist Party when he

Mussolini in uniform – a study in arrogance?

suggested that Italy should go to war. His newspaper was financed by the Allies and by Italian industrialists. When Italy joined the Allies in 1915, Mussolini went to war, but in 1917 was released from the army, having suffered an injury in grenade practice.

Backing for the Fascists

The Fascists failed to get any seats in the 1919 elections. Mussolini blamed the extreme anti-clericalism of some of his followers and moderated his programme. In 1921 the Fascists gained 35 seats in parliament. The election was fought against a background of street battles between **Blackshirts** and Communists. Mussolini claimed that Fascists stood for law and order and were victims of unprovoked attacks. They gained financial backing from leading bankers and from top industrialists like Pirelli, the tyre manufacturer, and Agnelli of Fiat.

The political system was unstable. Italy had five governments between 1918 and 1922. Violence between the Fascists and Communists continued. The showdown came in August 1922. A general strike was broken by Blackshirt squads. The government seemed unwilling to stop the spread of Fascist violence and were incapable of protecting the rights of all citizens.

March on Rome

> 'Either we are allowed to govern, or we will seize power by marching on Rome.'
>
> (Mussolini)

Mussolini and the Blackshirts march on Rome, 1922

This ultimatum dismayed the Italian government. The Fascist leaders assembled about 25,000 of the Blackshirt Squadristi, north and east of Rome. The Prime Minister, **Luigi Facta,** asked **King Victor Emmanuel III** to impose martial law and sanction the use of the army. When his request was refused, Facta resigned. The King called on Mussolini to form a coalition government. Some of the Blackshirts did march on Rome. Mussolini himself arrived by train on 30 October 1922, in time for the victory parade on the following day.

The Fascist Party

Benito Mussolini played the part of war hero, but returned to his job as newspaper editor. He criticised the government's war effort. By 1918 he was calling for a dictator, 'a man who is ruthless and intelligent enough to make a clean sweep'. He criticised the 'mutilated peace' for not giving Italy the promised lands.

In March 1919, Mussolini founded the ***Fascio di Combattimento,*** a group of fighters. The symbol of the new **Fascist Party** was the fasces or bundle of rods carried by officials in ancient Rome. The rods had an axe in their midst and were symbols of authority and punishment. The Fascists also adopted the Roman salute used by D'Annunzio and his followers. Wearing black shirts, carrying skull-embroidered flags, singing patriotic songs and shouting nationalist slogans, they marched against the Socialists and Communists. By 1922 a quarter of a million had joined the Fascist ***Squadristi.*** By provoking disorder, they hoped that the people would be willing to accept the authoritarianism which Mussolini preached as a cure for the country's ills.

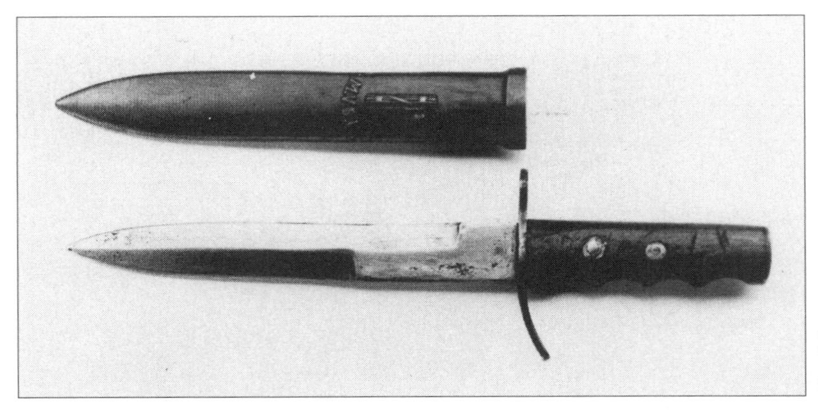

A fascist dagger. The sheath bears the *fasces* symbol.

Mussolini's followers in Italy, the Nazis in Germany, the Falangists in Spain, the members of the Leagues in France and Mosley's supporters in Britain were all Fascists. They were all anti-communist and to a greater or lesser degree exhibited most of the following characteristics:

The Fascist Programme

- Nationalism: Fascists were more than nationalistic – they were extremely or ultra nationalistic. The Italians were anxious to wipe out the shame of Caporetto (see chapter 11) and to gain the promised lands denied to them at Versailles. They sought more colonies and hoped to restore the glorious days of the Holy Roman Empire. This extreme nationalism was also evident in Nazi Germany with its policy of *Lebensraum* (living space).
- Racism: From extreme nationalism grew racism, a belief in the superiority of one's own race. This reached extremes in Germany with the persecution of Jews and non-Aryans. Mussolini was prevailed upon by Hitler to pursue a racist programme, but it was not followed with great vigour in Italy.
- Totalitarianism: Mussolini defined Italy's totalitarian regime as 'everything within the state, nothing outside the state, nothing against the state'. The individual had to exist and work completely for the benefit of the state and not vice-versa. In Italy and Germany this led to one-party government and the death of democracy.
- Anti-Communism: The post-war years saw strikes and demonstrations organised by the working classes, the growing strength of trade unions, terrorist acts carried out by Communists and the emergence of a Communist state in Russia. Fascist parties grew up throughout Europe in opposition to Communism. They gained the support of capitalists, industrialists and people of property.
- Cult of the Leader: Fascism offered an alternative to the system of parliamentary democracy which failed to solve the post-war social and economic problems. The Fascists offered a different system of government with strong rule and one all-powerful leader. The cult of the leader was emphasised by an elaborate propaganda programme.

Mussolini's Italy

The Road to Dictatorship

In 1922 Benito Mussolini became Italy's youngest Prime Minister, aged 39. However, out of a total of 535 deputies, the Fascists numbered only 35. Mussolini, in the early stages, moved carefully. There were only four Fascist ministers in his coalition government.

The first step towards dictatorship was the **Acerbo Law** of 1923. This law reserved two-thirds of the seats in parliament for the party which gained most votes in future elections. In the 1924 elections, Fascist thugs used violence to intimidate the opposition parties and to frighten voters. This ensured that the Fascists won a majority of seats.

Mussolini decided that he should personally control parliament. As well as being Prime Minister he was his own Foreign Minister. He also held the important post of Minister for the Interior. This meant that he controlled the police. The legalised Blackshirts were renamed the **Volunteer Militia for National Security** and were paid by the state. They swore an oath of allegiance to Mussolini rather than to the King. The **Fascist Grand Council,** formed as a consultative body, was nominated by Mussolini. The King and Constitution were retained to give an appearance of democracy, but Italy was well on the road to dictatorship.

The Murder of Matteotti

Up to 1924, some politicians were still brave enough to condemn Mussolini's dictatorial tactics. **Giacomo Matteotti,** a respected Socialist politician, wrote a book entitled ***The Fascists Exposed.*** He attempted to have the 1924 elections declared invalid. He had compiled a file of Fascist crimes to be published abroad. In May 1924 he attacked Mussolini in a speech in parliament. Mussolini called it 'outrageously provocative' and said it deserved 'some more concrete reply' than a speech in return. Ten days later some Fascist thugs took their leader at his word. Matteotti was kidnapped and murdered.

The murder was traced to two extreme right-wing Fascists. They were imprisoned, but were released after a year. Opposition deputies accused Mussolini of complicity in the murder. Over 100 deputies withdrew from the Chamber of Deputies in protest. In the long term this was a mistake, because it meant that there was now no opposition voice in parliament. For a while Mussolini felt that he would not survive the crisis.

The King refused to censure the Prime Minister and Mussolini soon turned the incident to his advantage. He recovered his courage and addressed the Chamber on 3 January 1925:

> 'I assume full responsibility for all that has happened. If Fascism has been nothing more than castor oil and the rubber truncheon, instead of being the proud passion of the best part of Italian youth, then I am to blame! Gentlemen, Italy wants peace, tranquillity, calm in which to work. We will give her this by means of love if possible, but by force if necessary.'

Il Duce – Cult of Leader

All portraits of Mussolini after 1925 were entitled ***Il Duce*** (The Leader). The cult of Mussolini was fostered mainly through his newspaper *Il Popolo d'Italia*, then run by his brother Arnaldo. It set an example to other publications by writing *DUCE* in capital letters.

Fascist Blackshirts await the arrival of Mussolini

In January 1926 parliament granted Mussolini the power to rule by decree. This marked the end of Mussolini's semi-democratic rule and his emergence as a full-blown dictator. He issued about 100,000 decrees in the next twenty years. Italy became a one-party state. Strict censorship ensured that the people heard only what their leader wanted them to hear. He was skilfully presented to them as Italy's man of destiny. Slogans such as *'Il Duce la sempre ragione'* ('The Leader is always right') were scrawled across buildings all over the country.

Schoolteachers had to take an oath of loyalty to the Fascist regime. Only committed Fascists could teach history, regarded as a most important subject. Of course, facts were altered to present Mussolini as a Superman, the saviour of his country. He posed as a 'man of culture' who wanted to bring about an artistic revolution. After 1931, all public servants had to join the Party. Children at the age of eight were expected to join the ***Balilla***, the Fascist Youth Movement. The symbols of the movement were the rifle and the book. Children were taught to repeat the Fascist creed: 'I believe in the genius of Mussolini.'

The problem for Mussolini was that he began to believe his own propaganda. He became convinced that he was infallible:

> 'On my grave I want this epitaph: "Here lies one of the most intelligent animals ever to appear on the surface of the earth".'

This extraordinary arrogance eventually led to Mussolini's downfall.

Repression

The Fascist Government set up machinery to deal with enemies of the State. In 1926, a secret police force called **OVRA** was formed to deal with those who opposed the regime. A special court tried people accused of crimes against the State. It had power to sentence people to death for treason. Despite the emphasis on law and order, crimes against opponents of the Fascists went unpunished. Fascist thugs were given a free hand.

Concentration camps or penal settlements were built on islands off the coast of Italy for political prisoners. However, the treatment of prisoners was far less severe than in Nazi Germany, or in Stalin's Soviet Union. Between 1927 and 1940 only ten people were executed. Less than 4,000 people were sent to concentration camps.

The persecution of Jews did not take place until the last years of the Fascist regime and then it was at Hitler's prompting. Mussolini told a member of his party that he did not 'believe in the least in this stupid anti-Semitic theory'. The **Race Law** of 1938 classified a Jew as anyone with a Jewish parent. They were banned from owning large businesses, buildings, or estates in the country. They could not join the Fascist party, were not allowed to fight in the army and were barred from working for the government.

The Economy

Mussolini was fortunate in his first Minister for Finance, De Stefani, who was a liberal economist and held his post from 1922 to 1925. De Stefani introduced a simpler tax system and attracted capital to Italy from abroad by offering foreign businesses exemptions from taxation. He believed that, wherever possible, the government should not interfere in business. He reduced government expenditure and achieved a budget surplus for the first time since the war. Trade revived and unemployment fell from 500,000 to 122,000.

Some industrialists pressed Mussolini for state aid and, at their behest, he dismissed De Stefani in 1925. *Il Duce* now took a more direct interest in the economy. He believed that the economy should be subject to considerations of national prestige. He revalued the *lira* by 10% in 1926, thereby making Italian exports dearer. He cut the salaries of public employees and raised tariffs on imports. The economy became severely depressed and unemployment exceeded a million in 1932. Mussolini stubbornly refused to devalue the *lira* until 1936.

The Italian government tried to implement a policy of national self-sufficiency. As a result of the 'Battle for Grain', wheat imports were reduced to a quarter of their previous level. However, other crops were neglected and the Fascist regime never achieved its aim that the country should become self-sufficient in food.

After 1932, the government adopted a system of state capitalism. Big businesses, such as shipbuilding, aviation and the automobile industry, benefited from grants and government orders. By 1940, the state held a 20 per cent interest in Italian industry, a very high percentage at that time.

Church-State Relations

In order to create a fully totalitarian state, Mussolini sought to reach an understanding with the Catholic Church. Since the Unification of Italy in 1870, the Church had withheld recognition from the Kingdom of Italy. Mussolini had personally abandoned Catholicism early in life, but he knew that an agreement with the Church would increase his prestige, not just in Italy but throughout the world.

Pope Pius XI saw Fascism as a bulwark against Communism and in 1922 appealed to Catholics not to resist the Fascists. Relations between Church and State gradually improved and in 1929 the **Lateran Treaty and Concordat** was signed. Its main terms were:

- The Church recognised the Kingdom of Italy and agreed to seek parliamentary approval of Church appointments.
- The Church supported the Fascist anti-Communist policies.
- The State recognised the independence of the Vatican City. It was to have its own army, police, courts, postal system and independent radio station.
- The Pope gave up all claims on Italian territory. Generous compensation was paid to the Vatican for the loss of the Papal states.
- Catholicism became Italy's official religion and the Church was given control of religious education in schools.

Good relations between the Pope and Mussolini did not last. The Pope objected to restrictions placed on Catholic organisations and newspapers and on religious instruction in schools. He published an encyclical (letter) condemning 'the monopoly of youth' by the *Balilla* and a 'pagan worship of state'. Mussolini knew that the Church had a strong hold on the hearts and minds of most Italians and he modified his attitude.

The Pope has often been criticised for his support of 'the man sent by Providence', his description of Mussolini in 1922. However, he showed more courage than most European statesmen. He declared the invasion of Abyssinia in 1935 'an unjust war' and condemned the anti-Jewish laws of 1938. The neutrality of the Vatican City state was respected by all powers during World War II.

The Corporate State

> 'We control the political forces, we control the moral forces, we control the economic forces. Thus we are in the midst of the corporate Fascist state.'
>
> (Mussolini)

Corporations were semi-state bodies set up in order to achieve the economic recovery of Italy. Mussolini dissolved trade unions which he regarded as communist in outlook. Every profession and industry set up its own **Corporation.** Each had employer delegates, employee delegates and three Fascist Party members. Strikes and lock-outs were forbidden and problems would be solved through negotiation. In 1939 a **Central Committee of Corporations** replaced the Chamber of Deputies.

In practice, Corporations favoured the employers. Between 1925 and 1938 wages on average fell by 10 per cent. Living standards generally remained among the lowest in Europe. Workers benefited from improved schemes of insurance, covering unemployment, sickness and accident. Corporations never became the 'fascist institutions par excellence' that Mussolini had hoped. A costly and often corrupt bureaucracy failed to prevent economic collapse during the war.

Reforms

Despite its serious defects the Fascist regime had some considerable achievements:

- Mussolini reformed Italy's transport system. He insisted that the 'trains run on time'. The building of bridges, canals and major road systems called *Autostrada* improved communications throughout the country.
- A team of engineers drained the **Pontine Marshes,** swampland that lay between Rome and Naples. People from the poorest regions were settled on this and other reclaimed land.
- Mussolini had some successes in his economic policies. He announced the **'Battle for Grain'** which doubled grain production between 1922 and 1939. Industry and agriculture were both encouraged to modernise.
- Mussolini's government improved educational standards. The school-leaving age was raised, new schools were built and higher standards of attainment were set. Between 1922 and 1939, the number of secondary schools increased by 120 per cent. Physical fitness became a priority. It was unfortunate that the Fascists should have spoiled the reforms by using education for indoctrination.
- Mussolini also had a measure of success in dealing with the **Sicilian Mafia.** He had in 1927 suffered the humiliation of travelling round the island under the protection of the local Mafia boss, rather than under the protection of the police. The ruthless methods of the Fascists curbed the powers of the Mafia. Many gangsters emigrated to the U.S.A.

The bus-station at Littoria, a town built on the Pontine Marshes

A distinguished Italian writer, Alberto Moravia, observed years after that 'if Mussolini had had a foreign policy as clever as his domestic one, perhaps he'd be *Duce* today'. It certainly was his disastrous foreign adventures that caused his downfall.

Foreign Policy

Early successes

'I will make Italy great, respected, feared'.

(Mussolini)

Mussolini saw relations with other countries as a means of getting prestige for Italy. He acted as his own foreign minister up to 1936 and then gave the post to his son-in-law, **Ciano.** He believed that Italy should dominate neighbouring states in Southern Europe and North Africa and that the Mediterranean should become an 'Italian lake'. **Fiume** was restored to Italy in 1924 after successful negotiations with Yugoslavia. Mussolini also gained prestige and popularity among nationalists by his aggressive handling of the **Corfu** incident (see chapter 13). He defied the League of Nations and forced Greece to pay 50 million lire in compensation for the death of four Italians on Greek territory. Italy also established a protectorate over **Albania** in 1926.

THE WAR SALAD
Mussolini "Let me see – they say 'A miser for the vinegar; a spendthrift for the oil, and a madman to stir it.' But – is the oil going to hold out?"

Moderate Policies

While Mussolini was establishing his power in Italy, during the 1920s, his foreign policy remained moderate. In 1925, he signed the **Treaty of Locarno** (see chapter 18), thereby accepting the division of Europe agreed at Versailles. In 1928, he signed the **Kellogg-Briand Pact,** thereby agreeing to renounce war as an instrument of policy. References to an Italian Empire seemed, at this stage, mere talk, mainly for the benefit of ardent Italian nationalists.

Imperialist Expansion

In 1925, Mussolini announced that his objective was to 'found an empire' and to 'win glory and power'. His visit to **Libya** in 1926 provoked a long and costly war with rebels who would not accept Italian rule of their country.

In 1935 Italy launched a sudden attack on **Abyssinia** (see chapter 13). The League of Nations imposed only partial sanctions on Italy. These had little effect, but Mussolini withdrew his country from the League in 1937. Victor Emmanuel was declared **Emperor** of the new Italian Empire.

Relations with Hitler

Mussolini was worried about Hitler's aggression in Europe and opposed the idea of union between Germany and Austria in 1934: a weak Austria acted as a buffer state between Italy and Germany. Italian troops were moved to the Austrian border in 1934 and the idea of union between Germany and Austria was postponed. In 1935 Italy, Britain and France joined the **Stresa Front** and condemned German rearmament.

Hitler and Mussolini were both Fascists and the Italian invasion of Abyssinia won Hitler's admiration. The Führer used the crisis to occupy the Rhineland. The Germans defied the League's sanctions and continued to trade with Italy. Relations continued to improve and the **Rome-Berlin Axis** was signed in 1936. In 1937 Italy joined Germany and Japan in the **Anti-Comintern Pact.** The Italians were not yet ready for war and delayed signing a full military alliance, the **Pact of Steel,** until 1939.

Mussolini and Hitler cooperated in support of their fellow Fascist, Franco, in the Spanish Civil War (see chapter 21). Over 70,000 Italian troops were committed to the war. Mussolini was strongly anti-Communist. He also saw the war as an opportunity of displaying Italy's growing military strength (to the world). The succession of wars that Mussolini fought since 1926 in Libya, Abyssinia and Spain meant that Italy was militarily and economically exhausted before World War II started.

Hitler allowed Mussolini to think he had 'saved Europe' at **Munich** in 1938. Italy was not ready for war and Mussolini gloried in the role of peacemaker. His solution was to give in to Hitler's demands and sacrifice Czechoslovakia (see chapter 18).

In September 1939, Mussolini was forced to break his Pact of Steel Agreement when he announced that Italy was not involved in the war. The Italian army was not

Hitler and Mussolini in Rome, 1938

ready for battle. Mussolini declared war in June 1940, as he then considered a German victory certain. Germany had to send troops to help the Italians in **Greece** and **North Africa.** Hitler later said that the Italian fiasco in Greece cost him the war, as it forced him to delay the invasion of Russia.

End of Fascism

Mussolini deposed

World War II cost Italy her empire and Mussolini his life. In 1941 Italy lost Abyssinia and Italian East Africa. The Italian North African army surrendered at **Tunis** in May 1943. The Allies invaded **Sicily** and the first air-raid on **Rome** occurred two months later. This led the Fascist Grand Council and the King to remove Mussolini from his position.

Marshal Badoglio took over as Prime Minister of Italy in July 1943. Mussolini was arrested and kept in confinement in a mountain resort. Hitler needed his Fascist colleague to act as a puppet leader in a German-dominated northern Italy. Mussolini was rescued by a German commando unit in a daring raid. He set up a rebel Fascist government at **Salo** in Northern Italy. This led to eighteen months of bloody civil war with Badoglio's supporters and the Allies.

Final Days

Rome fell to the Allies in June 1944. Mussolini's position became hopeless. He refused to surrender, as he was a mere tool of the Germans, and he feared a trial as a war criminal. Disguised as a German soldier he joined a retreating German convoy in

April 1945. He was captured by Italian partisans. On April 28, 1945, Mussolini and his mistress were executed. Their bodies were left hanging upside down in a street in **Milan,** objects of derision by a disillusioned and hostile public.

Mussolini had said that Fascism would die with him. It was one of the few accurate predictions that he had made!

Assessment

Despite some minor successes and reforms, the Fascist regime in Italy ultimately proved a failure. In theory, the Corporate State was supposed to bring social justice and also real planning of the economy. In practice, it was a device through which the political dictatorship of the Fascist Party was exercised. The Corporations tended to favour the employers and to become corrupt instruments of state bureaucracy. The reduction in strikes was due to state repression, rather than to the Corporations.

Mussolini's economic policies proved a case of shadow rather than real substance. Intervention in Abyssinia, Spain and Albania proved too much of a drain on the country. By 1936 the budget was badly in deficit and Mussolini was forced to devalue the *lira*. The country had an inflation rate of 20 per cent in 1937. Italy had limited resources and a policy of self-sufficiency did not suit it. Real economic growth in the Fascist years was very small.

The Catholic Church and the monarchy both retained a large measure of independence in Italy. The power of the regime was limited, compared with Nazi Germany, or the Soviet Union under Stalin. Mussolini created a great army, but he did not have the resources to equip it properly. His pretensions were brutally exposed in 1939, when Italy was not prepared for war.

Mussolini's one great talent lay in propaganda. This prevented the inefficiency of his regime from being appreciated and remedied. Eventually he began to believe in the image he had created. It was ironic that his single great skill contributed to his ruin and to the downfall of Fascism in Italy.

Chapter 16

The Weimar Republic, 1919-1933

Germany in Crisis, 1918-1923

The Republic Proclaimed

Woodrow Wilson insisted that the Kaiser would have to go, before the Allies would enter into peace negotiations with Germany. William II abdicated on 9 November, 1918. He was given protection by the Dutch and lived in Holland until 1940.

On the day of the Kaiser's abdication, a republic was proclaimed in place of the Second Reich. The acting Chancellor of the new **Provisional German Government** was **Friedrich Ebert**, leader of the **Social Democratic Party**. He had at one time been a saddle-maker and became a union organiser. It was Ebert's government which signed the armistice with the Allies on 11 November, 1918, and brought the war to an end.

The Spartacist Rising

The Social Democrats took over a land of misery, refugees, homelessness and starvation. Many discontented people had joined the Communist Party, founded in 1918 by **Karl Liebknecht** and **Rosa Luxemburg**. They were known as the **Spartacists**, named after Spartacus who had led a great revolt of slaves in ancient Rome.

March of communist children in protest against fascist violence

The Spartacists were calling for nationalisation of industry and the abolition of the army. They attacked the Chancellory Building in Berlin in December, 1918. Riots and street fighting in January 1919 were threatening the very existence of the new government. Ebert was forced to bargain with the ***Freikorps*** (Free Corps). These were irregular forces, mainly ex-army officers and soldiers. They were a conservative but undisciplined body, committed to right-wing policies and law and order at any cost. The Spartacist revolt was crushed by the Freikorps with great brutality. Karl Liebknecht and Rosa Luxemburg were captured and murdered. Post-war politics in Germany were off to a bloody start.

The Constitution

In January 1919, elections were held for a National Assembly. The Social Democrats were the largest party and the majority of those elected favoured a republic. On 9 February the new National Assembly met in the town of **Weimar**, a cultural centre in south-east Germany. It could not meet in riot-torn Berlin and wished to demonstrate a break with the past. Weimar, the home of Goethe and Schiller, signalled a move away from the old military authoritarianism.

The Weimar Constitution gave its name to the new Germany, which became known as **The Weimar Republic**. It was liberal and democratic. The President, elected by popular vote, would hold office for seven years. He would appoint the Chancellor, who would select a Cabinet to govern the country. The Cabinet would be answerable to the lower house, **the Reichstag**. The second house, **the Reichsrat**, had limited legislative power.

The Republic was a federal State, in which great freedom was given to the local parliaments. Civil and religious liberties were guaranteed and women were given the vote. Important issues could be referred to the people in a referendum. In time of national emergency, the President could assume special powers and suspend civil liberties. Ebert became the first President of the Weimar Republic

The Humiliating Treaty

> 'Today in the Hall of Mirrors of Versailles the disgraceful Treaty is being signed. Never forget it!... Today German honour is dragged to the grave. Never forget it! The German people with unceasing labour will push forward to reconquer that place among the nations of the world to which they are entitled. There will be vengeance for the shame of 1918.'
>
> (*Deutsche Zeitung* (German News) 28 June, 1919)

The Allies presented the terms of the peace treaty to the National Assembly. The Germans were angered by the loss of territory, the guilt clause and the demand for huge reparations (see chapter 12). It is untrue that the Germans accepted the terms tamely, as Hitler later suggested. In protest, they scuttled their fleet at Scapa Flow, the British naval base, rather than hand it over to the Allies.

The German deputies disputed the terms bitterly. However, the country was in no position to resume the war, as the blockade remained in place. Representatives of Ebert's government signed the Treaty of Versailles on 28 June, 1919.

The Kapp Putsch

The extreme nationalists could not bring themselves to believe that Germany had lost the war. Some had greeted the returning troops in 1918 as heroes. They now blamed the government for accepting the humiliating peace terms. The legend grew up that Germany had not lost the war, but had been stabbed in the back by the 'November Criminals', who signed the Armistice in November 1918.

These extreme nationalists now combined with the *Freikorps*, who had tasted blood and power during the Spartacist revolt. In March 1920, **Wolfgang Kapp**, a politician and rabid nationalist, led a group of ex-army officers and *Freikorps* in an attempted *coup d'état*. They marched on Berlin and forced Ebert's government to flee. The regular army, the Reichswehr, refused to act against the rebels and the Republic seemed doomed.

Ebert and the Minister for Defence, **Noske**, called on the workers to defend the Republic. A general strike led to the breakdown of public services and commercial life. Kapp found himself isolated and fled to Sweden. His putsch had been thwarted by the workers. It was significant that some units of the *Freikorps* wore a swastika badge when they marched on Berlin.

Invasion of the Ruhr

In 1921, the Reparations Commission fixed £6,600 million as the sum the Germans would have to pay in compensation to the Allies for damage caused in the Great War. It was to be paid partly in money and partly in goods. Most of it would come from the Ruhr, the industrial centre of Germany.

At the end of 1922, the Germans fell behind with their repayments. **Raymond Poincaré**, the new French Premier, decided to act. The rebuilding of his war-damaged country was costing a great deal of money. The French taxpayer would have to be penalised even more unless the German payments continued. French and Belgian troops occupied the Ruhr in January 1923.

At first the French thought they could make the Germans work for them. However, the German government and people were united against a common enemy. German miners and factory-workers in the Ruhr went on strike. They refused to co-operate with the invaders. The French and Belgians brought in their own workers and tried to cut off the Ruhr from the rest of Germany. They deported the leaders of the passive resistance. Violence flared and people were killed on both sides.

Germans load baskets of paper money onto wagons during inflation.

Inflation

The Weimar government backed those who were on strike in the Ruhr by paying their wages. The crippling cost of the war, the pressure of the Reparation demands and the problem in the Ruhr all combined to cause massive inflation. The government responded by printing billions of banknotes. Germany lacked the real wealth to back these notes. The *mark* soon became almost worthless.

The drop in the value of the *mark* is best seen in the number of marks needed to buy **one dollar's worth** of goods.

Year	No. of *marks* needed to buy one dollar's worth of goods
1919:	8.9
1921:	70
1922 (Jan):	192
1922 (Aug):	1,000
1923 (Jan):	18,000
1923 (July):	160,000
1923 (Aug):	1,000,000
1923 (Nov):	4,200,000,000

The inflation was felt particularly by the German middle-classes whose savings became valueless. Skilled workers, those with modest nest eggs, people living on pensions and those with investments were severely hit. Their loyalties to the Republic and their life savings were blown away like leaves in the wind.

The Stresemann Era, 1923-1929

Fulfilment

Gustav Stresemann was an industrialist and was leader of the People's Party in the Reichstag. He was originally a right-winger and a supporter of *Weltpolitik* (see chapter 9). Above all, Stresemann was a realist and he came to believe that unless democracy survived, Germany would soon become a dictatorship.

In the Reichstag, Stresemann came to the fore when he bitterly attacked the occupation of the Ruhr. He gained support from the Socialists when he condemned businesses who were using the inflation for their own benefit. He was elected Chancellor, heading a coalition government, in August 1923. Inter-party disputes led to the fall of his government in November of that year. He was Foreign Minister from 1923 until 1929. His influence was such that these years have become known as the 'Stresemann Era'.

Gustav Stresemann followed a policy of fulfilment. He believed that if Germany tried to fulfil the terms of the Versailles Treaty, she would become 'acceptable and respectable'. The Allies would treat her with sympathy and would cancel some of the harsher terms of the settlement.

Economic stability

When he took office, Stresemann ordered an end to the policy of passive resistance in the Ruhr, as it had been depriving the Government of vital revenue. He abolished the

Gustav Stresemann, Foreign Minister 1923-29

old *mark* and introduced a new currency, the *Rentenmark*. Tax reforms and the reduction of Government expenditure helped the economic situation. The Government announced that Germany would resume the payment of reparations. However, Stresemann also opened negotiations with the Allies on the question of his country's payments.

Dawes Plan

A committee under the American banker, **Charles Dawes**, drew up a more reasonable schedule of reparations in 1924, based on Germany's ability to pay. An American loan of $800 million was granted to Germany. It has been calculated that foreign loans granted to Germany between 1924 and 1929 exceeded the amount paid in reparations. The German economy quickly recovered and the country enjoyed a brief period of prosperity and political stability.

Rearmament

Stresemann did all in his power to make Germany acceptable to other countries. However, at heart he was an ardent nationalist and he allowed the army to find ways of getting around the military clauses of the Treaty of Versailles. Secret army units were disguised as labour battalions and heavy industry cooperated in producing arms.

Germany and the U.S.S.R. were 'outcast' nations, the former guilty aggressors and the latter, Communists. The Germans helped with the reorganisation of the Red Army. In return, the Russians provided facilities for the production of German tanks and aircraft. The **Treaty of Rapallo in 1922** established formal relations between the two countries. Stresemann built on this Treaty during the 1920s and re-established his country among the community of nations.

The Locarno Pact

Stresemann's policies gradually convinced Britain and France that Germany was no longer a threat to the peace of Europe. **The Locarno Pact of 1925** was another triumph for Stresemann. By the terms of this agreement:

- The frontiers between Germany, France and Belgium were accepted as final.
- The Rhineland was to remain demilitarised, and Allied troops would be gradually withdrawn from the area.
- Germany would not attempt to alter her eastern frontiers by force. Disputes with Czechoslovakia and Poland about frontiers would be submitted to arbitration.

The Locarno Pact was also signed by Britain and Italy as guarantors. Europe relaxed and old enmities seemed to be disappearing. Germany was invited to become a member of the **League of Nations** in 1926. She signed the **Kellogg-Briand Pact** in 1928, outlawing war as a means of settling differences.

The Young Plan

Stresemann continued to press for more concessions. He demanded a final settlement of the reparations issue. The Allies set up a committee to investigate the matter. It was headed by an American businessman, **Owen Young**. In 1929, Stresemann accepted the Young Plan. It recommended that the amount Germany owed should be reduced to £2,000 million and that the period of repayment should be extended to 59 years.

Death of Stresemann

Ebert, the first President of the Weimar Republic, had died in 1925. His place was taken by **Field Marshal Paul von Hindenburg**, hero of World War I. In 1925, he was 77 years old and though at heart a monarchist, he was loyal to the Republic during his term of office.

On 3 October, 1929, Stresemann died suddenly from a heart attack. He had been responsible for saving his country and making her respectable among the nations of Europe. However, many Germans were not satisfied with the gradual erosion of Versailles achieved by Stresemann. Within a few years, German foreign policy was to follow a very different direction.

A street brawl between rival political groups, 1929

The Fall of Weimar

The Depression

The **Wall Street Crash** occurred on 29 October, 1929, less than a month after the death of Stresemann. American money soon stopped flowing into Germany. Short-term loans were called in. Factories and businesses closed. Unemployment soared.

Year	Unemployed
1929	1.3 million
1930	3.0 million
1931	4.3 million
1932	6.0 million

The government response was an attempt to cut state expenditure, including unemployment benefits. The government did not survive the crisis.

Brüning

The new Chancellor, **Heinrich Brüning**, 1930-32, could not command a majority in the Reichstag for his austerity programme. He was forced to rely on Article 48 of the Constitution, which allowed rule by presidential decree. Violence increased and the parties of the extreme left and extreme right gained massive support, as people lost faith in the democratic system. The **Nazi Party**, led by **Adolf Hitler** increased their seats in the Reichstag from 13 to 107 in the 1930 elections (see chapter 17 for details of the rise of Nazism.)

Hindenburg was re-elected President in 1932, but Hitler came a close second. Brüning was beginning to lose control of the Reichstag and was dismissed by Hindenburg in 1932.

Von Papen

Hindenburg appointed **Franz von Papen** as Chancellor. He had even less support than Brüning in the Reichstag and formed a government from outside Parliament.

Hitler with President von Hindenburg

Election poster in support of Hitler – 'one people, one leader, one YES'.

Election poster 1932 – 'We want work and bread! Vote Hitler!'

This 'Cabinet of Barons' was extremely right-wing and was forced to rely on the President signing emergency decrees.

The Nazis won 230 seats in the July 1932 elections and became the largest group in the Reichstag. After another inconclusive election and serious street rioting, Von Papen was forced to resign. An interim government could not rely on the support of the Reichstag and the way was open for the Nazis to make a bid for power. (see chapter 17 for details.) On 30 January 1933, Hitler became Chancellor of Germany. His aim was to destroy the Republic. On March 23, the **Enabling Act** made Hitler sole ruler of Germany. The Weimar Republic was dead.

von Hindenburg dies:
Hitler appoints himself Führer

Reasons for failure

- The Weimar Republic was born out of Germany's defeat in World War I. It accepted the humiliating Treaty of Versailles and was a symbol of national disgrace. Not enough Germans supported it or wished to keep it alive.
- Germans had little experience of democracy. They were used to the authoritarian leadership of kings, emperors, chancellors and generals.

- The German political parties grew up as representatives of sectional interests. They failed to develop policies which would appeal to the nation as a whole. Every government between 1919 and 1933 was a weak coalition, as no party could achieve an overall majority.
- The officer class in the German army failed to support the Weimar Republic. They resented the loss of the privileges which they had enjoyed under the Kaisers. As the government could not rely on the army, political parties created their own paramilitary sections. These did much to cause violence and intimidation and led to the de-stabilisation of the Republic.
- Other important groups, such as the educators, the judiciary, the industrialists, financiers and landowners, were generally hostile to the Republic. They looked back to the old days or forward to an authoritarian government of the right which would make Germany great again.
- The inflation and depression, 1929-1933, weakened the country and the lawful government. It enabled the reactionary forces to come to the fore. President Hindenburg was becoming senile and was too easily persuaded to accept these forces.
- The determination of Adolf Hitler was the final blow. He destroyed the Republic by outlining policies that appealed to many Germans. Instead of unemployment, depression and despair, he offered employment, stability and hope of greatness.

Cultural Achievements

Reasons for Innovation

Germany's defeat and the horrors of war led to a reaction against pre-war idealism. Post-war political and economic upheaval accelerated the trend. Artists felt that a new country and a new morality were needed.

The old restrictions associated with an authoritarian empire were swept away. Germany became the cultural centre of the new Europe, just as France had been before 1914. This was clearly seen in the fields of architecture, literature, theatre and cinema.

Architecture

Walter Gropius founded the *Bauhaus* Institute of Architecture, which sought to unite art, architecture and design. Gropius abandoned the old decorative and ornamental styles and concentrated on functional and useful aspects. Basic forms and straight lines became the order of the day. Steel frameworks, austere pillars and the use of glass were features of the *Bauhaus* school.

The Bauhaus also influenced design in furniture, sculpture, painting, textiles and pottery. Artists wished to serve people in the modern age. Hitler and the Nazis had little time for these designs, as they wished to recapture the old days of German glory. Many artists and architects emigrated during the Nazi regime and spread their ideas in America and other western countries.

Literature

Some writers used the new freedom from censorship to depict the horror of war. **Eric Remarque** (see chapter 11) produced a remarkable picture of a soldier's life in World War I in his famous novel *All Quiet on the Western Front*. **Oswald Spengler's** *The Decline of the West* depicted the decline of Western civilisation.

Thomas Mann in *The Magic Mountain* was concerned with the decline of capitalism. He cast a cold eye on German twentieth-century life in his allegorical tale *Dr Faustus*. Mann received the Nobel prize for literature in 1929, but was among those who left the country in the early thirties.

Existentialism, a philosophy which was later more closely associated with the French writers Sartre and Camus, was a predominantly German phenomenon between the wars. **Martin Heidegger** asserted that man must not attempt to flee from the reality of life. The subjective and personal must be placed above the logical, objective and intellectual.

Theatre

Theatre flourished in Weimar Germany, especially in Berlin. This was the time for experimentation. Even stage design was influenced by the *Bauhaus* school. Out went the cluttered stage, as the audience were encouraged to use their imaginations. New techniques added to the excitement. Expressionism aimed at creating emotional reaction in theatre audiences.

Bertolt Brecht criticised capitalism and appeared to praise communism in such works as *Mother Courage* and *The Beggar's Opera*. Other famous playwrights were **Heinrich Mann** and **Hermann Sudermann**.

Cinema

As in other European countries, Hollywood dominated the German commercial cinema. At home the expressionists of the 1920s experimented with lighting, acting and sets.

The most famous expressionist film was *The Cabinet of Dr Caligari*. The film *Western Front* realistically depicted the horrors of war. One of the earliest sound films was *The Blue Angel* with Marlene Dietrich, who later starred in many Hollywood roles.

Chapter 17

Germany under Hitler, 1933-1939

The Road to Dictatorship

Hitler's early life

Adolf Hitler was born in 1889 in **Braunau**, a small town in Austria. His father was a customs official and the family was not poor, contrary to Hitler's later claims. At school he showed some ability at art, but he was a lazy student. He failed his examinations and left school at sixteen. He blamed his teachers for his failure. Hitler was to seek out many other scapegoats in later life.

Following the death of his parents, Hitler moved to **Vienna**, where he lived for four years. He applied for a place in the Art Academy but failed the entrance examinations. He had no job but scraped out a living doing odd jobs and designing postcards. During this time he showed an interest in art, opera and politics.

It was while he lived in Vienna that Hitler formed the ideas which he developed in later years. He came in contact with German nationalism and began to hate people of foreign races.

The War Years

Hitler left Austria in 1913 and moved to **Munich**, a centre of German cultural life. He volunteered to join the German army at the outbreak of war in 1914. He was a brave soldier and received the Iron Cross First Class for bravery. Corporal Hitler said that war was 'the greatest of all experiences' and was bitter when Germany surrendered in 1918.

The Nazi Party

Adolf Hitler stayed on in the army after the war. He was asked to spy on political parties, in order to find out if they were dangerous. In September 1919, he attended a meeting of the newly-founded **German Workers' Party** in Munich. Hitler liked the ideas of the founders and joined the party. Before long he became its leader and renamed it the **National Socialist German Workers Party**, or **Nazi Party** for short.

Beer Hall Putsch

The Nazis formed their own small army, the **SA**, *(Sturm Abteilung)* known as the **Brownshirts**, to protect their meetings. Hitler made the hooked cross, the swastika, the emblem of the party. He tried to copy Mussolini and adopted the brown shirt as the uniform of his followers. He built up the party and encouraged his followers to call him ***Führer*** (Leader) The Weimar government was in difficulty in 1923 (see chapter 16) and Hitler decided it was time to make a bid for power.

In imitation of Mussolini's march on Rome, Hitler decided to organise a march on Berlin. On 9 November 1923, Hitler issued this announcement in Munich:

> 'Proclamation to the German people! The Government of the November Criminals in Berlin has today been deposed. A provisional German National Government has been formed. This consists of General Ludendorff, Adolf Hitler...'

Ludendorff, the heroic leader of the German army during World War I, had been attracted to the party by its nationalism. He had been disgusted when Stresemann called off the passive resistance in the Ruhr. The general and the ex-corporal led their band of about 2,000 brown-shirted Nazi **Storm-Troopers** through Munich at midday. The police and the army had been alerted and the way was barred. Shots rang out and 14 Nazis and three policemen were killed. Hitler and Ludendorff were arrested and charged with treason.

Mein Kampf

Ludendorff was acquitted, but Hitler saw his trial as an opportunity to make political speeches:

> 'I nourish the proud hope that one day the hour will come when these rough companies will grow into battalions, the battalions to regiments, the regiments to divisions...that the old flags will wave again.'

Hitler entertains friends in Landsberg prison, 1924. Rudolf Hess is second from right.

Hitler received the minimum sentence of five years. He served only nine months in comfortable surroundings in **Landsberg Prison.** While there, his assistant, Rudolf Hess, wrote down his leader's thoughts in a book which became Hitler's autobiography. It was titled ***Mein Kampf*** (My Struggle). It is long, repetitive and boring, but Hitler's political theories and the Nazi Party policies are to be found in this book.

Nazi Party Policies

Nationalism

All German-speaking people should be united. This meant that those living in Austria and in other areas outside Germany should become citizens of the Fatherland.

Socialism

Germany's largest industries and department stores should be nationalised. Unearned incomes and ground rents should be abolished, while profits made during the war should be confiscated.

Master Race

The best races are pure ones which have not interbred with others. Pure-bred Germans, typically tall and blonde, belonged to the **Aryan** race. They should keep themselves pure, in order to become the *Herrenvolk,* or master race.

Non-Aryans, such as Jews and Slavs, are inferior to the master race. These enslaved races should work for the benefit of the Aryans. The Jews are involved in a plot to take control of the world. They helped bring about Germany's defeat in the war and must be destroyed.

Living Space

Germans need more living space or ***lebensraum***. They must expand by taking over countries of the inferior races to the east of Germany, for example Poland and Russia. They will have to use force, if necessary, to take over these lands.

The Treaty

The Treaty of Versailles was signed by those who betrayed Germany. It is unfair and must be cancelled. The territory taken away from Germany must be returned. France must be destroyed.

Germany cannot achieve these aims unless she is strong. Conscription is needed to build up the army. The youth of Germany need disciplined training. Rearmament is vital.

Communism
Communism, the political system in Russia, is dangerous. It must be completely destroyed.

Totalitarianism
The interest of the state must come before the interests of the individual.

The Führer Principle
Nothing can be achieved unless Germany is ruled by a single, strong leader who has great power, a *'Führer'*.

The Lean Years

When he came out of prison in December 1924, Hitler found that the country had changed. The French troops had gone and the country enjoyed the stability and relative prosperity of the 'Stresemann Era' (see chapter 16). Hitler had learned a lesson after the failure of 1923:

> 'Instead of working to achieve power by armed coup, we shall have to hold our noses and enter the Reichstag...Sooner or later we shall have a majority and after that, we shall have Germany.'
>
> (Karl Ludecke, *I knew Hitler*, 1938)

Electoral success

For the next five years the Nazis built up the party and waited. The death of Stresemann, the depression, unemployment and the instability of the Weimar government (see chapter 16) gave the Nazis their opportunity. They had only 12 deputies in the Reichstag in 1928. In 1930, a year after the Wall Street Crash, they had 107 seats. This figure was increased to 230 in 1932.

The increase in the vote for the Nazi Party was due to Hitler's propaganda. He promised jobs to the unemployed, more land to the poorer farmers, government contracts to business chiefs and rearmament to the army. Instead of weakness and indecision, he offered strong leadership and decisive action.

Hitler becomes Chancellor

After 1930, President Hindenburg was forced to come to terms with the Nazis. He had come to rely increasingly on the advice of General **Kurt von Schleicher** of the German army. Von Schleicher recommended the dismissal of the moderate Chancellor, Brüning, and his replacement by the aristocrat, **Franz von Papen**.

Von Papen posed as champion of law and order. The German army was used against Prussian Socialists. The Chancellor lifted the ban which Brüning's government has imposed on Hitler's private armies. Von Papen failed to gain the confidence of the Reichstag. President Hindenburg agreed to the appointment of von Schleicher as Chancellor in December 1932.

Von Schleicher also failed to get the support of the Reichstag. He urged the President to dissolve the parliament and give the Chancellor powers to govern without it. Hindenburg refused, as he feared that civil war might result, and von Sleicher resigned.

Von Papen now urged the President to appoint Hitler as Chancellor, with von Papen himself as Vice-Chancellor. Hindenburg despised Hitler and the Nazis. However, due to continuing instability, he felt he had no choices left. On 30 January 1933, a new government was formed. Hitler was Chancellor in a cabinet of three Nazis and ten conservatives. Von Papen and the right wing felt that they would be able to control Hitler. They quickly learned that they had made a serious error.

One Supreme Leader

New Elections

> 'We will become Reichstag deputies in order to cripple...Weimar...with its own type of machinery. If democracy is so stupid as to provide us with free tickets... that is their own affair...We come as enemies! We come like the wolf which breaks into the sheepfold.'
>
> (Joseph Goebbels)

Hitler had no intention of upholding democracy. His first act was to call for another election. In his own mind it was to be the last. He controlled the police, the armed services and his private armies. During the run-up to the elections, political meetings of the Communists and Socialists were broken up. People were encouraged to associate chaos and the collapse of law and order with the left-wing parties.

Reichstag Fire

One week before the election the Reichstag building went up in flames. A Dutch Communist, **Marinus van der Lubbe**, admitted that he set the fire as an act of protest. The event suited the Nazis very well. Hitler said that it was the start of a Communist plot to take over the country.

He persuaded President Hindenburg to sign a law for the **Protection of the People and the State**. This law banned the Communists from campaigning in the election. Their newspapers were closed down and four thousand of them were thrown into prison. In the 1933 election, the Nazis won 288 seats, but failed to gain an overall majority.

March 1933, members of the SA, the Nazis' private army, terrorize Communists in Berlin

Enabling Act

Hitler wished to be able to act without the permission of the President or parliament. He therefore proposed the **Enabling Bill** on 23 March 1933. The parliament meeting was surrounded by **Storm-Troopers**. Some deputies were prevented from entering, while others were intimidated. Only 94 Socialists voted against the bill which gave Hitler power to suspend the constitution and rule by decree.

Hitler used the Enabling Act to put Nazi officials in charge of local governments. All political parties other than the National Socialists (Nazis) were banned. The Weimar Republic was dead. In the new one-party totalitarian state, democracy could not survive.

The Storm-Troopers (SA)

Hitler had used the SA, his brown-shirted Storm-Troopers, in his bid for power. These were the thugs who had smashed the Communists in the 1933 elections. By 1934, the ranks of the SA had swollen to over 2 million. Many of them leaned towards the left in politics and were disappointed that most of the socialist policies of the Nazis had not been implemented. These policies had been included in order to attract the working class and small traders. When Hitler began to get help from leading industrialists and landowners, he tried to quietly forget about his socialist programme.

The leader of the SA was **Ernst Röhm**. He wanted to become a general and thought that the two million Storm-Troopers should be formed into a people's army under his command. He began to criticise Hitler, calling him a 'swine' a 'traitor' and an 'ignorant world war corporal'. The SA were beginning to call for a 'second revolution'.

The regular German army, the ***Wehrmacht***, was in a weak position due to the restrictions placed on it by the Treaty of Versailles. Army leaders hoped that Hitler would introduce conscription and rearm the forces. Two of Hitler's ministers, **Himmler** and **Göring**, as well as the German High Command, wanted to see an end to the power wielded by the upstart, Röhm, and the SA.

The SS

Hitler relied for protection on the **SS**, an elite paramilitary bodyguard. This black-shirted force came under the command of **Heinrich Himmler**. This ruthless leader saw his duties as extending beyond protection to the elimination of all opposition to Hitler and the Nazis.

The Night of the Long Knives

Himmler and Göring passed on information to Hitler about the behaviour of the SA and Röhm's ambitions. Hitler saw an opportunity of finding favour with the *Wehrmacht* command, while at the same time destroying the threat from the SA. On 30 June 1934, the SS butchered 150 people at Stradelheim Prison. Röhm and about

Heinrich Himmler – in charge of Hitler's SS

250 others were murdered. Most were members of the SA, but Hitler and Himmler also included a number of citizens whom they considered enemies of the Nazis. Hitler justified his actions in a chilling speech to the Reichstag:

> 'I ordered the leaders of the guilty shot. If someone asked me why we did not use the regular courts, I would reply: at that moment I was responsible for the German nation... It was I alone who, during those twenty-four hours, was the Supreme Court of Justice of the German People!'

The Führer

On 2 August 1934, President Hindenburg died. Hitler decided to unite the offices of Chancellor and President. He also declared himself supreme commander of the armed forces and called himself *Der Führer* (The Leader). All members of the armed forces now swore on oath of loyalty to him alone:

> 'I swear by God this sacred oath, that I will give unconditional obedience to the *Führer* of the German Reich and *Volk*, Adolf Hitler, the Supreme Commander of the *Wehrmacht*, and, as a brave soldier, will be ready at any time to set down my life for this oath.'

In a plebiscite, 90% of those who voted ratified Hitler in his position as Führer.

End of Rise to PWR.

Nazi Germany

Nazi Persecution

The Jews

> 'The Jew...is and remains a parasite, a sponger.'
>
> (Hitler)

> 'The Jew...is the real cause of our loss in the Great War'.
>
> (Goebbels)

Just to be a Jew became a crime in Germany. It is calculated that about 600,000 Jews lived there in 1933. They owned many shops and businesses and were prominent in banking, law and medicine. However, there were many poor Jews also. Jews were persecuted for no other reasons than race, religion and envy.

When they came to power in 1933, the Nazis engaged in a campaign of violence against the Jews – window-smashing, looting of shops and assaults. These eased for a time after condemnation abroad, especially in the United States. However, in April 1933, the Jews were expelled from the Civil Service and excluded from universities.

Demonstration outside Jewish-owned shop, 1938. Poster at right says 'Germans do not buy from Jews'.

They could not work as journalists. The **Nuremberg Laws** of 1935, 'recognising that purity of blood is essential for the survival of the German race' stated:

- Marriages between Jews and Germans are forbidden. Those marriages which have already been contracted in contravention of this law are declared null and void.
- Relations between Jews and Germans outside of marriage are forbidden.

Jews were deprived of their citizenship and forced to wear the **Star of David** so that they could be publicly identified. In 1936 the anti-Jewish campaign was eased, as the Nazis tried to present a benign face during the Olympic Games held in Berlin. However, Hitler was annoyed when the American negro athlete, Jesse Owens, won four gold medals. The Chancellor would not present the prize to Owens, whom he perceived as a representative of an inferior race!

An American writer who lived in Germany during the 1930s described how the Jews were treated:

> 'Over the doors of the grocery and butchery shops, the bakeries and the dairies were signs, "Jews not admitted". Pharmacies would not sell them drugs or medicine...Hotels would not give them a night's lodgings. And always, wherever they went, were the taunting signs "Jews Enter This Place at their own Risk"...At a

sharp bend on the road near Ludwigshafen was a sign "Drive Carefully! Sharp Curve! Jews 75 Miles an Hour!"

(W.L. Shirer, *The Rise and Fall of the Third Reich*)

Many Jews feared for their lives, or were not prepared to endure the persecution. **Albert Einstein**, the physicist, was among those forced to emigrate. On 9 November 1938, the Nazis launched a terrorist campaign against the Jews. It became known as ***Kristallnacht*** (Crystal Night) due to the amount of glass broken. Shops were wrecked, synagogues were burned and about 90 Jews were killed. The campaign was an act of revenge against the murder, on 7 November, of a German diplomat in Paris by a Polish Jew.

In the years which followed, conditions for Jews became even worse. During World War Two, Hitler decided on his **'final solution'** of the Jewish problem. This involved the mass murder of Jews from Germany and the occupied territories.

Concentration camps

The first concentration camp was opened near the village of **Dachau**, outside Munich, in March 1933. The prisoners in the earliest camps were Communists and Socialists. Catholic and Protestant clergy, gypsies, tramps, other 'undesirables' and 'enemies of the state' were also confined.

The concentration camps were run by a branch of Himmler's SS, the **Death's Head Units**, who wore skull and crossbone badges on their uniforms. Prisoners lived in overcrowded dormitory blocks, under a system of strict discipline. They wore prison uniforms and their heads were closely shaven. Prisoners were beaten and tortured. Many died from epidemic diseases or ill-treatment. Others were shot:

'Physical punishment consisted of whipping, frequent kicking (abdomen and groin), slaps in the face, shooting, or wounding with the bayonet. These prisoners were forced to stare for hours into glaring lights, to kneel for hours and so on.'

(Bruno Bettelheim, *The Informed Heart*, Moem, 1960)

In later years the camps became more closely associated with the Jews. Acting on Hitler's orders, Himmler drew up plans for the **'final solution'** of the **'Jewish problem'**. During the war, mass extermination camps were built at **Auschwitz**, **Treblinka**, **Belsen** and **Buchenwald**. Jews were rounded up in Germany and in the conquered territories (France, Holland, Poland, etc.) and transported to these camps. They were poisoned in gas chambers which were disguised as showers. After the war it was established that some six million Jews had lost their lives in the Nazi genocide.

Religious persecution

Hitler was brought up a Catholic, but soon abandoned all religion. He admired the organisation and ceremonial of the Churches, but regarded Christianity as fit only for slaves. Nazism should become the new religion of a superior people.

The Catholic Church was impressed by Hitler's opposition to Communism. A **Concordat** (agreement) was signed between the Vatican and the German State in 1933. The Church was allowed to manage its own affairs and the Catholic education system was protected. The Catholic Bishops took an oath of allegiance to the German Reich.

The Catholic Church soon realised that it had signed away its independence. In the following years many priests were arrested, monasteries were closed, Catholic organisations banned and publications censored. In 1937, **Pope Pius XI** condemned the evils of Nazism in an encyclical letter, ***'With Burning Anxiety'.*** This led to further hostility between the State and the Catholic Church.

At first, some Protestant Church leaders also seemed to welcome the Nazis. Hitler decided to merge the many Protestant Churches into one **National Church,** the ***Reichskirche.*** Among its rules were the following:

— 'In the National Reich Church... only national "Orators of the Reich" will be allowed to speak.
— The National Reich Church demands an immediate stop to the printing and sale of the Bible in Germany.
— On the altars there must be nothing but ***Mein Kampf,*** and to the left of this a sword.'

Many Protestant churchmen resisted Hitler's plans. Some 800 were arrested and deprived of state pensions. Pastor **Martin Niemöller** led a group of Protestant churchmen who rejected the authority of the *Reichskirche.* He was arrested after preaching an anti-Nazi sermon and kept in solitary confinement for the next seven years.

In hindsight it can be said that the churches failed to react strongly enough against the Nazis. However, the churches have traditionally tried to avoid politics and to live within the system, rather than seek confrontation and risk persecution of their members. Some individual bishops and priests spoke out. The Protestant pastor, **Dietrich Bonhoeffer,** was hanged for his anti-Nazi activities. The Jesuit priest, **Alfred Delp,** was executed for conspiring against Hitler.

Life in Nazi Germany

The Economy

There were 6 million people unemployed when Hitler came to power in 1933. He had promised 'work for all' in the election campaigns and he was spectacularly successful in keeping his promise. By 1936, the number was down to about 2.5 million. It was negligible by 1939.

There were a number of reasons for the fall in unemployment:

- Hitler's economic policies were based on the principle of **autarchy**, that is, self-sufficiency in food and essential raw materials. He was well aware that Germany had lost the Great War because she was unable to survive the Allied blockade.
- Self-sufficiency helped create new jobs. The policy was reasonably successful in industry, but less so in agriculture. Many left the land for better-paid jobs in the cities. The Finance Minister, **Schacht**, wanted Hitler to divert money from armaments to agriculture. This led to the famous 'guns or butter' controversy. In the long term, Hitler felt that only the capture of the rich lands to the east would completely solve Germany's agricultural problems.
- World recession had bottomed out by 1933. German industry flourished and consumer goods became more plentiful. This helped to build up national morale. However, when war broke out in 1939, Germany did not have adequate supplies of oil, copper, or rubber. Self-sufficiency had not been achieved!
- The introduction of conscription in 1935 and the expansion of the armed forces gave employment to thousands of young men.
- Rearmament created thousands of jobs. The steel, chemical and coal industries prospered.

Hitler opens an Autobahn

- Many public works were started. Great public buildings would symbolise the achievements of the Third Reich. The road-building programme which created the famous *autobahnen* (motorways) employed 200,000 men.
- Hitler ordered that a 'people's car', the ***Volkswagen***, must be built at a price everyone could afford. Eventually over 1 million people worked in the motor industry.
- The vast state bureaucracy and the Nazi Party itself were sources of many jobs.
- Workers were sent where there was a shortage of labour. Wages and prices were strictly controlled, so that factory owners could keep down their costs. Trade unions were banned and workers had to join the Nazi-controlled Labour Front. Strikes were forbidden and trade union activists were imprisoned.
- It is also sadly true that the imprisonment of Jews, Communists and trade unionists in concentration camps also reduced unemployment, as their jobs were taken up by Nazis.

The creation of jobs and the stimulation of the economy were major factors in Hitler's popularity. Most people enjoyed a rise in living standards. The loss of democracy and the sacrifice of individual freedom seemed worthwhile.

Propaganda

One of the most important posts in the Nazi government was the **Ministry of National Enlightenment and Propaganda**. The Minister, **Joseph Goebbels**, aimed to make Germans believe in Nazi ideas and to be loyal to Hitler and the Party:

> 'The essence of propaganda consists in winning people over to an idea so sincerely, so vitally, that in the end they will succumb to it utterly and can never escape from it.'

Goebbels made sure that newspapers printed only stories favourable to the government. Newspapers which printed stories he had not approved were closed down. He also made use of the State-controlled radio to inform the German people. Strict censorship was used to ban information or entertainment which the government considered harmful. The Führer was presented as a kind man whose pleasures were few and simple. Goebbels suggested that good Germans should demonstrate their loyalty to their leader by putting out flags in his honour.

The most spectacular form of propaganda was the mass rally. The most famous of these rallies was held in August each year at **Nuremberg** and lasted for a week. Hundreds of thousands watched army parades and gymnastic displays. They listened to massed choirs, brass bands and speeches. An eyewitness described the 1935 rally:

'Twelve huge SA bands played military marches with beautiful precision and terrifying power. Behind the bands, on the field itself, solid squares of uniformed men were ranged in strict military order, thousands strong....A blare of trumpets rent the air and a hundred thousand people leaped to their feet. All eyes were turned towards the stand, awaiting the approach of the *Führer.* There was a low rumble of excitement and then the crowd burst into a tremendous ovation, the *"Heils"* swelling until they were like the roar of a mighty cataract.'

Hitler had developed into a highly effective speaker. He learned how to catch the attention of his audience and how to read their minds. Words such as 'smash' 'violent', 'hatred' and 'power' worked his mass audience into a frenzy. Hitler sometimes seemed to lose all control and by the time his speech ended, he was on the verge of hysteria. His listeners loved the speeches and the spectacle. The rallies played no small part in the *Führer's* appeal to Germans.

Hitler addresses a rally at Nuremberg, September 1935.

Hitler Youth greet their Führer, 1935.

Youth Movements

Hitler ensured that young people were loyal to him and to the Nazi Party. Textbooks were rewritten to paint a good picture of the Nazis. Teachers had to belong to the **German Teachers League** and had to put across Nazi ideas in their lessons.

Young people had to belong to youth organisations which taught them loyalty to Hitler and trained them in military skills. By 1939, some six million young German boys belonged to the **Hitler Youth Movement**, the ***Hitlerjugend***. Girls were told to join the ***Bund Deutsche Mädchen***, the **German Girls League**. They were taught to honour the Nazi state and to become fit mothers of soldiers of the Third Reich.
Hitler spoke of the purpose of the youth organisations:

> 'The weak must be chiselled away. I want young men and women who can suffer pain. A young German must be swift as a greyhound, as tough as leather and as hard as Krupp's steel.'

The role of women

The Nazi leaders said that the job of women was to bear as many children as possible. Hitler said that women should stick to 'the three Ks' *Kinder, Kirche und Küche* (Children, Church and Cooking). Women found themselves being forced to stay at home and many doctors, civil servants, lawyers and teachers lost their jobs.

The Totalitarian State

Between 1933 and 1939, the Nazis turned Germany into a Totalitarian State. The State controlled the political system, the workers, education and the youth movements and dominated the Churches. The courts and the judicial system were in Nazi hands. Individuals had to yield their freedom to the state machine.

Hitler was in control of the ***Wehrmacht***, the German army, after the 'Night of the Long Knives'. From 1934 onwards, Himmler's SS, kept a close eye on all political activity in the state. In 1936 Hitler gave Himmler overall control of the German police force. Himmler appointed SS officers to key positions in the police service. In particular, his chief assistant, **Reinhard Heydrich**, controlled the secret police, the ***Gestapo***, the most feared and ruthless instrument of State policy.

Some historians have asked why more Germans did not oppose Hitler. Those who spoke up paid with their lives. In the Nazi police state, spies were everywhere and people soon saw the wisdom of keeping quiet. It must also be noted that many patriotic Germans saw Hitler as their saviour. He stood up to those who had humiliated their country. He provided jobs and hope. The propaganda machine soon convinced those with doubts. By 1939 not many Germans were opposed to Hitler and the Nazis.

Chapter 18

Hitler's Foreign Policy: the Road to War

Undoing the Treaty

Hitler's aims

> 'The Treaty of Versailles is engraved on the minds and hearts of the German people and burned into them. Sixty million people will find their souls aflame with a feeling of rage and shame. Then the people will join in a common cry: "We will have arms again".'
>
> (Hitler)

The first aim of Hitler's foreign policy was the destruction of the hated treaty, together with its guilt clause and its restrictions on Germany.

Hitler's second aim was to unite all Germans in a single country. This *Grossdeutschland* theory of German unity had long been the objective of German nationalists in both the German and Austrian Empires. There were millions of Germans living in Austria, Czechoslovakia and Poland. These countries would have to be brought under his control.

His third aim was to provide Germany with *Lebensraum*, or 'living space'. In order to provide Germans with sufficient food and raw materials, he would take over the rich lands and resources of Poland and Russia. While achieving his third aim, Hitler would also destroy Communism.

These three aims could only be achieved if Germany greatly increased her armed forces, and rearmed.

Rearmament

The League of Nations **Disarmament Conference** was in process when Hitler became Chancellor in 1933. He took the high moral ground by challenging the other powers to disarm as Germany had done. France was still very worried about her security and refused. In October 1933, Hitler withdrew from the Disarmament Conference and the League of Nations.

Hitler's action meant that Germany was going to rearm. Göring had already begun to secretly form a German air force. In March 1935, Hitler spoke openly about

HITLER'S FOREIGN POLICY 1933-1939

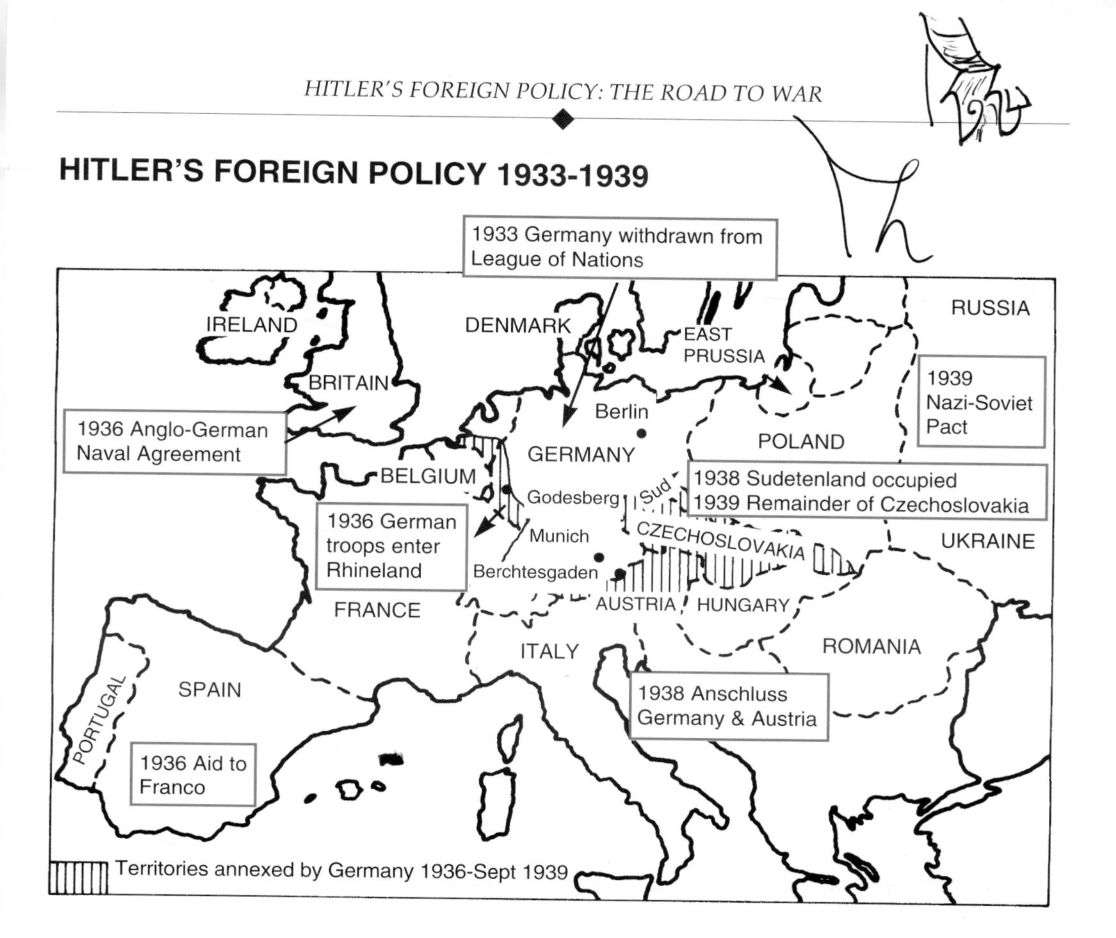

the *Luftwaffe*. Later that month he announced plans to conscript an army of over half a million men. He assured British diplomats that his purpose was to make his country 'safe against the Bolsheviks'.

Pact with Poland

In January 1934, Germany signed a ten-year non-aggression pact with Poland. Hitler adopted the role of peaceful statesman by accepting the **Polish Corridor** and the borders between the two countries. His main purpose in signing the agreement was to break the alliance between France and Poland. In one stroke he weakened the security of both countries.

The Stresa Front

Austrian Nazis hoped for union between Austria and Germany. They were encouraged by the Germans. In July 1934, the Austrian Chancellor, **Dollfuss**, was murdered and it seemed as if Hitler's troops would take over the country. The Italian dictator, Mussolini, was at this time wary of Hitler and he rushed troops to the **Brenner Pass**. Hitler backed down and disassociated himself from the failed attempt to seize power.

The events in Austria and worries about Hitler's foreign policy led to closer co-operation between Britain, France and Italy. Representatives from these countries met at Stresa in Northern Italy. In what became known as **'The Stresa Front'**, they condemned German rearmament and expressed concern about Hitler's aggressive foreign policies. However, the Stresa Front quickly collapsed after the Italian invasion of Abyssinia and Britain's Naval Pact with Germany.

The Saar

The Saar had been placed under the control of the League of Nations under the terms of the Treaty of Versailles (see chapter 12). In a plebiscite in January 1935, the inhabitants of the region voted by nine to one to reunite with Germany. This encouraged Hitler to pursue his aim of uniting all Germans in a single country.

The Anglo-German Naval Pact, 1935

Hitler said that the German navy should be one third the size of the British Royal Navy. Britain did not wish to enter into another naval race with Germany, as had happened before World War I. Some British politicians also had sympathy with German demands. The **Anglo-German Naval Pact** was signed in June 1935. This limited the size of the German fleet to 35 per cent of the British fleet. However, there was no limit to the number of submarines Germany could build. The French Premier, **Laval**, was loud in his protest:

> 'A question which affects all the signatories of the Treaty of Versailles has been treated more or less as a private matter between Germany and Great Britain'.

The Anglo-German Naval Pact was very significant. It showed Hitler that the Treaty of Versailles was no longer sacred. He was now aware that the British could be manipulated and that Britain and France no longer spoke with one voice.

The Rhineland 1936

The Treaty of Versailles forbade Germany from stationing troops or building fortifications within 30 miles of the River Rhine. Hitler found this situation intolerable.

On 7 March 1936, small detachments of German troops entered the demilitarised zone in the Rhineland. The French army, far stronger than the Germans, made no effort to stop them. They were not prepared to act without British support, because they still believed in the principle of collective security, laid down in the Charter of the League of Nations. However, most Britons agreed with the observation of a Conservative politician, Lord Lothian, that the Germans were 'only going in to their own back garden'.

German soldiers enter the Rhineland, 1936.

Hitler admitted to his personal interpreter that he was taking a gamble:

'The forty-eight hours after the march into the Rhineland were the most nerve-racking of my life. If the French had then marched into the Rhineland, we would have had to withdraw with our tails between our legs, for the military resources at our disposal would have been wholly inadequate for even a moderate resistance.

Hitler's Interpreter, Dr Paul Schmidt (Heinemann, 1951)

The failure of the British and French to oppose Hitler's march into the Rhineland in 1936 was to have enormous significance. Hitler had gambled and won. He no longer believed in Anglo-French solidarity. He could continue to gamble until somebody called his bluff.

Rome-Berlin Axis

Hitler had always admired Mussolini's fascist policies. He hoped to forge links between Italy and Germany. An opportunity to show support for his fellow dictator arose in 1935 when Italy invaded Abyssinia. Hitler completely ignored League of Nations sanctions against Italy. He also assured Mussolini that he would not seek to bring into the Reich the German-speaking people of the Tyrol in Northern Italy. He accepted the Brenner Pass as the border between Austria and Italy.

This new understanding between Hitler and Mussolini led to the signing of the **Rome-Berlin Axis** in October 1936. Later Germany, Japan and Italy signed the **Anti-Comintern Pact**. This was an agreement between the three countries to join together to fight the advance of Communism. The first joint military action by Hitler and

Mussolini was in support of Franco in the Spanish Civil War, 1936-1939. Hitler used the war to test the Condor Air Legion and to draw Germany and Italy closer together. Finally, in May 1939, Hitler and Mussolini formed a military alliance known as the **Pact of Steel**.

Anschluss

Hitler was born in Austria. The vast majority of the Austrians spoke German, but the Treaty of Versailles forbade *Anschluss* (Union) between the two countries. Hitler felt that Austria should naturally be the first acquisition of the new German Empire.

In February 1938, at a meeting with the Austrian Chancellor, **Schuschnigg**, Hitler demanded the appointment of a number of Nazis to the Cabinet and freedom for all Nazi prisoners. Failure to comply would mean invasion. Schuschnigg was bullied and threatened by Hitler:

> 'Who knows, perhaps you will find me one morning in Vienna like a spring storm. Then you will go through something!... The SA and the legion will come in after the troops...wreaking vengeance.'

The Austrian Chancellor had little choice. He signed the agreement in order to avoid civil war. A leading Nazi, **Arthur Seyss-Inquart**, was appointed to the key post of Minister for the Interior. When he returned to Austria, Schuschnigg decided on one last bold attempt to save his country from the Nazis. He called for a

Nazi poster for plebiscite on the Anchluss: *'The whole people says Yes! on 10 April.'*

referendum on Austrian freedom, on a formula of words which were likely to meet with success:

> 'Are you in favour of a free and German, independent and social, a Christian and united Austria?'

Hitler saw this as a threat to his plans for *Anschluss* and moved troops to the border. Schuschnigg asked France, Italy and Britain to protect his country, but all three refused. He resigned and was replaced as Chancellor by Seyss-Inquart. The plebiscite was cancelled and the new Chancellor sent a pre-arranged telegram asking for the German army to be sent in to 'establish peace and order... and to prevent bloodshed'.

On 12 March 1938, the German army marched into Austria. They were greeted by cheering crowds, Nazi salutes and swastikas. Hitler himself returned to Vienna as hero and conqueror, some 25 years after he had quietly left the city. A month later, over 99 per cent of those who voted approved of the union between Germany and Austria. Hitler said:

> 'This is the proudest hour of my life'.

The Destruction of Czechoslovakia

The Sudetenland

Czechoslovakia was created as a new country in 1919. An area known as the **Sudetenland**, within the borders of Czechoslovakia, contained three million German speakers. Here were the factories and the mines, the power-stations and the fertile farm lands, and the **Skoda arms works**, biggest in Europe.

German troops enter the Sudetenland, October 1938

Konrad Henlein

Most Germans living in the Sudetenland supported the Sudeten German Party, led by **Konrad Henlein**. With Hitler's support Henlein began to stir up trouble. He accused the Czechs of discrimination against the German minority. He organised riots and demonstrations and in 1938 began to agitate for self-government for the Sudetenland.

Concessions were made to the Germans in the Sudetenland. However, Hitler was not satisfied and demanded justice for what he called these 'tortured creatures'. **Ribbentrop**, the German Foreign Minister, greatly encouraged the Sudeten German Party in a speech in Berlin in March 1938:

> 'The Sudeten Germans must realise that they are backed up by a nation of 75 million, which will not tolerate a continued suppression of the Sudeten Germans by the Czechoslovak Government'.

Appeasement

Neville Chamberlain had been elected Prime Minister of Britain in 1937. He took a strong personal interest in foreign affairs. He believed in a policy of appeasement and felt that concessions should be made to Hitler. Chamberlain felt that another

Chamberlain flew to see Hitler in his mountain retreat near Berchtesgaden. Angel of hope!

horrific war must be avoided. If Germany completed the process of unifying all Germans and re-establishing the security of her frontiers, Hitler would have no further reason to threaten the peace of Europe.

Peace Talks

During the summer of 1938 the situation in Czechoslovakia deteriorated. Twelve German divisions massed on the Czech border. The Czech government mobilised half a million troops. The French had treaty obligations to Czechoslovakia and consulted with Britain. The Soviet Union proposed collective action to support the Czechs, but neither Britain nor France trusted Stalin. Communism was still the major enemy.

On 12 September 1938, Hitler bitterly attacked Czechoslovakia in an address at Nuremberg. He warned other countries not to protect 'the oppressors of Germans'. Three days later Chamberlain flew to meet Hitler at his Alpine home at **Berchtesgaden** in Bavaria. He was bluntly told by the German Chancellor that Czechoslovakia must give up the Sudetenland.

The British Prime Minister flew home and consulted with France. The two nations drew up a plan and the Czech government reluctantly agreed to concede the Sudetenland to Germany. Chamberlain then flew to meet Hitler at **Godesberg** in the Rhine Valley. He was bitterly disappointed to find that Hitler now had new demands on Czech territory. He could not yield to these demands and war now seemed very close. Chamberlain addressed the people of Britain:

> 'How horrible, fantastic, incredible it is that we should be digging trenches and trying on gas-masks here because of a quarrel in a faraway country between people of whom we know nothing'.

The Munich Agreement

The Italian dictator, **Mussolini**, called a conference between himself, the French Prime Minister, **Edouard Daladier**, **Hitler** and **Chamberlain**. They met at Munich, in South Germany, on 29 September 1938. Neither Czechoslovakia nor her eastern ally, the Soviet Union, was invited to attend. The Czech delegation waited in a Munich hotel until they were informed of their fate.

Hitler repeated his earlier demands on Czech territory. These were conceded by Daladier and Chamberlain, in return for a guarantee from Hitler that this was his 'last territorial demand in Europe'. Polish and Hungarian demands on Czech territory were also to be met. The Czechs were told that they must accept the agreement within twelve hours, or they could expect no help from Britain or France. When informed of the fate of his country, **President Benes** of Czechoslovakia resigned. Germany took over the Sudetenland on 1 October 1938.

Mussolini and Hitler, Munich 1938. Göring, Ciano and Hess are immediately behind.

Results of Munich

Neville Chamberlain was greeted as a hero on his return to London. At the airport he showed the piece of paper that he and Hitler had signed in a separate agreement at Munich:

> 'I had another talk with the German Chancellor, Herr Hitler, and here is the paper which bears his name upon it as well as mine. We regard the agreement signed last night and the Anglo-German Treaty as symbolic of the desire of our two peoples never to go to war with one another again.'

That evening in Downing Street, Chamberlain said, to great applause, that he had brought back 'peace and honour...I believe it is peace in our time.'

Chamberlain was congratulated by the King and by most Conservatives. **Clement Attlee**, leader of the Labour Party, felt that Czechoslovakia had been betrayed. The Conservative politician, **Winston Churchill**, said that the agreement was a disaster for Britain and France.

In the longer term, the **Munich Agreement** had a number of results:

- Hitler now believed that neither Britain nor France would fight against Germany. He continued to gamble with German expansion.
- The Soviet Union was convinced that collective security to guarantee existing borders was abandoned. She would have to look after her own interests.
- It gave Britain and France some time and space in which to rearm.

Neville Chamberlain on arrival at Munich, 29 September 1938.

- The acquisition of the Skoda arms works was of major benefit to Hitler during World War II.

The Fate of Czechoslovakia

On 15 March 1939, **President Hacha** of Czechoslovakia was summoned to Berlin. He was told that disorder in his country was a threat to Germany and must be stopped. He was threatened and bullied and forced to sign an agreement asking the Führer to 'protect' his people.

Within four days, a German army of occupation was in control of the remainder of Czechoslovakia. A new puppet government was appointed. Britain and France protested, but it was clear that the policy of appeasement had failed.

Count-down to War

The Polish Corridor

Britain and France now awoke to the dangers posed by Hitler. They both ordered rearmament. For the first time in her history, Britain introduced compulsory military service in peace-time. The situation became more tense when Hitler took the predominantly German city of **Memel** from Lithuania, in March 1939. In April, Mussolini ordered the Italian army to seize Albania.

Hitler had begun to make demands on Poland about the free city of **Danzig** and the **Polish Corridor**. This was a strip of territory which Poland gained at the Treaty of Versailles in order to gain access to the Baltic at Danzig. It was mainly German-speaking. On 31 March Chamberlain gave the Poles a pledge:

> 'Any action that clearly threatened Polish independence, and which the Polish government accordingly considered it vital to resist with their national forces, His Majesty's Government would feel themselves bound at once to lend the Polish government all support in their power'.

It was obvious that a line had now been drawn by the western powers against Hitler. It was, however, surprising that Chamberlain's pledge left the Polish government free to decide when aid should be granted. It was a blank cheque issued by Britain to Poland.

The German-Soviet Pact

Hitler immediately tore up German agreements with Poland (1934) and Britain (1935). The Soviet Union talked about collective action with Britain and France. However, both of these countries were reluctant to form an alliance with a Communist power. They delayed making a decision until it was too late.

The Soviets were afraid that the western powers would let them face Hitler alone. The USSR needed time and the new Foreign Minister, **Molotov**, began to sound out the Germans for some sort of deal. Germany wished to avoid war on two fronts, if France and Britain moved to protect Poland.

On 23 August 1939 the world was told the stunning news that Germany and the Soviet Union had signed a ten-year **Non-Aggression Pact**. This was referred to as the **Ribbentrop-Molotov Pact**, after the two foreign ministers who negotiated the terms. Secret clauses in the Agreement allowed the Soviets to take over the Baltic States and arranged for the partition of Poland between Germany and the Soviet Union.

The Invasion of Poland

Hitler now felt that he was ready to pursue his **'final demand'** that the Polish frontier should be rectified. A series of incidents were faked on the frontier. At **Gleiwitz** twelve German criminals were killed by SS men dressed in Polish uniforms and left lying near a German Radio Station. The Germans said they would punish this wicked 'Polish' attack.

At 4.45 a.m. on 1 September 1939, German forces invaded Poland. The British and French governments demanded their withdrawal. On 3 September Chamberlain broadcast to the British nation:

> 'This morning the British Ambassador in Berlin handed the German Government a final note stating that unless we heard from them by eleven o'clock that they were prepared at once to withdraw their troops from Poland, a state of war would exist between us. I have to tell you now that no such undertaking has been received and that consequently this country is at war with Germany'.

France declared war on Germany on the same evening. World War II had begun.

Chapter 19

France between the Wars, 1918-1939

Post-War France

The Lost Generation

France lost 1,390,000 men in World War I. This represented about fifteen per cent of all French males between the ages of eighteen and twenty-seven. Almost three million had been wounded. About 750,000 of these were the *mutilés*, the hopelessly maimed soldiers.

This 'lost generation' left France with long-term problems. The country was seriously short of workers. She was forced to employ immigrant workers, mainly Spaniards, Belgians, Italians and Poles. Many French citizens resented the presence of these labourers and there was little integration between them and the native population.

The loss of a generation of men meant that France's birthrate fell. Her population did not rise between the wars from its 1914 level of 40 million. During the same period the population of Germany rose from 60 million to over 80 million. In 1939, France had only half the normal number of eighteen- to twenty-one-year-olds. Many believed that another war must be avoided at all costs.

Material Damage

Northern and Eastern France suffered great devastation in World War I. Towns and cities were destroyed and three and a quarter million hectares of her rich agricultural land had been blasted. Roads, railways, factories, homes and mines were ruined.

War devastation – France after World War I

Before the war, France was a creditor nation. By 1919 she had accumulated a national debt of 175 billion francs due to the massive borrowing needed to fund her war effort. French investors had lost heavily overseas. The new Bolshevik government in Russia claimed that it was not responsible for debts incurred under the Tsars and another 100 million francs were lost to France.

A New Government

In 1919, a group of mainly right-wing parties, the *Bloc National*, formed a government. It had the support of big business and contained many ex-servicemen. The government was strongly anti-communist and was led by the ageing **Clemenceau**, now known as *'père la-victoire'*, the father of victory.

The new government's first task was to re-build the war-torn country and to help the ravaged economy. Clemenceau could not afford to antagonise his supporters by increasing taxation. Instead, France continued to borrow heavily, in expectation of the 'magic cure' of reparations.

In 1919 the French presented a 'bill' of 200,000 million francs to the treaty-makers. This was not only to compensate for damage to goods and property, but also contained an estimate to cover pensions for war widows and the disabled.

Wilson and Lloyd George considered such a figure outrageous. A **Reparations Commission** was set up to investigate the matter. It found that Germany could not afford to pay the sum demanded by the French. In 1921, the Commission decided on reparations of £6,600 million. Even this figure was reviewed and lowered on a number of occasions over the next ten years (see chapter 12). It has been estimated that France received only a fraction of the cost of re-building the devastated areas.

The Twenties: Crisis and Recovery

Bloc National, 1920-1924

Raymond Poincaré had been President since 1913. His term of office came to an end in 1920. Clemenceau hoped to take his place. However, it was felt that he was too lenient on Germany during the negotiations at Versailles and he failed in his bid. He resigned as Premier in 1920.

Clemenceau was succeeded by **Paul Deschanel** and six months later by **Alexandre Millerand**. The Chamber of Deputies was mainly Catholic and good relations were re-established between France and the Pope. St Joan of Arc was canonised and her feast day became a national holiday.

Poincaré became premier of the ***Bloc National* Government** in 1922. He was the first ex-president to head a government. He was anti-German and when his policy of occupation of the Ruhr (see chapter 18) failed, he had to increase taxation at home. He also ran into difficulties when he tried to prevent civil servants from having the right to strike. Many people resented the attempts to give the Catholic Church more influence in education. Inflation and a series of strikes further weakened the government and there was a swing to the left in the 1924 elections.

Cartel des Gauches, 1924-1926

The Socialists and the Radicals combined in an electoral alliance called the ***Cartel des Gauches*** *(Alliance of the Left)* and formed a government under **Edouard Herriot**. The Premier was a **Radical** but was opposed to the Socialist policy of State interference. The two parties were united in their opposition to the Catholic Church. However, the government failed in its attempts to introduce the **Law of Separation of Church and State** and other anti-clerical measures in Alsace-Lorraine.

The Cartel also failed to deal with the financial crisis. It was unable to impose taxes high enough to bring in social reforms and to balance the budget. The government continued to borrow and inflation worsened. In 1926 the franc was only worth about a third of its 1924 value. The Cartel collapsed in 1926.

Union Nationale, 1926-1929

The new **Government of National Unity** was a strong centre-based coalition. It was again led by Raymond Poincaré. Due to the financial crisis, he was given the power to rule by decree. He took measures to stabilise the franc and expand the economy. Germany resumed its payment of Reparations thanks to the new conciliatory policies of the French and German governments.

The economy recovered and France enjoyed a boom period. She became Europe's leading car manufacturer through her ***Renault, Peugeot and Citroën*** factories. New chemical and engineering enterprises flourished. French clothes and perfumes were sought after by the rich and famous. The tourist industry grew and Paris, the 'City of Light', became the home of international artists and writers. France's economy was now in such a strong position that the Wall Street Crash of 1929 had no immediate effect.

The Thirties: A Dismal Record

The Depression

The Depression which began in the U.S.A. in October 1929 soon hit other countries like Britain and Germany. However, France seemed unaffected throughout 1930 and most of 1931. When Britain devalued the pound in 1931, France did not follow. Her products soon became overpriced on an already depressed world market.

André Tardieu had succeeded Poincaré as Prime Minister in 1929 and tried to implement a five-year plan which would modernise industry and agriculture. The Chamber of Deputies failed to support his programme of state investment in large public works schemes.

Unemployment rose sharply and the voters turned again to the left in the 1932 elections. Six governments of the left were in power between 1932 and 1934. They

French Hunger March, 1933: *du pain et du travail* (bread and work!)

attempted to introduce a policy of deflation. However, they could not get agreement in the Chamber either to raise taxes or make significant cuts in government expenditure. They were also subject to attacks from the extreme right-wingers, who were now beginning to flex their muscles.

The Leagues

Democracy took away power from the élite and gave it to the herd.

This was the opinion of **Charles Maurras**, who founded ***Action Française***. This was the oldest of the anti-parliamentary leagues. It called for a restoration of the monarchy, was fiercely nationalistic and blamed the Jews for the depression and most of the country's ills.

Other leagues were ***Jeunesses Patriotes, Solidarité Française*** and ***Croix-de-Feu***. These right-wing fascist groups took part in menacing street demonstrations against the Third Republic. They wanted 'France for the French' and attacked the left-wing governments. They considered these governments too weak to deal with the depression or with the threat of Communism. As in Germany and Italy, France was in danger of falling to fascism.

Wall posters in Paris, 1934. *'On étouffe l'Affaire Stavisky'* – "The Stavisky Affair is being hushed up'.

The Stavisky Affair

The Fascist Leagues were boosted in 1933 by the Stavisky Scandal. It seemed to confirm their claims that the democratic government was corrupt. **Serge Stavisky** was a financier of Russian-Jewish origins. He was a swindler who used his friends in high places to cover up his shady deals. Through his contacts, one of his trials was postponed nineteen times between 1926 and 1933.

In December 1933, a fraudulent scheme to float a loan of 200 million francs' worth of bonds collapsed. Before the police could arrest Stavisky, he was found dead, apparently by suicide. Right-wing newspapers alleged that he had been murdered to protect his friends in high places, some of them government ministers. The Prime Minister, **Pierre Chautemps**, was brother-in-law of the prosecutor who had delayed Stavisky's trial. Chautemps was forced to resign on 27 January 1934 and was succeeded by **Edouard Daladier**.

Place de la Concorde

> 'Your Parliament is rotten. Your politicians corrupt. Your country betrayed by the slime of scandal. Your security menaced.'

This was an appeal by the right-wing newspaper, *Solidarité Française,* on 6 February 1934 urging the French people to demonstrate against the government. Other newspapers also joined in the appeal as Daladier's Government made its first appearance. That evening several thousand members of the leagues converged on the *Place de la Concorde,* the great square just across the Seine from the Chamber. They had been told to 'chase the robbers' from Parliament.

Serious rioting occurred and fifteen people were killed in clashes between the police and members of the leagues. It was the bloodiest street-fighting in Paris since the days of the Commune. The Government stood firm. Daladier was held partly responsible for the trouble and forced to resign. However, the Third Republic again survived.

Riots in Paris, February 6, 1934

The Popular Front

The parties of the left were horrified and frightened by the events of 1934. The **Communists, Socialists** and **Radicals** came to believe that they ought to form a united front against the Right. The Communist Party in Moscow had, up to this time, opposed co-operation between the Communists and Socialists. Now they urged the parties of the Left to unite against the fascist dangers of Hitler and Nazi Germany.

On Bastille Day, 14 July, 1935, a third of a million people – Communists (led by **Maurice Thorez**), Socialists (led by **Léon Blum**) and Radicals (led by **Daladier**) – marched through Paris. They showed the people that they could give massive and united support to the Third Republic. For the 1936 elections, they adopted the slogan **'Bread, Peace and Liberty'**.

Léon Blum

> 'Your coming to power, Monsieur le Président du Conseil, is undoubtedly an historic occasion. For the first time, this ancient Roman Gallic country will be governed by a Jew.'

These were the bitter words of a Fascist deputy to the new Prime Minister, Léon Blum, on 6 June 1936, two days after he had been elected Prime Minister. The **Popular Front**, of Communists, Socialists and Radicals, had won the election with 380 supporters against 230 for the parties of the Right.

Blum, in power, continued his attack on the monied classes. He singled out for special mention the 'two hundred families', the richest in France, who through the Bank of France, exercised great power behind the scenes of government.

The victory of the Popular Front had raised the hopes of many trade unionists that the factories would be handed over to them. Large-scale strikes took place as workers demanded the reforms promised to them during the elections. Blum called a conference of employers and workers at **Hotel Matignon**.

The **Matignon Agreement** was a triumph for Blum. He won considerable concessions from the employers – a rise of 12 per cent in wages, holidays with pay and the promise of a 40-hour working week. The agreement brought peace for some time. Blum introduced other social reforms and banned the leagues.

In a short time inflation went out of control. Blum, who was a moderate, was under attack from both right and left. The right still controlled the Senate and when it failed to support his financial proposals the Prime Minister resigned in June 1937. By April 1938 the Popular Front Government had collapsed.

Government of National Defence, 1938-1939

The worsening international situation called for a strong figure. Like his predecessors, Clemenceau and Poincaré, **Edouard Daladier** was seen as such a figure. He was elected Prime Minister of a **Government of National Defence** in 1938.

Confidence now returned. Unlike Blum, Daladier was given financial powers by parliament. He devalued the franc by 10 per cent. He followed this up with a lengthening of the working week. In protest, a general strike was called for 30 November 1938. It was a dismal failure. The monied class now believed that the left-wing threat was dead and began to bring back their money into France. By 1939 the economic situation had stabilised. However, France's foreign policies since World War I looked as if they were doomed to end in failure.

Foreign Policy

The Search for Security

At Versailles, Clemenceau failed to get more than a fraction of the amount demanded in reparations from Germany (see chapter 13). When the United States failed to accept the Treaty of Versailles and the League of Nations, France lost her guarantee of security from Britain and America against German aggression. This meant that she had to rely on Britain alone. However, British self-interest came before any commitments to France.

In 1923 Poincaré ordered French troops to occupy the **Ruhr** in collaboration with the Belgians. This was done in order to force Germany to pay reparations. In 1924 the Dawes Plan reduced German reparations and enabled France to withdraw from the Ruhr. However, France was hardly comforted when she saw that the plan would help Germany's recovery.

Aristide Briand, Foreign Minister 1926-32

Aristide Briand, Foreign Minister from 1926 to 1932, pursued a policy of reconciliation towards Germany. He was helped in this by the friendly attitude of the Weimar Foreign Minister, Gustav Stresemann (see chapter 18). Briand signed the **Locarno Pact** in 1925. This confirmed the existing borders between France and Germany and accepted the demilitarised zone of the Rhineland. France welcomed Germany into the League of Nations in 1926.

In 1928 France and the United States were responsible for drawing up the **Kellogg-Briand Pact** which outlawed war as a means of settling disputes. Sixty-five countries signed the pact.

Failure to Act

Hitler was not at first recognised by the French as a serious threat. Appeasement had its advocates in France. After all, Hitler was the sworn enemy of the greater threat, Communism. Soon Hitler's aggressive policies began to frighten some French politicians. Germany's withdrawal from the Disarmament Conference and the League of Nations convinced France of the need for allies. In 1935, she signed the **Franco-Soviet Pact**. However, the French Foreign Minister, **Laval**, was unwilling to extend this agreement into a military alliance against Hitler.

In 1935 France, Britain and Italy reaffirmed the Locarno Pact and condemned Hitler's rearmament plans in an understanding known as the **Stresa Front**. This Front quickly fell apart when Mussolini invaded Abyssinia and Britain signed the **Anglo-German Naval Pact**. Later in 1935, the French Foreign Minister, Pierre Laval,

Fascist march, Paris 1936

met the British Foreign Secretary, Sir Samuel Hoare, to discuss Mussolini's invasion. The **Hoare-Laval Pact** recognised Italian rights in Abyssinia. Laval came under fierce attack from Blum and the left and was forced to resign.

Léon Blum's **Popular Front Government** took no action against Hitler when Germany marched into the Rhineland in 1936 (see chapter 18). It seems that French military intelligence over-estimated the strength of the German army. France was still clinging on to the idea of **collective security**, that is, joint action with Britain against the Germans. Britain, however, no longer supported the idea.

Non-intervention

Blum, against his better judgement, backed the British policy of **non-intervention** in the Spanish Civil War, 1936-1939 (see chapter 21). He had been anxious to help the Popular Front Government in Spain and sent some assistance in 1936. However, Britain soon convinced him of the advantages of neutrality.

Blum's Government was in the process of collapse when Hitler took over **Austria** in March 1938 and was in no position to intervene. In any event, the French were again discouraged from taking any action by the British, who regarded Austria as Germany's 'back-yard'.

Another war looms

France did not honour its commitments to **Czechoslovakia** in 1938. Daladier signed the Munich Agreement, but was deeply unhappy about the whole episode. His foreign minister, **Georges Bonnet**, was a pacifist who believed in peace at any cost. Daladier also knew that he would not get the backing of the French Right for action against Hitler. The fear of war, and the feeling on the Right that France had more to fear from Communism than from Hitler, had created a mood of pacifism in the country. When Daladier returned home from Munich in 1938, he was received as a hero.

France was less enthusiastic than Britain in making a stand against Hitler over the invasion of Poland. Many Frenchmen did not want to 'die for Danzig'. It was with reluctance that France declared war on Germany on 3 September 1939.

Maginot Line Mentality

The French were not anxious to get involved in another horrible war so soon after 1918. They adopted a defensive strategy based on the **Maginot Line**. This was a series of fortifications and fortresses constructed from the Swiss to the Belgian borders. They were named after the Minister of War, **André Maginot**. The Line was not extended north-westwards because of the cost and because it would offend the Belgians.

The Maginot Line gave the French a comfortable feeling of security. However, a young colonel, **Charles de Gaulle**, condemned the **'Maginot Line mentality'**. He said that modern tank and air power made fixed fortifications like the Maginot Line obsolete. His military studies were published in France. He advocated a mechanised army with quick-moving tanks supported by aerial bombardment. His ideas were ignored in France, but studied with interest in Germany.

When Germany invaded France in 1940, France was militarily unprepared to withstand the onslaught. The Germans simply skirted around the Maginot Line, advancing through Belgium and the Ardennes. France had not achieved the security she had sought since 1918 and her foreign policy was in ruins.

Chapter 20

Britain 1918-1939

The post-War Years

Post-War problems

World War I took the lives of three-quarters of a million British men and wounded one and a half million more. In 1914, Britain had been one of the world's most powerful nations, immensely rich from trade. In 1918 she had a post-war debt of £7,500 million, owed mainly to America. She had sold many of her overseas investments to finance the war. Customers abroad, neglected by Britain, turned elsewhere. Some never returned, while many began building up their own industries. Britain found it impossible to regain her place as a leading world power.

These economic difficulties made it hard for the government to find money for social reforms and expensive welfare measures. However, the **Education Act** of 1918 stated that all those who left school at 14 should continue part-time education until they were 18. Bright children from poorer backgrounds were given more free places in secondary schools. **The Representation of the People Act**, 1918, gave the vote to all men over 21 and women over 30.

Armistice Day, London, 11.11.1918

The Coupon Election

There had been no election in Britain since 1910. The Liberal leader, **Herbert Asquith**, was Prime Minister when the country went to war in 1914. His government was criticised for its conduct of the war and, in 1915, Asquith gave in to pressure from the Conservatives and formed a **National Government**. His Liberal colleague, **Lloyd George**, was more acceptable to the Tories and, with their support, he took over from Asquith in December 1916. This led to a split in the Liberal Party.

In December 1918, Lloyd George called an election in order to gain a mandate from the people for the Paris Peace Conference (see chapter 12). He promised to 'make Germany pay'. At home he promised to make Britain 'a land fit for heroes to live in'. Lloyd George's wing of the Liberals joined the Conservatives in insisting that this could be achieved by a continuation of **coalition government** between the two parties. They gained an overwhelming victory over Labour and the Asquith wing of the Liberals.

Lloyd George's Government, 1918-1922

As Prime Minister from 1918 to 1922, Lloyd George devoted most of his time to foreign affairs. At home, 4 million men were released from the army within a year. They found jobs reasonably easy to obtain, for 1919-20 was a period of post-war boom. Industry was working all-out to make up the leeway caused by the war.

Labour Party election posters in the 1920s

An **Unemployment Insurance Act** of 1920 extended unemployment assistance to most workers and raised the benefit. A new **Housing and Town Planning Act**, 1919, helped to make up the arrears in house-building caused by the war.

Depression

By the middle of 1920, the boom had turned into a slump. By 1921 unemployment had reached 2 million, as exports fell by 50 per cent. The government had lifted war-time price controls. Inflation grew rapidly, but wages failed to rise in line.

Strike fever, which had hit Britain between 1910 and 1914, now returned. The police, railway workers and coal-miners were all involved in disputes. The government appointed **Sir Eric Geddes** as chairman of a committee set up to enforce cutbacks. His proposals, known as the **'Geddes Axe'**, took a hundred million pounds out of housing, education and social services.

Lloyd George, as Prime Minister of a government which enforced these cut-backs, was becoming very unpopular. It was rumoured in 1922 that he was involved in conferring honours in exchange for contributions to his party fund. Many Conservatives now felt that they no longer needed Lloyd George and his Coalition Liberals. When they withdrew their support, the Prime Minister resigned. His achievements had been immense. Winston Churchill paid him this tribute:

> 'When the English history of the first quarter of the twentieth century is written, it will be seen that the greater part of our fortunes in peace and war were shaped by this one man.'
>
> (Churchill)

Britain in the Twenties

Baldwin's First Government, 1922-1924

The Conservatives won the election of 1922 and **Bonar Law** became Prime Minister. He died from cancer in 1923 and was replaced by **Stanley Baldwin**. The new Chancellor of the Exchequer was **Neville Chamberlain**. Baldwin adopted a policy of protecting British industry from foreign competition by putting tariffs on all imported goods except foodstuffs. The Conservatives traditionally said that Britain's position as a world-power was based on free trade. Protection met with great opposition, much of it from within the Conservative Party. The Prime Minister fought an election on the issue. The Conservatives lost a number of seats and Baldwin was forced to resign in 1924.

The First Labour Government, 1924

> 'Today 23 years ago, dear Grandmama died. I wonder what she would have thought of a Labour Government.'
>
> (Entry in King George V's diary, 22 January 1924)

Stanley and Mrs Baldwin

Queen Victoria could hardly have foreseen that a Labour Government would rule Britain within a quarter of a century of her death. With support from the Liberals, the new Labour Government was headed by **Ramsay MacDonald**. Born into a poor family in Scotland in 1866, MacDonald became Secretary of the Labour Party in 1900, an M.P. in 1906 and Leader of the Party in 1911. He resigned his position as Leader in 1914, as he opposed the war, but was re-elected in 1922.

The government was committed to free trade and abolished various tariffs. Labour had promised in their election campaign that they would introduce socialist policies. The party emphasised the importance of State support for housing and education. MacDonald was in favour of moderate policies. In any case, he depended on Liberal support and would not have got their backing for any extreme socialist measures. The most successful domestic reform passed was the **Housing Act** introduced by the Minister of Health, **John Wheatley**. It led to the clearance of slums and the building of over 2 million council houses by 1939.

The Red Scare

MacDonald made a treaty with the U.S.S.R. in 1924. In August of that year, the government did not proceed with a charge of sedition against the Communist newspaper, ***The Workers' Weekly***. The Liberals withdrew their support and an election was called.

Four days before the election, a letter was published in British newspapers, supposedly from **Zinoviev**, head of the Comintern in Moscow. It gave instructions on methods of creating a revolution in Britain. The **'Red Letter'** was probably a forgery, as the original has never been produced. The British electorate was, at this

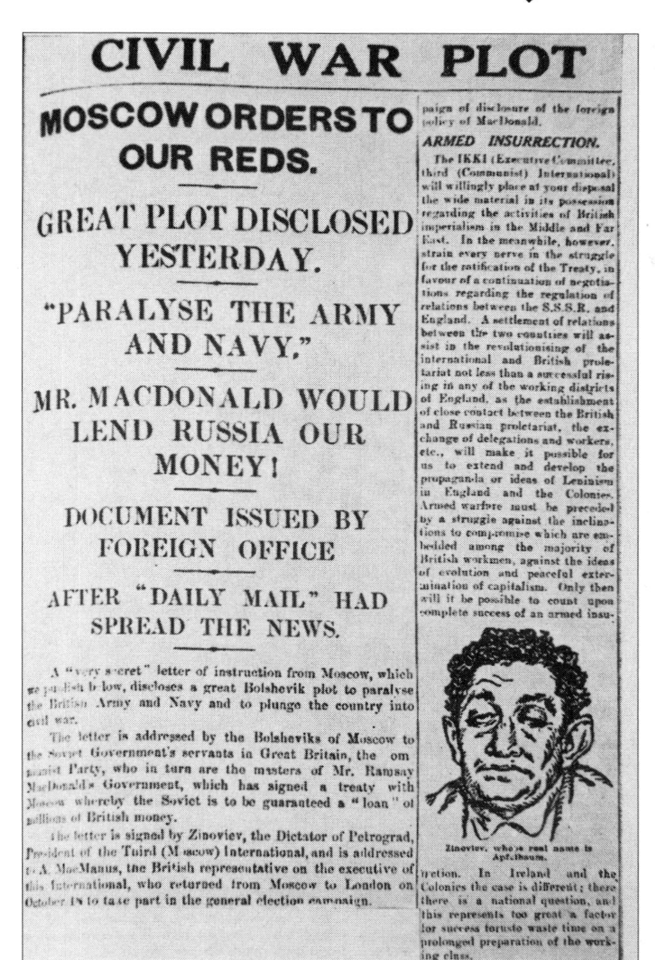

CIVIL WAR PLOT

MOSCOW ORDERS TO OUR REDS.

GREAT PLOT DISCLOSED YESTERDAY.

"PARALYSE THE ARMY AND NAVY."

MR. MACDONALD WOULD LEND RUSSIA OUR MONEY!

DOCUMENT ISSUED BY FOREIGN OFFICE

AFTER "DAILY MAIL" HAD SPREAD THE NEWS.

A "very secret" letter of instruction from Moscow, which we publish below, discloses a great Bolshevik plot to paralyse the British Army and Navy and to plunge the country into civil war.

The letter is addressed by the Bolsheviks of Moscow to the Soviet Government's servants in Great Britain, the Communist Party, who in turn are the masters of Mr. Ramsay MacDonald's Government, which has signed a treaty with Moscow whereby the Soviet is to be guaranteed a "loan" of millions of British money.

The letter is signed by Zinoviev, the Dictator of Petrograd, President of the Third (Moscow) International, and is addressed to A. MacManus, the British representative on the executive of this International, who returned from Moscow to London on October 18 to take part in the general election campaign.

...paign of disclosure of the foreign policy of MacDonald.

ARMED INSURRECTION.

The IKKI (Executive Committee, third (Communist) International) will willingly place at your disposal the wide material in its possession regarding the activities of British imperialism in the Middle and Far East. In the meanwhile, however, strain every nerve in the struggle for the ratification of the Treaty, in favour of a continuation of negotiations regarding the regulation of relations between the S.S.S.R. and England. A settlement of relations between the two countries will assist in the revolutionising of the international and British proletariat not less than a successful rising in any of the working districts of England, as the establishment of close contact between the British and Russian proletariat, the exchange of delegations and workers, etc., will make it possible for us to extend and develop the propaganda or ideas of Leninism in England and the Colonies. Armed warfare must be preceded by a struggle against the inclinations to compromise which are embedded among the majority of British workmen, against the ideas of evolution and peaceful extermination of capitalism. Only then will it be possible to count upon complete success of an armed insurrection. In Ireland and the Colonies the case is different; there there is a national question, and this represents too great a factor for success for us to waste time on a prolonged preparation of the working class.

Zinoviev, whose real name is Apfelbaum.

The 'Red Letter' Scare, 1924

time, scared by the Bolshevik Revolution in Russia and the threat of Communism. The 'Red Letter' helped defeat MacDonald and the Conservatives returned to power with a large majority.

Baldwin's Second Government, 1924-1929

Stanley Baldwin became Prime Minister in the Conservative Government of 1924. He had abandoned ideas of protection and the economy improved gradually. **Winston Churchill** became Chancellor of the Exchequer. He put Britain back on the gold standard. Gold became obtainable for paper money, on demand, at the banks. The value of the pound increased and prices declined. This was therefore a time of relative prosperity for the middle classes and those on stable incomes, but that was of little consolation to more than a million people who remained unemployed. The stronger pound made British exports dearer. The employers cut workers' wages in order to compete. This led to unrest.

In 1925 Neville Chamberlain introduced the **Widows, Orphans and Old Age Pensions Act**. He abolished the Poor Law Unions and Guardians set up in 1834. The **Central Electricity Board** became responsible for a national system of electrical distribution, while the **British Broadcasting Corporatio**n (BBC) became a public body. In 1928 the vote was extended to all women over the age of 21. Only one major event disturbed the calm during the period 1924-1929. That was the industrial unrest of 1926.

The General Strike 1926

The employers in the coal industry had failed to modernise. The pound was also overvalued and when the French occupation of the Ruhr (see chapter 19) ended, British coal prices collapsed. The mine owners tried to impose wage cuts and longer hours on their labour force.

The miners refused to budge and the government set up an enquiry under **Sir Herbert Samuel**. Neither the unions nor the owners would accept his report, which recommended a reorganisation of the industry and a temporary cut in wages. The mine-owners locked out the workers on 1 May, 1926. The Trades Union Congress called a General Strike for 4 May in support of the miners.

'Not a minute on the day, not a penny off the pay'

(A.J. Cook, Miners' Union Leader, 1925)

Most workers responded to the call for a General Strike: power supplies, transport, and the building, printing and manufacturing industries halted. Churchill, in the *British Gazette*, which he had printed during the dispute, presented the strike as an attempt to overthrow the government. He antagonised the workers by referring to

Military convoys protect food supplies, General Strike 1926.

them as 'the enemy'. The T.U.C. leaders were uneasy about the strike, especially when Baldwin condemned to it as unconstitutional. They also feared that union assets would be seized if the strike was declared illegal.

The government held firm and organised emergency supplies. Volunteers who saw the strikers as revolutionaries and communists kept essential services going. The T.U.C. called off the strike on 12 May. The miners felt betrayed and stayed out for another six months, after which they returned to work on worse terms than they had rejected earlier.

The General Strike seriously weakened the unions. They were accused of 'unconditional surrender' and 'betrayal' by the miners. Union membership dropped from 8 million to 5 million. In 1927, the government brought in the **Trade Union and Trade Disputes Act**, which, in effect, banned general strikes. Most workers now turned to political action and support for the Labour Party increased. Ramsay MacDonald had played no part in the strike. When it was all over he wrote in the *Socialist Review*:

> 'The General Strike is a weapon that cannot be wielded for industrial purposes. It is clumsy and ineffectual...... I hope that the result will be a thorough reconsideration of trade union tactics. If the wonderful unity in the strike...... would be shown in politics, Labour could solve the mining and similar difficulties through the ballot box.'
>
> (*The Life and Times of Ernest Bevin*, A. Bullock, Heinemann, 1967)

The Hungry Thirties

MacDonald's Second Labour Government, 1929-1931

The general election of 1929 brought Labour back to power, with the support of the Liberals. The government was in power for only a few months when the Wall Street Crash occurred. The cautious MacDonald decided on severe cuts in public spending as the only solution. By the end of 1931, unemployment had risen to almost 3 million and Britain was in crisis.

A cabinet committee under **Sir George May** proposed cuts in unemployment assistance and in workers' wages. The cabinet split on the proposals and MacDonald announced the resignation of his government.

National Governments, (1931-1940)

Due to the economic crisis, Britain was ruled for the remainder of the 1930s by National Governments. These were coalitions and were dominated by the Conservatives. **MacDonald** was asked by King George V to form a temporary National Government in 1931 and to remain on as Prime Minister. He accepted and was expelled from the Labour Party.

Labour Government 1929-31
Centre front: Ramsey MacDonald, P.M.

BRITAIN BETWEEN THE WARS

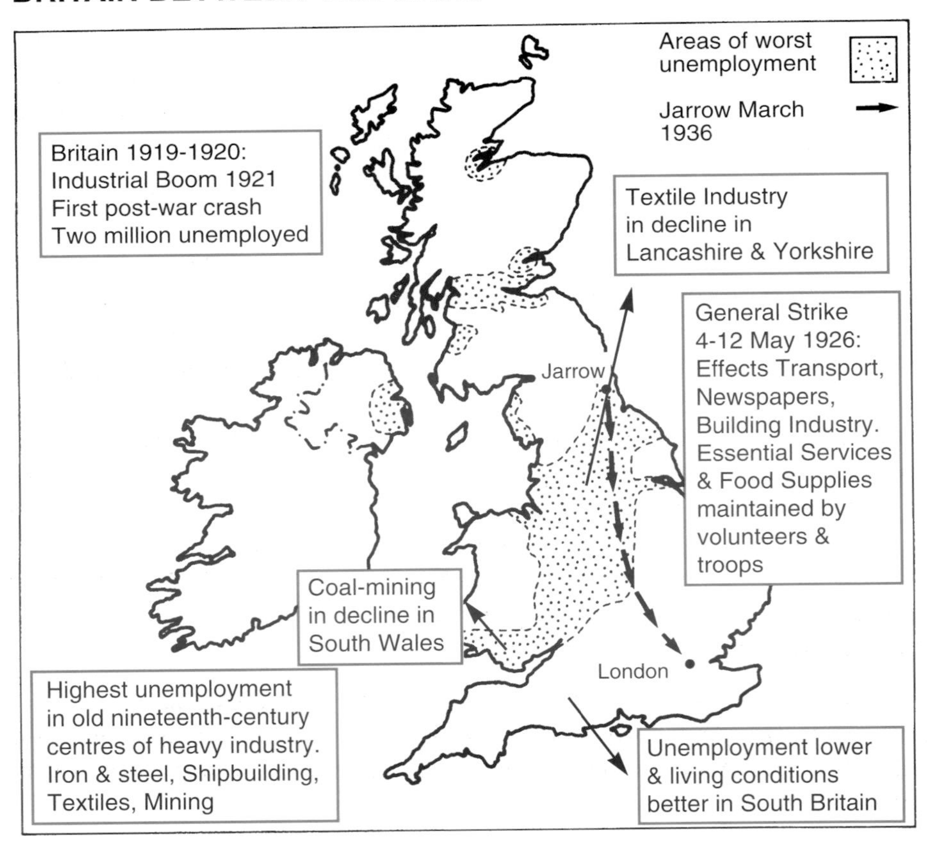

The new government won a massive majority in the 1931 election. It was composed of Conservatives, Liberals and a few Labour members and lasted until 1935. **Baldwin** replaced MacDonald as Prime Minister from 1935 until 1937. The third National Government of the 1930s was presided over by **Neville Chamberlain** from 1937 until 1940. Among the measures brought in by these governments to fight the **depression** (often called **the slump**) were the following:

- The gold standard was abandoned. This led to a devaluation of the pound by 10 per cent and British exports became cheaper.
- A duty of 10 per cent was placed on all imports except those from the Commonwealth countries.
- Unemployment benefits were more strictly means-tested.
- Public employees' salaries were cut by 10 per cent.
- Income tax was increased from 22.5 pence to 25 pence in the pound.
- Interest rates were kept at a low level.
- The farming and building industries received supports and subsidies and both enjoyed a period of revival.

By the mid-thirties, Britain had recovered from the worst of the recession. However, unemployment remained stubbornly high and had climbed back to almost 2 million by 1938. The national governments had to deal with this problem, together with the threat from Nazi Germany abroad, and an abdication crisis at home.

Unemployment

Britain suffered from the problem of unemployment right through the years between the wars, 1918-1939. Those working in the old-established and heavier industries suffered most. These were mainly coal-mining, iron and steel and textiles. They were situated mainly in what became known as the 'depressed areas' of Scotland, Wales and the North of England.

Since the depression affected most countries in the world, there was a general decline in shipping. Some shipbuilding towns, such as Jarrow on the Tyne, were so severely affected that unemployment levels reached almost seventy per cent. This led to the **'Jarrow Crusade'**, a march of the Jarrow unemployed to London in 1936.

The greatest damage done by unemployment was the despair it brought to certain areas. Some men were unemployed for years. They lost hope and they became as depressed as the surroundings they lived in:

> 'When a man fell out of work he would, on the first day, dress in his Sunday suit with collar and tie. He shaved, pinned on his ex-serviceman's badge and, head held high, lined up at the Labour Exchange. He kept up his spirits by joking with his mates...

Jarrow march of the unemployed to London, 1936

As the weeks passed the unemployed man changed. He stopped dressing up. He acquired a characteristic slouch...... He left the stubble on his cheeks.

As the months passed his hands grew white – softer and whiter than those of his wife.'

(*The Twentieth Century World*, J. Ray and J. Hagerty, 1991. Stanley Thornes)

The writer and socialist, **George Orwell**, described such a man in Wigan in the mid-1930s:

'He was standing there as motionless as a statue, cap pulled over his eyes, gaze fixed on the pavement, hands in pockets, shoulders hunched, the bitter wind blowing his thin trousers against his legs. Waste paper and dust blew about him.'

(*The Road to Wigan Pier*, G. Orwell, 1937)

New industries

But all was not gloom. Although the old industries were in decline, new ones flourished: the manufacture of radio and electrical goods and the production of motor vehicles brought new prosperity to many areas in the Midlands and South-East. The writer, **J.B. Priestley**, described a different England in his ***English Journey***, published in 1936:

'...... the northern entrance to London, where the smooth wide road passes, between miles of semi-detached bungalows, all with their little garages, their wireless sets, their periodicals about film stars, their swimming costumes and tennis rackets and dancing shoes.'

After 1936, Britain turned to rearmament and this provided some jobs. Yet, by 1939, there were still nearly a million and a half out of work. The war brought a period of boom and work was found for all.

The Abdication Crisis, 1936

In January, 1936, George V died and his eldest son succeeded as **Edward VIII**. The King, who was still a bachelor, wished to marry an American, **Mrs Wallis Simpson**. She had been twice married and was soon to be divorced for the second time. On top of all this, Mrs Simpson was a 'commoner'.

The King thought that his private life should be kept separate from his public life. Despite the monarch's personal popularity, the Prime Minister, Mr Baldwin, opposed the marriage. The Archbishop of Canterbury felt that such a marriage would be contrary to the teaching of the Church of England, and Edward was head of that Church.

The King then proposed that he should be allowed to marry Mrs Simpson, but that she would not become Queen. Baldwin and the dominion governments rejected this proposal. Edward decided to abdicate, rather than give up the woman he loved. He was succeeded by his brother who ruled as **George VI**. Edward and Mrs Simpson married and lived out their lives in exile near Paris.

Foreign Policy

Britain's role in Europe has been examined in detail in chapters 12, 13 and 18. The principal aspects of this policy can be summarised as follows:

The Re-settlement of Europe

Lloyd George had fought the 1919 election with the slogan 'Make Germany pay'. However, at Versailles he was not as severe as the French in his demands. (see chapter 12). He ensured that strict naval restrictions were placed on Germany. Many League of Nations mandates were administered by Britain.

During the 1920s, Britain began to feel that the Treaty of Versailles had been too severe on Germany. Ramsay MacDonald's Labour Government of 1924/5 supported the **Dawes Plan** to reduce Germany's reparations, and also helped to solve the problems posed by France's occupation of the Ruhr. Britain worked towards the maintenance of peace and stability in Europe. She did not want to be dragged into another conflict on the continent. Her main concerns were the solution of domestic problems and the development of relations with the Commonwealth.

Britain acted as a guarantor of the **Treaty of Locarno**, 1926, by which the borders decided at the Treaty of Versailles were recognised.

The British looked after their own interests and did not give strong support to the policy of **collective security** advocated by the League of Nations. They were willing to compromise with Mussolini rather than risk war. The **Hoare-Laval Pact** (see chapter 19) was rejected by the British public, as it rewarded aggression by suggesting that Italy should be allowed retain two-thirds of Abyssinia.

Anti-Communism

Another central pillar of Britain's policy between the wars was her total opposition to Communism. British forces were sent to Russia in 1919, to oppose the Bolsheviks. However, Britain was war-weary, the intervention was only half-hearted and the forces were quickly withdrawn.

British politicians during the 1930's preferred to compromise with Hitler than to debate with Stalin. Soviet attempts to set up a solid military alliance to oppose the Nazis were treated with some disdain. Britain could only blame herself when Stalin turned to Hitler and accepted the Nazi-Soviet Pact of 1939 (see chapter 18).

When France and the Nationalists rebelled against the lawful government of Republicans, Socialists and Communists in Spain in 1936 (see chapter 21), Britain was to the forefront in establishing a policy of international non-intervention. Many Britons felt that, if the Communists had rebelled, Britain would have become involved.

The Commonwealth

Britain's colonial empire gradually evolved into the British Commonwealth. At various Imperial Conferences, Dominion States like Canada, Australia and Ireland looked for a greater say in their own affairs. This was finally granted by the **Statute of Westminster, 1931**, which gave the Dominions equality, with the power to repeal laws passed by the British Parliament. During the inter-war years, other colonies, like India, sought independence. This was not granted until 1947.

Britain's attempts to protect her industries led to duties being placed on a wide range of imports in 1932. This provoked protests from the dominion states within the British Commonwealth. At the Imperial Economic Conference, held in Ottawa later in 1932, a system of Imperial preferences was worked out between Britain and the dominions. This led to an increase in trade between Britain and the Commonwealth.

Appeasement

The policy of appeasement is associated mainly with Neville Chamberlain. As soon as he became Prime Minister, Chamberlain set out to achieve conciliation with Germany. Diplomats who were sympathetic to the Nazis were appointed to Berlin. The Foreign Secretary, **Anthony Eden**, stood for the idea of collective security and the upholding of the League of Nations. He resigned in protest at the Prime Minister's policy of appeasement.

Hitler's seizure of Austria shocked most people in Britain, but many British politicians argued that the Nazis were justified in absorbing those of Germanic race. Chamberlain (see chapter 18) played a leading role in the **Munich Peace Conference**

'Peace in our time' Chamberlain returns from Munich, 1938.

and the betrayal of Czechoslovakia. The British Prime Minister seems not to have realised, until too late, the brutal and immoral nature of Hitler's regime. He could not stop Hitler by granting concessions.

It was clear that Britain's twin policies of self-interest and appeasement had failed completely by the end of the 1930s. Chamberlain recognised this in a speech in Parliament in 1939:

> 'Everything that I have worked for, everything that I have hoped for, everything that I have believed in during my public life, has crashed into ruins.'

Declaration of WWII: Britain declares war on Germany, September 3, 1939 – crowds in Downing Street hear the announcement.

Chapter 21

Franco and the Spanish Civil War

Political Background

A country in decline

In the 18th and 19th centuries, the once proud country of Spain went into decline. She lost her great empire which had been built up in the 16th and 17th centuries. The Spanish people blamed the monarchy for this disaster. In 1868 Queen Isabella II was deposed and the **First Spanish Republic** was established.

The Republic proved unstable and in 1875, after a military rising, a constitutional monarchy was proclaimed under Isabella's son, Alfonso XII. After the king's death in 1885, his wife, Maria, took over as regent. Her son, Alfonso XIII, succeeded to the throne in 1902.

At the beginning of the 20th century, it seemed that Spain was on the road to true democracy. The King ruled with the help of the ***Cortes***, the Spanish parliament, which had an upper and lower house. All adult males could vote in general elections.

However, Spain was not without serious problems. In the Spanish-American War, which began in 1898, she lost her remaining colonies of Cuba, the Philippines and Puerto Rico. The Spaniards tried to restore their pride by seizing Morocco. The Moroccans refused to submit to Spanish rule and caused endless trouble for the colonists. In 1921 the Spanish army suffered a major defeat in Morocco.

After World War I, in which Spain remained neutral, the country was beset by poverty and unemployment. Trade unions grew in strength, strikes were commonplace, and serious rioting occurred. Between 1918 and 1923 Spain had twelve governments. Many called for the setting up of a new republic. Others, including the King, favoured stronger government.

Primo de Rivera

In 1923, **General Primo de Rivera** seized power with the help of the army and the King, **Alfonso XIII.** De Rivera was an authoritarian who, during his seven years in power, brought strict army discipline into Spanish life. The *Cortes* was closed down and all political parties were banned. His slogan was 'Country, Religion and Monarchy'.

During the 1920s, Europe's economies improved. Rivera modernised Spain by improving the roads, irrigating dry areas, re-equipping railways, bringing electricity to rural areas and introducing an efficient telephone system. He ended the war in Morocco and at home gave Spain law and order.

By 1930, world depression had hit Spain and her economy was in difficulty. Rivera had created many enemies through his curbs on freedom. He had lost the confidence of the army when he changed the system of promotion. Above all, he did not get on well with the King. Alfonso forced Rivera to resign and the general fled abroad.

The King held municipal elections in 1931. The Republicans won a majority and Alfonso went into exile in France. Rivera's rule had ended in failure. However, many conservative Spaniards were later to look back fondly to the period of stability between 1923 and 1930.

A Divided Country

Regionalism

We who are as good as you swear to you who are not better than we,
to accept you as our king and sovereign lord
provided you accept all our liberties and laws.

This very limited promise of loyalty made by the people of Aragon to the King of Spain shows the independent and rebellious attitude of the people in the regions towards those who wanted a united country. **Separatism** in Spain was due partly to geography, partly to history. Large mountain ranges cut across the country, dividing it into smaller areas. Many people in these smaller areas (regions) wanted self-rule and had no allegiance to the Spanish nation.

Basque Separatists were Catholic and conservative. They wished to preserve their language, culture and unique way of life. They resented having to take orders from Madrid and obtained a measure of independence in 1936. They set up a **Basque Republic** and were prepared to defend it with their lives.

Catalonia traditionally had its own ***Cortes*** or Parliament. This province was the industrial centre of Spain, accounting for almost 80 per cent of the country's output. Catalans also had a strong sense of a separate cultural and historical identity. Because of industrialisation, communist and anarchist movements flourished, especially in the capital, Barcelona. The province sought a separate Catalan Republic and was granted a **Statute of Autonomy** in 1932. This gave Catalonia a limited degree of self-rule.

Agrarian Problems

Spain was mainly an agricultural country. In many areas huge estates – ***latifundia*** – developed. These were owned by rich families who hired landless labourers and often left large areas uncultivated. The poor labourers were paid a pittance by the landlords. There were about 2,500,000 *braceros*, or landless labourers, in Spain in the 1930s.

Some provinces, for example Catalonia and Valencia, had no *latifundia*. They suffered from the problem of ***minifundia,*** farms too small and miserable to support a family. The poverty and hunger for land often led to agrarian violence. The communists promised land to the peasants and the problem of inequitable land distribution was one of the main causes of the Civil War.

The Church

In the 1930's, the Catholic Church was in a more powerful position in Spain than in any other country in Europe. She owned much land and property and generally took the side of the wealthy in disputes with the poor. As an old traditional institution she had become very conservative and opposed all forms of liberalism, socialism and republicanism.

Church outside Madrid, destroyed by communists, 1936.

Except in the Basque region, the Catholic clergy became identified with the upper classes. This meant that most workers became anti-clerical and often joined communist and anarchist groups. They burned churches, the 'temples of the rich', and were very willing to execute the hated clerics during the Civil War. The Catholic Church had become a symbol of division in Spain.

The Army

The army, like the Church, had held on to its privileged position. It regarded itself as protector of the old way of life. Its privileges had come from the days when Spain was a great colonial country with a powerful military machine. Army officers opposed change within the army and the country. The aristocracy expected that it would continue to provide Spain with its officer class and resented moves by any government which attempted to change this situation.

At home, the army was used to suppress the restless peasants. A local armed force, the ***Civil Guard***, dealt mercilessly with disturbances. In the Spain of the 1930s the army was ready to support the right-wing political parties. The immediate cause of the Civil War was the decision by army generals to oppose a left-wing government which seemed unable to uphold law and order.

Left v Right

Spain did not escape from the clash of fascism and communism seen in other European countries during the 1920s and 1930s.

The main conservative party was the CEDA or Catholic Party. It turned against the Republic due to the attacks on the Church and by 1933 was the strongest opponent of the government. Its leader was **Gil Robles** who strongly opposed the democratic system. He visited Germany in 1933 and expressed his admiration for the Nazis.

The *Falange Española,* a Spanish fascist party, was founded in 1932 by **José Antonio de Rivera,** son of the former dictator. The party's motto, 'Arise! Spain. One Great and Free' appealed to nationalists. De Rivera was executed by Republicans in 1936.

Calvo Sotelo became leader of the Nationalist Bloc in 1934. This Bloc was a coalition of Monarchists, Fascists and right-wing independents, opposed to Republicans, Communists and left-wing groups. Sotelo had been a minister in Primo de Rivera's government and went into exile in France in 1930. The Nationalist Bloc rejected the basic principles of democracy. It set out to destroy trade unionism. Sotelo's assassination in 1936 was an immediate cause of the Civil War.

Republicans in Spain were in favour of a constitutional democracy. Some Republicans were conservative in outlook. The left wing of the movement was led by **Manuel Azana.** The Socialist Party wished to change Spanish society through peaceful and democratic means. **Largo Caballero** was the main socialist leader.

Manuel Azana, President of the Republic.

'La Pasionaria' (Dolores Ibarruri), fiery communist orator

On the extreme left were the Communists and the Anarchists. The most influential figure in the Spanish Communist Party was **Dolores Ibarruri**, known as ***La Pasionaria***, the 'passion flower'. She was a member of the *Cortes* and proved a great orator. Always dressed in black, she moved her listeners through her deeply sincere Communist faith, her courage and her fiery speeches.

The Anarchists wanted to abolish both private property and organised religion. Their aim was to destroy the government. In their ideal world nobody would give orders and the country would be an association of independent local districts or societies. The Spanish Anarchists advocated the overthrow of capitalism through strikes and violence. The best known Anarchist leader during the Civil War was **Buenventura Durruti**.

The Spanish Civil War was a struggle between right and left. Those who supported the monarchy, the propertied classes, the Church and the army were prepared to sacrifice everything to ensure victory over the Republicans, Socialists, Communists and Anarchists.

Left v	**Right**
Republicans: Manuel Azana Socialists: Largo Caballero Anarchists: Buenventura Durruti Communists: Dolores Ibarruri *Supported by –* Working class, trade unions, landless labourers, liberals, intellectuals, separatists, International Brigades, Stalin.	CEDA (Catholic Party): Gil Robles Falange Espanola (Fascists): Jose Antonio de Rivero Nationalist Bloc: Calvo Sotelo *Supported by –* Catholic Church, army, monarchists, landlords, propertied classes, big business, Hitler, Mussolini, Salazar, O'Duffy.

The Second Republic

The First Spanish Republic had been established in 1868 but lasted for only seven years. The Second Republic was proclaimed in 1931. **Prime Minister Manuel Azana** led a government of Left Republicans and Socialists. A new democratic constitution was introduced.

The Church had to hand over to the state much of its property and its control of education. Government payment of priests was stopped and some monastic orders were abolished. While these measures pleased the liberals, they frightened many of the more conservative supporters of the Republic.

Spain's major problem, the hopeless situation of the landless peasant, was not tackled with zeal. **The Agrarian Law** of 1932 applied to only some regions and affected only 10 per cent of the *latifundia*. Very few peasants got land. The world-wide slump was biting deep and the government became more unpopular. The 1933 elections returned a right-wing government.

Under the new government, most of Azana's reforms were repealed. The CEDA, a Catholic Confederation of Conservative Parties, was included in the government in 1934. Socialists and Republicans feared that a military dictatorship would be set up. An all-out strike was called and serious disturbances occurred in the mining districts of Asturias. These were put down with terrible cruelty and much bloodshed by a young general, **Francisco Franco**.

The Popular Front, 1936

The left-wing Republicans, Socialists, Communists and Anarchists now joined together in a **Popular Front** of anti-fascist groups. They won a majority in the general election of 1936. The Popular Front government led by **Azana**, tried to introduce reforms and to divide the large estates among the peasants. However, many people began to seize land for themselves. Widespread disorder occurred and churches were burned. Hatred of Church authorities led to the murder of a number of priests. Between March and July, 1936, over 100,000 peasant families were granted land. The government ordered the closure of all church schools. The building workers in Madrid went on strike and disorder continued.

In 1936, the Popular Front Government became worried about the possibility of an army revolt. Suspect generals were transferred. General Mola was sent to Navarre in the north of Spain. General Franco was moved to the Canary Islands. There he brooded over the collapse of law and order and the attacks on the Catholic Church.

Nationalists began to believe that the government was unable or unwilling to protect property and religion. Army generals met in secret and decided to stage a *coup* against the government of this 'Godless Republic'. They hoped to link up with the Falangists, the Monarchists and CEDA. The leader of the Monarchists in the *Cortes*, **Calvo Sotelo**, became involved in the proposed *coup*. Sotelo was assassinated on 12 July, 1936. The army generals felt it was time to act.

The Spanish Civil War began five days later, on 17 July, 1936. The revolt was led by **General Sanjurjo**. After his death in 1936 and the death of **General Mola** in 1937, **General Francisco Franco** became the leader of the Nationalists.

The Civil War

Franco: profile

> Spaniards! To whomever feels a holy love of Spain, to those who in the ranks of the army and navy have made professions of faith in the service of the country, to those who have sworn to defend her from her enemies, the Nation calls you to her defence.

These were the opening lines of the manifesto issued by General Franco to the Spanish people on 19 July, 1936 before he flew from the Canary Islands to Morocco, at the beginning of the Spanish Civil War.

Francisco Franco was born in 1892 in El Ferrol, Northern Spain. His distinguished military career began in Morocco. As second-in-command of the Spanish Foreign Legion he helped to secure Morocco for Spain in a campaign between 1920 and 1925. In 1926 Franco became Europe's youngest brigadier-general. He was in charge of the suppression of the revolt of the miners of the Asturias in 1934. In 1935 Franco was appointed Chief of Staff, the top post in the army. Early in 1936 he was transferred to the Canary Islands, as the government became worried about the loyalty of the army.

Franco decided to take part in the army revolt in July 1936. The 'enemies' he referred to in his manifesto included members of the Popular Front government. He

Franco is proclaimed Head of the Spanish State, October 1936.

saw their victory in the 1936 elections as a prelude to a Communist take-over of power.

The First Phase

Franco's first problem was to ferry his **Moroccan army** to mainland Spain. He appealed to Hitler for help. The German dictator sent twenty transport planes to airlift Franco's elite African troops. This army proved immediately successful and soon captured **Seville**. Before long much of southern Spain was under his control. He then moved north and hoped to link up with the other nationalist forces.

The first setback for the army rebels was the death of the nominated leader, General Sanjurjo, in a plane crash at the beginning of the war. In northern Spain, General Mola proved successful. The Nationalists soon captured **Pamplona, Burgos, Valladolid** and **Salamanca**. However, they failed to capture the major cities of Barcelona, Valencia, Madrid and Toledo, in eastern and central Spain.

The Siege of Toledo

Franco had marched north from Seville and hoped to capture **Madrid**. However, he was told that nationalist troops were in difficulty in the ancient city of **Toledo.** There **General Moscardo** had been forced by the Republicans to retreat within the stone fortress of **Alcazar** on the banks of the River Tagus.

For over two months the Nationalists held out. The new Republican Prime Minister, **Largo Caballero,** came from Madrid to supervise the siege. The defenders survived by eating dead mules. The Republicans mined under the fortress and set off explosives. Still Moscardo refused to surrender. On 27 September 1936, **General**

General Franco (centre) congratulates Colonel Moscardo, leader of the Alcazar garrison, on his resistance against the government siege, in the Alcazar after the relief.

Valera arrived with one of Franco's relief armies. The siege was lifted. The defence of the **Alcazar** was an inspiration to the Nationalists and Moscardo became a hero. Those Republicans who were captured were mercilessly slaughtered.

By the end of 1936, the Nationalist forces controlled western and northern Spain with the exception of the Basque region. The Republicans held Madrid and most of the territory to the east of the capital including Valencia and Barcelona. The failure of the Nationalists to capture Madrid meant that the war would drag on.

Jarama and Guadalajara

Early in 1937, the Nationalists decided to starve Madrid into submission. They attempted to cut off supplies from the Valencia road. Far some time they swept all before them. The Republicans, aided by some units of the International Brigades (see pages 273/4) made a stand in the **Jarama Valley.** Both sides suffered terrible casualties before the Nationalists withdrew. It was their first major defeat in battle.

'For them there would be no old age': Jarama, 1937

A month later, in March 1937, the Nationalists attacked again. They had the help of Mussolini's Italian motorised troops. Early successes were reversed when they tried to take the town of **Guadalajara** near Madrid. They were hampered by ice and snow. Then the light Italian cars and lorries were smashed by Russian-made tanks. Mussolini's troops scattered. The Nationalists had suffered another setback.

Guernica

Franco's motto was 'Spain, one great and free'. The Popular Front government had granted self-government to the **Basques** in 1936. Franco was determined to crush **Separatism** and saw the Basque region as an easy target, as it was cut off from other Republican areas.

Picasso's vision of the destruction of Guernica

The Nationalist commander, **General Mola**, began his attack at the end of March 1937. He had 50,000 well-equipped troops and was backed up by German aeroplanes. The Basques fought fiercely, but were driven back. On 26 April, 1937, the **German Condor Legion** dropped 100,000 pounds of bombs on the town of **Guernica,** spiritual capital of the Basque people. Over a thousand people were killed and many were injured.

The destruction of Guernica and the horror of war has been captured by **Pablo Picasso** in his famous painting called after the town. The bombing caused great horror around the world, especially in Britain and France. It damaged the Nationalist cause although Franco claimed that the destruction had been caused by retreating Communists and Republicans who had set fire to the town. However, at the Nuremberg Trials after World War II, the Luftwaffe commander, Herman Göring, admitted that the Germans had bombed Guernica.

The Battle of Teruel

The Nationalists failed to break through to Madrid during 1937. In December of that year, the Republicans tried to relieve the pressure on the capital by attacking the Nationalist stronghold of Teruel, east of the city. The battle was fought in the snow and ice of one of Spain's coldest winters ever. Led by the legendary ***El Campesino,*** the Republicans took Teruel after fierce fighting.

Franco was with the Nationalist troops who were besieging Madrid. He sent 80,000 soldiers to the relief of Teruel. They were accompanied by German and Italian aircraft and artillery. The town was reduced to ruins by shells and bombs. The Nationalists recaptured the town on 22 February 1938. They found the bodies of 10,000 Republicans under the debris of what had once been Teruel.

Teruel was the beginning of the end for the Republicans. Their morale was shattered and they were now in retreat. They were forced back across the Ebro. The International Brigades withdrew in November 1938. In January 1939, **Barcelona** fell to the Nationalists. Only Madrid stood between Franco and outright victory.

The Fall of Madrid

Franco did not want to take over a ruined capital. After allowing some bombing he did not permit the Condor Legion to destroy the city. He had also been diverted away from Madrid in order to relieve places like Toledo and Teruel. The Republicans put up a heroic defence of their headquarters. They were inspired by the Communist leader ***La Pasionaria.***

The Battle of Madrid had started on 8 November 1936. The Nationalists failed to capture the city in fierce hand-to-hand fighting. Casualties were heavy and both sides dug trenches. Stalemate resulted. Franco tried to starve Madrid into surrender. The besieged people suffered terrible hardships, but refused to yield.

Franco would accept nothing less than unconditional surrender. A military junta under **Colonel Casado** overthrew Prime Minister Negrin's government inside Madrid. Casado hoped to obtain better surrender terms from Franco. In this he failed. The only concession was that the officials of the junta were allowed to leave Spain.

THE SPANISH CIVIL WAR

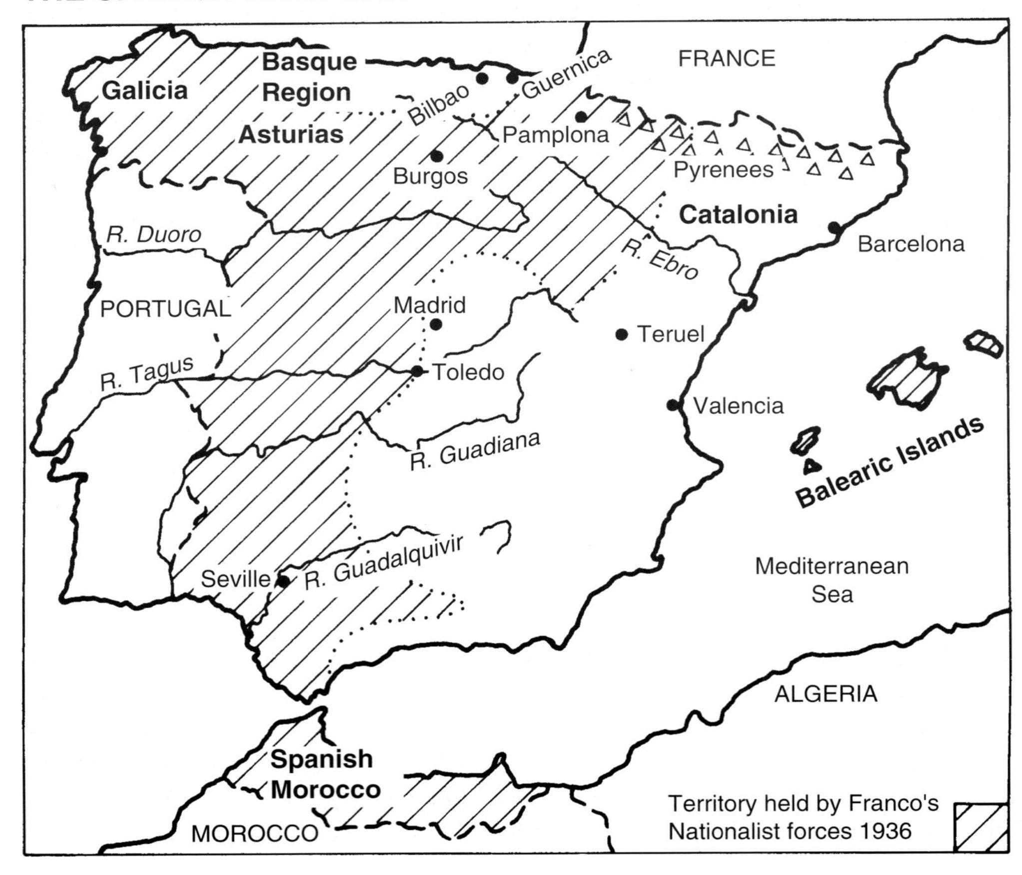

On 28 March 1939, Nationalist troops entered Madrid. The few remaining areas in Republican hands soon surrendered. The Spanish Civil War ended officially on 1 April 1939.

International involvement

Fascists for Franco

We must save Spain from Bolshevism.

(Hitler)

In Germany and Italy, sympathy lay with the fascist leader Franco. He was seen as the saviour of Spain against the scourge of Communism. Hitler sent 16,000 men and his **Condor Legion** of aircraft tanks and artillery to Spain when Franco appealed for help. This was an opportunity for Hitler to try out some of his forces, especially the newly-formed ***Luftwaffe***.

Mussolini sent over 50,000 men whom he called 'Volunteers'. The Italian navy also helped the Nationalist forces. The venture gave Mussolini a reason to claim that Italy was playing an important part in world affairs.

President Salazar of Portugal sent 20,000 troops to help Franco. Volunteers from other countries also fought for the Nationalists. Most of these Volunteers were convinced anti-Communists. They believed that the capitalist system, the Catholic Church, and democracy – as they understood the term – were all under attack. Among the Volunteers was a group of 'Blueshirts' from Ireland led by General O'Duffy. Help from the Fascist leaders strengthened the Nationalists greatly and was a major factor in their eventual triumph.

Stalin

The Republicans asked for help from the Soviet Union. Stalin supplied tanks, fighter planes and military advisers.He sent only limited supplies. He did not wish to unduly weaken his army, as he feared an attack by the Germans on his country. He sent just enough help to keep Germany and Italy tied down. Russian aid diminished later in the war, when the Nationalists blockaded the ports.

Non-intervention

In France, **Blum's** socialist government favoured the Republic and gave the government help with weapons and aeroplanes in the early part of the war. However, this aid soon stopped. Blum was afraid of splitting his country and of offending the British. Britain and France formed a **Non-Intervention Committee** and tried to prevent arms and munitions reaching either side. This weakened the Republican side, as it did little to curb the activities of Germany and Italy.

International Brigades

> On remote peninsulas,
> On sleepy plains, in the aberrant fishermen's islands
> Or the corrupt heart of a city
> The Volunteers
> Heard and migrated like gulls or the seeds of a flower.
>
> *Spain,* W. H. Auden

About 40,000 volunteers went to Spain to support the Republic. Most of these joined the **International Brigades.** Many were doctors, nurses, teachers, students and writers, who saw the war as a fight for freedom and democracy against the evil of fascism. **Josip Broz,** later Marshal Tito of Yugoslavia, was the chief organiser of these Brigades from a base in **Paris.**

The British poets, **Stephen Spender** and **W.H. Auden,** were among those who volunteered. **George Orwell** was another anti-fascist idealist. He described his experiences in the novel ***Homage to Catalonia.*** André Malraux, the Frenchman who commanded the Brigade's air squadrons, wrote ***Days of Hope.*** Perhaps the most famous novel to emerge from the struggle was ***For Whom the Bell Tolls,*** by the American writer and war correspondent, **Ernest Hemingway.**

The volunteers in the International Brigades fought with great courage. Their presence helped to prolong the war. However, they found it hard to overcome language problems and were not as disciplined or efficient as the trained troops against whom they fought. Those who survived withdrew at the end of 1938. ***La Pasionaria*** gave thanks and paid tribute:

> Today they are going away. Many of them, thousands of them, are staying here with the Spanish earth for their shroud, and all Spaniards remember them with the deepest feelings.

The Aftermath

A Costly War

The Spanish Civil War lasted a hundred days short of three years. Estimates of the casualties vary, but it is now believed that about half a million people lost their lives. The Republicans were guilty of about 20,000 executions, but the Nationalists were responsible for up to 10 times that number. The bitterness, horror and reprisals can only be accounted for by the absolute conflict between fascism and democracy, capitalism and communism, Catholicism and a 'Godless Republic'.

In his hour of victory, Franco showed no mercy. Thousands were executed after the war. Over 300,000 Republicans went into exile, mainly to France. Between 1939 and 1942 over a million were imprisoned, or forced to work in penal battalions,

Defeated Republicans are led across the frontier into France.

clearing up the war debris. Some of the prison terms were for up to thirty years. The unwillingness of the victors to forgive and forget meant that the bitterness of the Civil War has lasted in many areas up to the present day.

Reasons for Nationalists' Victory

- The Nationalists were very disciplined and united. They accepted Franco as sole commander. He showed a singular determination and utter ruthlessness in pursuit of victory.
- The Republicans, Socialists, Communists and Anarchists were never properly united, in ideology or as a fighting force. They lacked strict military discipline and did not have one supreme commander. They sometimes fought each other. This lack of unity seriously weakened the Republican cause.
- The Nationalists received more foreign aid than the Republicans. The decision by Britain and France to pursue a policy of non-intervention meant that the Republicans never had enough equipment to fight a successful war.

Franco's Spain

After the Civil War both town and country were devastated. The outbreak of World War II in 1939 meant that Spain could not rely on outside help towards reconstruction. She was in no condition to play a major part in World War II. Franco did allow a **Blue Division** of 18,000 Spanish Volunteers to fight against the Soviet Union in 1941, but he pointed out that this was part of the fight against Communism rather than against the Allies.

Spain was denied Marshall Aid given by the United States to war-damaged countries in Europe (see chapter 25). This meant that Spain remained economically backward. However, as the Cold War developed, Franco's regime began to gain international acceptance and respectability. In return for permission to set up four

U.S. bases in Spain, Franco received economic aid from the Americans. In 1955 Spain was permitted to join the United Nations. The 1960s saw the start of the boom in Spain's tourist industry.

After the Civil War, Franco set up a state which was 'pro-Spanish and pro-Catholic'. Divorce was forbidden, co-education was banned and control of education was placed in the hands of the Church. Strict censorship was enforced and trade unions were replaced by Fascist-type corporations which banned strikes. Spain was a totalitarian state in which Franco was head of state for life.

In 1969 Franco nominated **Juan Carlos**, grandson of the last king, Alfonso XIII, to succeed him. Franco died on 20 November 1975. He had curbed freedom, but had given Spain peace and stability for almost 40 years. There was a peaceful transition to Monarchy after his death. Under King Juan Carlos Spain has become a modern democratic state.

Chapter 22

Stalin's Russia, 1924-1939

The Struggle for Power

Stalin – A Profile

Joseph Djugashvili was born in 1879 in **Georgia** in the Caucasus. In 1913 he adopted the name Stalin which means 'man of steel'. He was the son of freed serfs. At the age of 15, Stalin was awarded a scholarship to a seminary for the training of Orthodox priests, at Tiflis in Georgia. According to his own account he was expelled 'for disseminating Marxist propaganda'.

Stalin then joined an underground revolutionary movement and organised a number of armed robberies to raise funds for the Marxist cause. He was imprisoned and was later exiled to Siberia. After his escape from exile he organised a Bolshevik group in Georgia. It was while attending a party conference in Finland, in 1905, that he first met Lenin.

Stalin kept in contact with Lenin and in 1912 was appointed to the **Bolshevik Central Committee.** Lenin saw that he was a good administrator and made him manager of the party's newly-founded newspaper, ***Pravda.*** Stalin was being watched by the Tsar's secret police and in 1914 he was again deported to Siberia. He was declared unfit for service in the army because of a deformed arm. When the Tsar abdicated after the February Revolution in 1917, Stalin was released and returned to Petrograd.

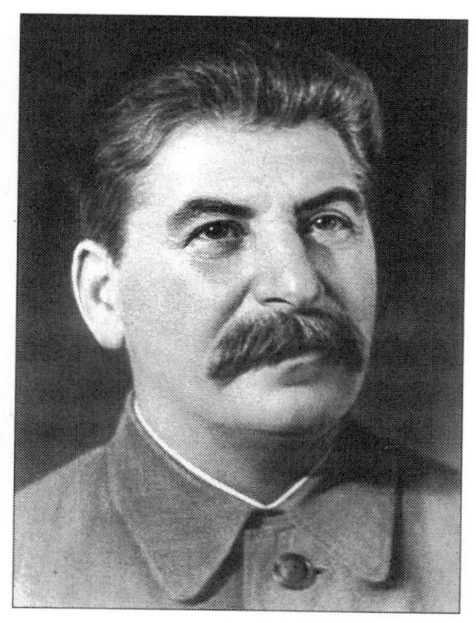

Joseph Stalin

Stalin played a minor role in the 1917 Revolution. He stayed close to Lenin and accepted the latter's policy of non-cooperation with Kerensky's Provisional Government. He was rewarded with the post of **Commissar for Nationalities**. This later gave him the support of many non-Russian Bolsheviks in the power struggle. Stalin also increased his power base by becoming **Commissar of the Workers' and Peasants' Inspectorate**. In 1922 he was appointed **General Secretary** of the Communist Party. He used his new position to nominate his own supporters to important positions in the Party.

Lenin's Testament

> Comrade Stalin, having become General Secretary, has concentrated enormous power in his hands and I am not sure that he always knows how to use that power with sufficient caution. Stalin is too rude, and this fault, entirely supportable in relation to us communists, becomes insupportable in the office of General Secretary. Therefore I propose to the comrades to think over the means of removing Stalin from that position and appointing to it another who is superior to Stalin, being more patient, more loyal, more polite and more attentive to comrades.
>
> (Lenin's Political Will)

Lenin suffered a stroke in 1922. Stalin acted as his personal assistant and hoped to become leader. Lenin became aware of Stalin's faults and before he died wrote his **'Political Will'**, a testament in which he warned the Party about his assistant. After Lenin's death in 1924 his will was read out to the Central Committee. One of the Party leaders, **Zinoviev**, saved the situation by saying that Stalin's behaviour had improved and that he had learned his lesson. Zinoviev did this because he feared that **Trotsky** would become dictator of the country.

The elimination of Trotsky

Trotsky had been the hero of the Civil War and was very popular with the Red Army. He was also an intellectual and believed in strict Marxist principles. The more cunning Stalin outwitted him by taking the role of organiser of Lenin's funeral and making the funeral oration. Trotsky was recognised by Lenin in his will as 'the most able man in the present Central Committee', but was admonished for his 'too far-reaching self-confidence'. It was this arrogance that led Lenin to withhold his complete blessing from Trotsky.

Stalin first turned **the right wing** of the party against his rival. Trotsky, a pure Marxist, was not enthusiastic about the New Economic Policy (N.E.P.) with its acceptance of limited capitalism. Stalin stood back and allowed an undignified squabble to develop between the Party leaders, **Zinoviev** and **Kamenev** and their perceived enemy, Trotsky. Stalin saw that Trotsky's policy of encouraging **'permanent revolution'** world-wide was neither realistic nor popular. He put forward his own slogan of **'Socialism in One Country'**. He said that the Russian revolution should be consolidated before revolutions elsewhere were attempted.

Trotsky and his wife, in exile

Stalin's intrigue undermined Trotsky's position. The latter lost his position as **War Commissar** in 1925 and in 1926 was removed from the Politburo. He was expelled from the Communist Party in 1927 and deported from the Soviet Union in 1929. Trotsky lived in exile in France and Norway before settling in Mexico. He continued to denounce Stalin in his writing. In 1940, as he sat in his study, an assassin drove a mountaineering pick into his head. The evidence points to the fact that the murderer was one of Stalin's agents.

Supreme Ruler

Stalin then proceeded to deal with his remaining rivals. **Zinoviev** was based in Leningrad and Stalin continued to centre all power in the **Kremlin** in Moscow. He then won the economic debate. Zinoviev wanted to concentrate on agriculture. Stalin argued that industrial development should be the main priority. Zinoviev was removed from his post as Secretary of the Leningrad Party Committee.

Only **Kamenev** remained as a rival to Stalin. He said that no single leader should take the place of Lenin. Stalin argued that a strong leader was absolutely necessary for party discipline and unity. He forced a vote on the issue in December 1927. Kamenev was defeated and dismissed from the Central Committee. Joseph Stalin was now the supreme leader of the Soviet Union.

Stalin's Economic Policies

A Planned Economy

> We are fifty or a hundred years behind the advanced countries. We must close this gap in ten years. Either we do it or they will crush us.
>
> (Stalin)

The new steelworks at Magnitogorsk in the Urals in 1930

Stalin believed that the only way to close the gap and save the country from attack was to develop Russia into a modern industrialised country. Russia had to build herself up in order to prove to the world that communism worked. This could only be done by setting strict targets for agriculture and industry.

The Five-Year Plans

In 1928 the NEP was replaced by the first Five-Year Plan. This was not a programme but a law which set down the rate of increase for industry and agriculture for the years 1928-1932. The emphasis was on fuel supplies and heavy industry. Coal, iron, steel and machine-building had to increase production by three hundred per cent. Factories that produced consumer goods like clothes and furniture were to develop at a much slower rate.

Changes in Industrial Production 1913-37

	1913	1927/9	1932	1937
Coal (millions of tonnes)	29.8	36.1	65.6	138.6
Steel (millions of tonnes)	4.4	4.1	6.0	18.1
Oil (millions of tonnes)	10.5	11.9	22.7	29.1

In the 1930s, new factories and towns sprang up all over the Soviet Union. A great iron and steel works was built in **Magnitogorsk** in the Urals. Europe's largest hydro-electric plant was constructed on the **River Dnieper**. Tractors and automobiles were produced in **Moscow** and **Gorky**. Coal was mined in the **Donetz Basin** and oil refined in the **Caucasus**.

The First Five-Year Plan was declared a success and was followed up by two further plans. In these there was more emphasis on transport and consumer goods. Great canals linked Moscow with the Volga, and the Baltic with the White Sea. Railway lines were extended into Central Asia and new roads were built. The famous Moscow Underground provided a cheap and efficient transport service for the capital. However, consumer goods were still very scarce, especially after 1937, when the country turned to armaments production.

The Stakhanov Movement

In 1935 **Alexei Stakhanov** produced 102 tons of coal in a six-hour shift. This was fourteen times the standard output. For this feat he was awarded the **'Order of Lenin'** and given the new title of **'Hero of Socialist Labour'.** Workers everywhere were urged to emulate his achievement and those who succeeded were given extra pay and free holidays. More recent evidence has shown that Stakhanov did not perform his great deed without help:

> Two assistants shored up the tunnel and removed the coal while Stakhanov worked at the face with hammer and pick. The event was moreover deliberately organised by the local party to meet Stalin's request for 'heroes'... Anyone who told the truth, the party warned, would be treated as a most dangerous enemy'.

(*The Independent*, 17 October, 1988)

Stakhanov explains how he cut over a hundred tonnes of coal in six hours.

Prizes and Punishments

The system of prizes and bonuses meant that Stalin was not adhering to the strict Marxist idea of equality. There were many who could not achieve their targets. Managers and workers whose performance was judged as poor were accused of being saboteurs. Punishment was severe. Some were imprisoned or shot. Safety standards were low and accidents were frequent.

Production targets could never have been met without the use of force. The White Sea Canal and the Moscow Underground were built largely by forced labour. The barbarous conditions in the labour camps are described vividly by **Alexander Solzhenitsyn** in ***The Gulag Archipelago***.

Despite these problems, full employment and improved medical care meant that the standard of living enjoyed by the average worker improved. This helped to make the shortage in consumer goods and the harsh discipline more tolerable. The major increase in industrial production was achieved at a time when the western world was suffering from severe depression and huge unemployment. Whatever the cost of the Five-Year Plans, there can be little doubt that they saved the Soviet Union from defeat in World War II.

Education

An important aspect of the Plans was education. There was no place for illiteracy. Many peasants had to learn to read and write in order to follow simple instructions. New schools, technical colleges and universities provided the country with a skilled labour force. Some specialists were brought in from abroad. More and more Soviet citizens graduated with engineering and technology degrees. The contribution of these educated people meant that by 1939 the Soviet Union was second only to the U.S. as an industrial power.

Collectivisation

> The way out is to turn the small and scattered peasant farms into large united farms, based on the common cultivation of the soil by means of a new and higher technique. The way out is to unite the small and dwarf peasant farms gradually but surely, not by pressure, but by example and persuasion, into large farms based on common, co-operative collective cultivation of the soil with the use of agricultural machines and tractors and scientific methods of intensive agriculture. There is no other way.
>
> (Speech by Stalin to the Fifteenth Party Congress, 1927)

Stalin's industrialisation policy required the transfer of 25 million workers from the land to the new factories. An agricultural revolution was needed to feed this labour force and to provide grain exports necessary for financing industrial development.

The local **collective farm** was called a ***Kolkhoz***. The setting up of the *Kolkhozes* meant the abolition of small peasant holdings. This was an attempt to replace private enterprise with communal farming. The peasant was allowed to keep some livestock and a small plot of land for himself. The collective had to produce a quota of crops and livestock for the central government. The *Kolkhoz* was only a step towards Stalin's policy of making the peasants work directly for the state.

The state farm was called the ***Sovkhoz***. It was owned by the state and worked by the peasants. It was a larger farm than the *Kolkhoz*, and was intended to act as a

Women harvesting on a Russian state farm

model for the collectives. The peasants who worked on state farms had no private land or livestock. The agricultural machines and tractors that Stalin had spoken about were always in short supply. However, Stalin wanted the Soviet Union to be self-sufficient in agriculture and was determined to overcome all opposition.

The Kulaks

Many peasants had done well under NEP. In particular the richer peasants, the ***Kulaks,*** were unwilling to see their land turned into common property. Some preferred to burn their homes and destroy their crops and cattle rather than hand them over to the state.

Stalin's first Five-Year Plan depended on the programme of collectivisation. He declared war on the better-off peasant farmers: 'We must smash the *Kulaks*, eliminate them as a class'. The Russian countryside became a battlefield. Soldiers moved in with machine guns. It is estimated that more than five million *Kulaks* were either executed or deported from their lands.

Results of Collectivisation

The peasants who were left had no choice but to join the collectives. Before long, 25 million one-family farms had been replaced by 300,000 collectives, each with an average of 80 families. After six years Russia had only half the number of animals she had before collectivisation began:

Number of animals in Russia		
	1928	1934
Horses	32.1 million	15.4 million
Cattle	60.1 million	33.5 million
Pigs	22.0 million	11.5 million
Sheep	97.3 million	30.9 million

The workers saw that most of their crops were taken away from the collective farms in government lorries. They spent as little time and energy as they could on these farms and hurried home to look after the plot that they had been allowed to keep. Yields from these small private plots were much higher than from the collectives. The chaos caused by collectivisation, together with bad harvests, led to a terrible famine in 1932 and 1933. It is thought that about ten million people died in the Soviet Union in these years as a result of the break-down of the agricultural system. It took a long time for the country to recover. Gradually production increased, but the targets set by the state planners were not achieved.

The Purges

Reasons for the Purges

Many of the 'Old Bolsheviks' who had joined the party before 1917, had become disillusioned with Stalin. They argued that he had moved away from pure Marxism and had substituted instead a personality cult. In the early 1930's, they began to feel that they had exchanged one form of dictatorship for another. These 'Old Bolsheviks' had attracted some younger communists to their cause. Stalin was aware through his secret police that plots against his rule were developing. He decided to act against his enemies.

As already mentioned, Trotsky continued to agitate against Stalin after he had been deported. He wrote pamphlets and contributed to journals. His supporters at home, known as **Trotskyites**, became prime targets for elimination.

Stalin suffered from insecurity. He was dull and lacked flair. He was jealous of intellectuals and of popular figures in the party, because they might prove a threat to him. To combat this, he initiated the **Cult of Stalin** when he took over in 1927. Towns and cities were called after him and statues erected in his honour. *Pravda* called him 'the wisest man of the age'. His jealousy and insecurity developed into paranoia. He saw enemies everywhere. Political rivals and army officers, anybody who might threaten his position, all suffered in the purges (Stalin's 'cleansing' or elimination of his enemies).

Stalin's wife apparently committed suicide in 1932. Opposition to his industrial and agricultural policies added to his isolation. He planned a terrible revenge on his enemies.

Kirov

Sergei Kirov was secretary of the party in Leningrad and a popular member of the Politburo. He was murdered in December 1934 in suspicious circumstances. Stalin may have been involved in the murder. He used the event as a justification for launching a campaign of terror. The secret police were ordered to round up anti-Stalinist Communists and put them on trial.

Thousands of suspects were arrested. Some were executed and many sent to labour camps. Among those who were imprisoned were Kamenev and Zinoviev. There was then a pause in the campaign of terror, but worse was soon to follow.

The Moscow Show Trials

> The Trotsky Fascist criminals who have made an attempt against the most precious thing of all, against the lives of our workers, have deserved their merciless punishment...This is the sentence of our court: *death to the enemies of the people.*
>
> (Moscow radio broadcast, 1936)

The three Moscow Show Trials of 1936, 1937 and 1938 involved major public or army figures. In August 1936, **Zinoviev, Kamenev** and fourteen others were found guilty of being 'agents of the German Polish Fascists' and executed. The most surprising aspect of these trials was that the accused stood meekly in the dock and confessed their guilt. They had either been beaten or brainwashed.

In later trials, anybody suspected of being a Trotskyite was executed. The top officers of the **Red Army** then came under attack. Seventy-five of the eighty members of the Supreme Military Council were executed, as well as all navy admirals. The Red Army entered World War II with few experienced military officers.

All members of Lenin's ***Politburo,*** except Stalin and Trotsky, were tried. The trials had been organised by a special secret police group, the **NKVD** under **Yezhov.** However, in 1939 Yezhov himself was executed. Stalin feared that he was becoming too powerful and needed to blame someone for the excesses. He accused the police chief of killing many innocent people.

At the Party Congress in 1939, Stalin declared that the purges had ended. It is estimated that over 1 million people had been put to death and a further 8 million

Kamenev and Zinoviev, Stalin's rivals for power

sent to labour camps. Fifty years later the purges were condemned outright by the Soviet leader, Mikhail Gorbachev. In a speech in 1987 Gorbachev spoke about the matter:

> We now know that the political accusations and repressive measures against a number of party leaders and statesmen, against many communists and non-party people, against economists, executives and military men, against scientists and cultural personalities were a result of deliberate falsification... The guilt of Stalin is enormous and unforgivable.
>
> (Mikhail Gorbachev)

Foreign Policy

Socialism in One Country

Stalin was suspicious of western powers. They were all capitalists and a threat to communism. In order to build socialism in one country, he needed a period of peace. He extended the terms of the **Treaty of Rapallo (1922)** with Weimar Germany. The Germans were given bases in the Soviet Union in order to build up their military power. This allowed them to get around the restrictions of the Versailles Treaty. In return, the Russians received German help in building armaments factories and in the area of defence.

In 1928 the Soviet Foreign Minister, **Litvinov**, called for total international disarmament. The U.S.S.R. signed the **Kellogg-Briand Pact**, which outlawed war. In 1933 the United States granted recognition to the Soviet State. Stalin had enough problems at home with the Five-Year Plans and collectivisation. He was happy to create the image abroad of Russia as a peace-loving country.

The Nazi Threat

During his first year in power, Hitler ended the co-operation between the German and Soviet armies. He made no secret of his anti-communism, his intention to expand to the east and of his contempt for all Slav people. Fear of the Nazis seems to have been the spur for the Soviet decision to join the **League of Nations** in 1934 and to sign a treaty with France in 1935.

Mutual suspicion remained between Stalin and the leaders of France and Britain, between communism and capitalism. Britain appeared reluctant to sign a defence pact with the U.S.S.R. in 1936. The policy of appeasement towards Germany in regard to Austria and Czechoslovakia led Stalin to believe that he would not receive help from the west if the Nazis attacked the Soviet Union.

Meanwhile Stalin sent help to the Republicans in the Spanish Civil War (see chapter 21). He seems to have sent enough to keep the Germans and Italians occupied while he built up his resources at home.

Stalin faced a stark choice in 1938 and 1939. He knew that the Anti-Comintern Pact of 1936 between Germany, Italy and Japan aimed to destroy communism and could involve him in a two-front war. He did not trust the western powers and was determined to avoid the mistakes of World War I. This time the Soviet Union would be last in and would not end up fighting for untrustworthy western allies.

During the 1930s, the Russians had favoured China with support in her ongoing disputes with Japan. In 1931 Japan invaded Mongolia. The Soviets regarded this as one of their spheres of influence. Stalin sent a large Russian army under **Marshal Zhukov** to fight the Japanese. Modern tanks and aeroplanes lent support. The Japanese were defeated by the middle of 1939. Russian forces had also gained some valuable experience.

The Nazi-Soviet Pact

The German seizure of non-German Czechoslovakia in March 1939 was seen by the Soviets as the real beginning of ***Lebensraum***, Hitler's expansion into Eastern Europe. Nazi threats against Poland added to Stalin's fears. The Soviet purges had shocked western opinion and negotiations with France and Britain were not easy. Stalin replaced Litvinov, his Foreign Minister, with **Molotov**. Secret negotiations between Molotov and the German Foreign Minister, **Ribbentrop**, began.

British cartoon, September 20, 1939 commenting on the Nazi-Soviet Pact

On 23 August 1939 the world was shocked by the signing of the **Molotov-Ribbentrop Pact**. This was a mutual non-aggression pact effective for ten years. Stalin gained both time and space by the signing of this agreement. A secret clause divided Poland between Germany and the Soviet Union. The Germans also recognised Stalin's interests in Lithuania, Latvia, Estonia and Finland.

Meanwhile Stalin ordered a complete retraining programme for the Red Army. He also built up his country's resources. Stalin hoped for more time, but when the Nazi invasion came, in June 1941, he was far more ready than he had been in 1939.

Reaction to the Pact

The Nazi-Soviet Pact was seen by the working class in the West as a betrayal, and by politicians as a cynical exercise. However, it probably saved Stalin and the Soviet Union. Stalin took over the territories allocated to him in the pact. In return, he supplied Hitler with enormous quantities of grain, oil and war materials. This was of huge benefit to the Germans during their successful campaigns up to mid-1941.

> The Soviet Union saved its skin at the cost of leaving the proletariat of the world without solutions, hopes, or help.
>
> (*Journals: 1934-1955*, Bertolt Brecht, Methuen, 1993)

Questions

ORDINARY LEVEL – D

Answer the following questions briefly. One or two sentences will be enough.

1 Explain the term 'fascism'?

2 State **one** important provision of the 1929 Concordat between Mussolini and the Papacy.

3 Why did the German mark become worthless in 1923?

4 Gustav Stresemann was an important figure in German history. Explain why.

5 Set down **one** important reason for the failure of the Weimar Republic.

6 How did the Reichstag fire of 1933 affect the career of Adolf Hitler?

7 Mention two ways in which Hitler tried to develop the German economy.

8 What was the 'Night of the Long Knives'?

9 What important decision was made at the Munich Conference, 1938?

10 What was the significance of the Nazi-Soviet Pact, 1939?

11 Describe one effect of the Stavisky scandal on France, 1934?

12 What was the Maginot Line?

13 The League of Nations gave Britain the mandate to administer Palestine. Describe the main problem which faced Britain in carrying out this task.

14 Give one reason why the General Strike of 1926 in Britain collapsed so quickly.

15 Set down one important cause of the Spanish Civil War.

16 What part did Nazi Germany play in the Spanish Civil War?

17 Give one reason for Franco's victory in the Spanish Civil War.

18 Why was Trotsky expelled from the U.S.S.R.?

19 What did Stalin mean by 'Socialism in one country'?

20 Explain briefly Stalin's policy of collectivisation.

Questions

ORDINARY LEVEL – E

Write a short paragraph on each of the following:

1. The characteristics of fascism
2. Gustav Stresemann
3. Hitler and the Jews
4. Hitler's use of propaganda
5. *Anschluss:* the union of Germany and Austria
6. Economic problems in Great Britain, 1919-1939
7. The Abdication Crisis in Great Britain, 1936
8. International involvement in the Spanish Civil War
9. Stalin's Five-Year Plans
10. The purges in the U.S.S.R. under Stalin.

ORDINARY LEVEL – F

Write a short essay on each of the following:

1. Benito Mussolini under each of the following headings:
   (i) His early career
   (ii) How he came to power
   (iii) How he ruled Italy
   (iv) A brief account of his foreign policy.
2. Hitler's foreign policies from his seizure of power, 1933, up to the outbreak of World War II, 1939
3. "The period 1920-1940 witnessed the decline and fall of the Third French Republic." Discuss.
4. The Spanish Civil War, 1936-1939, under each of the following headings:
   (i) The background to the war
   (ii) The course of the war
   (iii) World involvement in the war
   (iv) The results of the war.

Questions

HIGHER LEVEL

Write an essay on each of the following:

1 Outline the events which brought about the rise of Mussolini to power in Italy and discuss the main aspects of his domestic policy. (80)

2 Treat of the economic and political achievements of Gustav Stresemann in Germany during the period known as the 'Stresemann Era', 1923-1929. (80)

3 'The period 1919-1933 was, for Weimar Germany, a period of political instability and cultural achievement.' Discuss. (80)

4 'Between 1933 and 1939 Hitler and the Nazis turned Germany into a one-party totalitarian state.' How was this achieved? (80)

5 Treat of Hitler's handling of foreign affairs from 1933 up to the outbreak of World War II. (80)

6 'The period 1920-1940 witnessed the decline and fall of the Third French Republic'. Discuss. (80)

7 Discuss the economic and social policies of the Labour Party in Great Britain during the period 1924 to c. 1950. (80)

8 Treat of the origins of the Spanish Civil War and discuss international reaction to events in Spain, 1936-1939. (80)

9 Account for the victory of Franco in the Spanish Civil War and assess the significance of that victory for Spain. (80)

10 'From the death of Lenin, 1924, up to Hitler's invasion of Russia, 1941, Stalin consolidated his personal power and transformed the U.S.S.R.'. Discuss. (80)

SECTION FOUR

1939-c. 1970: THE SHAPING OF MODERN EUROPE

Chapter 23

World War II, 1939-1945

For now our linked-up globe has shrunk so small,
One Hitler in it means mad days for all.

Martyn Skinner, *Letters to Malaya*

Germany's run of victories

Invasion of Poland

At 4.45 a.m. on Friday, 1 September, 1939, Hitler's forces invaded Poland. The world now witnessed a new type of warfare called ***Blitzkrieg***, or lightning war. German *Stuka* dive bombers attacked Polish troops, while heavier bombers reduced industrial plants and fortifications to rubble. The ***Panzer*** (armoured and motorised) divisions followed up quickly, equipped with tanks and armoured cars. Behind came the infantry which overcame pockets of resistance.

The Polish army fought bravely, but their cavalry was no match for the German tanks. Within two days the German air force, the ***Luftwaffe***, had shattered the railway network and had destroyed the Polish air force on the ground. The French and British offered no help beyond declaring war on Germany. They were unprepared for war and the practical difficulties of intervening in a far-off country were quickly apparent. The city of **Warsaw** held out until 27 September when it was forced to surrender. The western part of Poland came under the 'protection' of the Third Reich.

Meanwhile Russia had invaded Poland from the east. Stalin was surprised at the speed of the German attack and moved to claim the territory promised to him in the 1939 pact with Germany. At the beginning of October, Poland was divided between Germany and the U.S.S.R.

The Winter War

Stalin now forced the three Baltic states of **Latvia**, **Estonia** and **Lithuania** to allow Soviet bases on their territories. In July 1940 these states were occupied by Stalin's forces and became part of the U.S.S.R.

Finland had once been under Tsarist rule. It included territory around Lake Ladoga near the city of Leningrad. Stalin was afraid that Hitler might double-cross him and invade the Soviet Union through Finland. For security reasons he

THE WAR IN EUROPE 1939-40

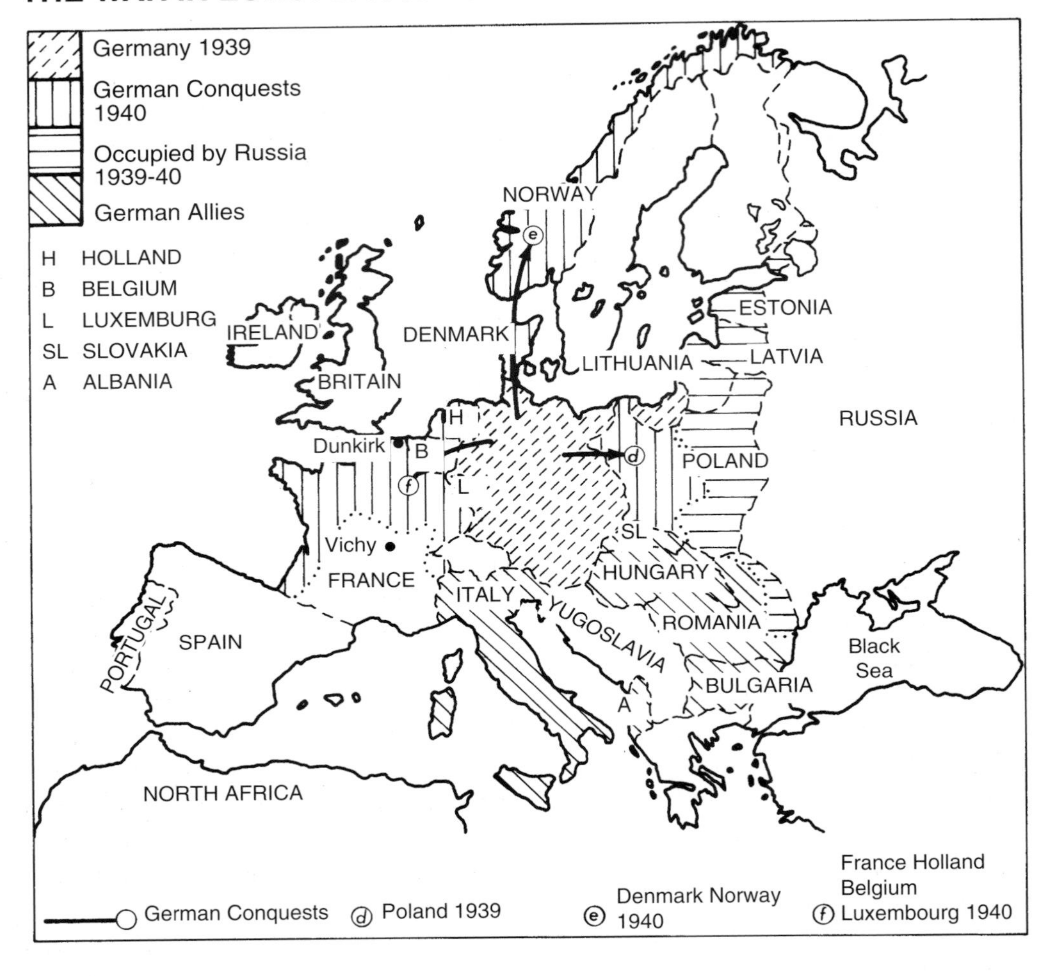

demanded the territory adjacent to Leningrad. When the Finns refused, he declared war on Finland on 30 November, 1939.

Stalin expected that the Red Army with its mass tank assaults would overrun the Finns in a few days. However, the forests, marshes and lakes of Finland were not suited to the Russian attack. Field Marshal Mannerheim's Finnish troops were familiar with the terrain and were better equipped for winter warfare. For over three months they held out against overwhelming odds. Finally in March 1940 the **Mannerheim Line** was breached and Finland sued for peace.

The Soviet Union received the area around Leningrad and territory to the north. The League of Nations expelled the U.S.S.R. The difficulties encountered by the Red Army aroused contempt in Germany. Hitler believed that his forces would have little difficulty when the time was right to invade Russia.

The Phoney War

'We'll hang out the washing on the Siegfried Line.'

The Popular English song of the time captured the mood of the winter-Spring of 1939-1940. The French felt safe behind their **Maginot Line**, a network of huge concrete fortifications along the border, with their guns pointed on Germany. Britain was concentrating on building up her forces. A British Expeditionary Force sailed to France but no fighting took place.

The German soldiers also waited for seven months behind their own barrier, the **Siegfried Line**. They called this period the *sitzkrieg* or sit-down war. Hitler wished to discuss the European question with Britain and France but was turned down. He then issued instructions for **'Operation Yellow'**, code-name for an attack on France through Holland and Belgium. The attack was postponed a number of times, due to poor weather. The newspapers began to refer to the 'Phoney War'. However, Hitler was now planning an attack on Scandinavia.

Scandinavia

Germany imported two-thirds of her iron ore from Scandinavia. The Allies were aware that if they cut the 'iron route', Germany's supplies of war materials would be severely curtailed. The British and French were considering capturing the Norwegian port of **Narvik** from which much of the ore was exported. They hesitated because such an action would violate Norwegian neutrality and provoke a German attack.

Hitler also noted Norway's weak military position and her lack of coastal defences. He held discussions on a Nazi invasion with **Vidkun Quisling**, head of the Norwegian Fascist Party. On April 9, 1940, German warships slipped into **Copenhagen** in Denmark. Landings also took place in **Oslo**, **Bergen** and other coastal towns in Norway. The Nazis received help from Norwegian sympathisers and Quisling became head of a puppet government in Norway.

Denmark surrendered almost immediately. After the first shock of invasion, the Norwegians began to fight back. France and Britain sent troops which gained footholds at Narvik and **Trondheim**. These were not provided with adequate air protection and were soon forced to withdraw. Germany now had control of the strategic fiords on Norway's coast and had secured her iron ore supplies.

Churchill

'Depart, I say, and let us have done with you. In the name of God, go.'

Oliver Cromwell's famous words were quoted in May 1940 by Leo Amery, a leading Conservative M.P. to **Neville Chamberlain** in the House of Commons. Chamberlain had suffered a major loss of prestige when his policy of appeasement failed. He was also subjected to constant criticism for failing to pursue the war with real vigour. The man of peace was replaced as Prime Minister by Winston Churchill. The new Prime Minister was a controversial figure. He had entered Parliament in

1900 as a Conservative. In 1906 he became a minister in a Liberal government. He was appointed First Lord of the Admiralty to take charge of the navy in 1914. He lost his position when his ill-fated plan to attack the Central Powers through Turkey failed (see chapter 11).

During the 1930s Churchill argued strongly against appeasement and for rearmament. He received little support in Parliament. However, his views were proved correct in September 1939, and he was again appointed First Lord of the Admiralty. In the 1940's, he proved a fine war leader, though he sometimes took an ill-informed part in military decisions. He was a great orator and immediately raised the morale of the British people. His words on taking office, on 10 May 1940, have become a symbol of sacrifice and defiance:

> I have nothing to offer but blood and toil and tears and sweat.
>
> (Churchill)

The Low Countries

On 10 May, 1940, the day Churchill became Prime Minister, German troops launched an attack on Belgium, Luxemburg and the Netherlands. The three invading armies were led by **Generals von Bock, von Leeb,** and **von Rundstedt**. *Blitzkrieg* methods were again used. The Allies expected that the Germans would concentrate their attack along the Belgian coast as in World War I. Instead, the main armoured divisions came through the poorly-guarded Ardennes, the hilly area of southern Belgium. France was now in danger.

The Dutch were defeated in less than a week. **Queen Wilhelmina** escaped to England and set up a government in exile. The bombing of Rotterdam had caused deep resentment among the Dutch. It meant that Germany had to use extra resources during the war to rule an unwilling people. On May 28, **King Leopold III** surrendered the Belgian army and became a prisoner.

Dunkirk

The German army which broke through the Ardennes came round behind the Allied forces. The **British Expeditionary Force** was caught in a trap. **General Gort** was forced to retreat. The Germans captured the Channel ports of **Boulogne** and **Calais**. The Allies were surrounded by German armoured divisions at **Dunkirk**, the only escape port on the northern French coast.

The German army halted about 20 miles from Dunkirk. It seems that Göring insisted that the mopping-up operation be left to the *Luftwaffe*. Churchill decided that the Allied troops should be evacuated. The RAF provided some air cover against attack. Rain and fog also gave the Allies a break from the *Luftwaffe* bombing. A fleet of Royal Navy ships assisted by privately-owned vessels succeeded in evacuating a quarter of a million British soldiers and over 100,000 Belgian and French troops from Dunkirk to Britain.

Dunkirk was a demoralising defeat for the Allies. Some 200 vessels were sunk. Over 40,000 prisoners were taken by the Germans, who also captured valuable military equipment. However, the ***Luftwaffe*** and ***Wehrmacht*** had failed to press home their advantage and the British army lived to fight another day. Indeed, Churchill made a morale-boosting speech which made Dunkirk seem almost like a victory:

> We shall fight on the beaches; we shall fight on the landing grounds; we shall fight in the fields and in the streets; we shall fight in the hills. We shall never surrender.

The Fall of France

There was now very little the French army could do to stop the German advance. **General Weygand** tried to form a new line of defence along the Aisne and the Somme but it soon crumbled. On 12 June the French declared Paris an open city rather than see it destroyed. Two days later the victorious German soldiers paraded down the Champs Elysées.

Premier **Paul Reynaud** had called for action by the RAF. He felt that Britain had let France down and resigned. His place was taken by **Marshal Pétain,** the hero of World War I. Pétain felt that the French position was hopeless and asked for an immediate armistice.

On 22 June, Hitler was present in **Compiègne** in the same railway carriage where the armistice with a defeated Germany had been signed in 1918. Now it was the French who agreed to an unconditional surrender. The country was to be occupied by the Germans, except about one third of its territory in the south. Pétain became head of a puppet government for this area, with headquarters in the town of **Vichy**. The Vichy government also ruled the French empire and retained control over the fleet. However, the British destroyed this fleet as it lay at anchor in the **Gulf of Oran,** in Algeria, lest it fall into German hands. A thousand French sailors were killed in the attack and Pétain broke off relations with Britain. Meanwhile **General Charles de Gaulle** had escaped to Britain. There he set up the **Free French Government** in London to continue the fight against the Nazis.

On 10 June, Mussolini, sensing a German victory, declared war on Britain and France. Italian troops invaded part of the southern coast of France. President Roosevelt of America declared that 'the hand that held the dagger stuck it in the back of its neighbour'. After the fall of France, Mussolini claimed **Nice** and part of **Savoy** as a reward for his efforts.

The Battle of Britain

A Defiant Leader

> I expect that the Battle of Britain is about to begin. Upon this battle depends the survival of Christian civilization. Upon it depends our own British life and the long continuity of our institutions and our Empire. The whole fury and might of the enemy must very soon be turned on us... Let us therefore brace ourselves to our duties, and so bear ourselves that, if

the British Empire and its Commonwealth last for a thousand years, men will still say, 'this was their finest hour'.

(Winston Churchill in The House of Commons, 18 June, 1940)

Operation Sea Lion

Hitler decided that a landing in Britain was possible 'provided that air superiority can be attained'. In mid-July he ordered preparations for **Operation Sea Lion**, the invasion of Britain in mid-September. In the meantime the *Luftwaffe* was set the task of defeating the British airforce, the RAF. The Germans had 1,900 bombers and 1,100 fighter planes compared to Britain's total of about 1,000 planes. Göring predicted that Germany would have mastery of the skies in four weeks.

The **Battle of Britain** opened on 10 July when German bombers attacked convoys in the Channel. It was not until the second week of August 1940 that the main attacks came. Almost every day for the next two months, hundreds of German bombers, escorted by fighter planes, flew to Britain from France. Göring at first tried to destroy the RAF and the fields from which it operated. Losses mounted on both sides, but Göring was beginning to gain the upper hand through sheer weight of numbers.

Spitfires in formation during the Battle of Britain

The Blitz

Production of British **Spitfires** and **Hurricanes** increased dramatically. Churchill decided to bomb Berlin on the night of 25 August and on several nights thereafter. Hitler and Göring resolved to retaliate. On 7 September, 300 bombers of the *Luftwaffe* attacked London. A thousand people were killed and damage was severe. The bombing of London and other cities continued for the next two months. About 60,000 people were killed during this period, known as the 'Blitz'.

Britain survives

Britain was saved by Hitler's decision to switch from bombing airfields to cities. The

Churchill visits bomb-damaged area of London during the Blitz, 1940.

German **Messerschmitts** were hindered by the necessity to return immediately to base after a mission, or risk running out of fuel. The RAF fighter-planes, the famous Spitfires, did not have such fuel problems and were easier to manoeuvre. RAF pilots who were shot down over Britain were able to bale out and most were soon ready to fly again. The British were also greatly helped by a new secret device, **radar**. This warned of approaching attacks and guided the RAF towards the *Luftwaffe*.

Operation Sea Lion was postponed until the spring of 1941. Air-raids continued for months against industrial towns and ports. By May 1941 the attacks had virtually stopped. The *Luftwaffe* was by then involved in the invasion of the Soviet Union. The Battle of Britain was the first turning point in the war against Hitler. Churchill paid the pilots of the Royal Air Force a famous tribute:

> 'Never in the field of human conflict was so much owed by so many to so few.'
>
> (Churchill)

The War in North Africa

Italians driven back

Mussolini, Hitler's ally, had a large army in Libya and Abyssinia. He now saw an opportunity of expanding his empire. The Italian leader felt that his country was strong enough to capture territory from the British and French in North Africa. He could then proceed to the Suez Canal itself.

In September 1940 the Italian commander, **Marshal Graziani**, advanced into Egypt. A British counter-attack under **General Wavell** drove the Italians back along the coast of North Africa and 130,000 Italian prisoners were taken. The British followed up into Libya and captured **Tobruk**. Abyssinia was reconquered by the Allies by April 1941 and Emperor Haile Salaisse returned to his throne.

NORTH AFRICA & ITALY 1942-43

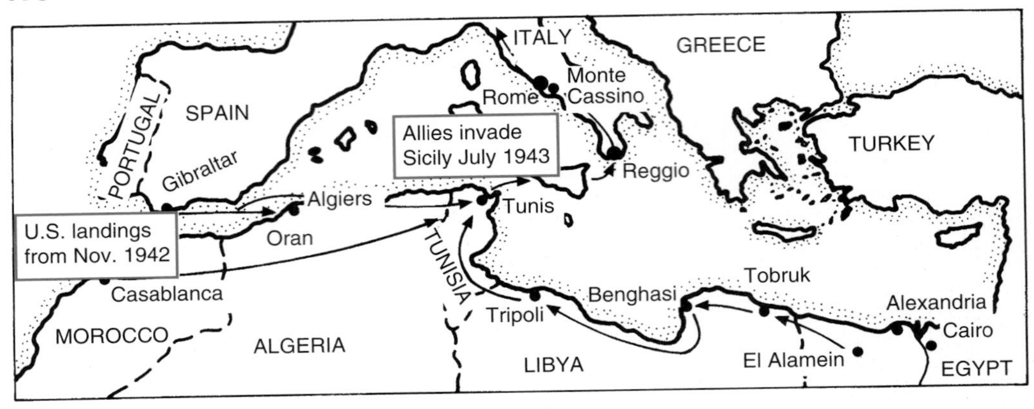

The Desert Fox

Wavell could not complete his task in North Africa because Churchill diverted some of his troops to fight the Italians and Germans in Greece. At the same time Hitler sent the elite and well-equipped *Afrika Korps* under **General Erwin Rommel** to help the Italians in Libya. Rommel was a very able general and became known as 'The Desert Fox'. He caught Wavell by surprise and quickly recaptured all the territory lost by the Italians.

Rommel now had to give up some of his troops for the invasion of Russia. Wavell was replaced by **General Auchinleck**. Rommel was unable to take Tobruk. For eighteen months the battle swung to and fro along the North African coast. In June 1942 Rommel finally succeeded in taking Tobruk. The British were driven back into Egypt and the Suez Canal was again threatened.

El Alamein

Churchill appointed **General Alexander** to take command in Africa. **General Montgomery** was given charge of the Eighth Army, later to become known as 'The Desert Rats'. Shipments of American tanks and bombers arrived to bolster the British. Hitler failed to send enough equipment to Rommel. German supplies were also curtailed by the bombing of Axis (German and Italian) convoys from the British base at Malta. By September, 1942, Rommel's *Afrika Korps* was outnumbered in tanks, men and guns by two to one.

Montgomery waited until Allied superiority was assured. On 23 October he opened the attack on German forces at **El Alamein**. He destroyed much of the German armour in the first few days. After twelve days Rommel's lines broke and he began his retreat. His army was pursued by a far superior British force. Hitler did not send much-needed supplies, as he was now in difficulty in Russia.

General Dwight D Eisenhower

Operation Torch

Meanwhile on 8 November 1942, a strong Anglo-American force under **General Dwight Eisenhower** landed in Morocco and Algeria. These territories were under the control of the Vichy government. The Vichy forces quickly came over to the Allied side. The Germans now placed Vichy France under direct German control.

The *Afrika Korps* was caught in a pincer movement. Tripoli fell and Tunis was surrounded. Rommel's genius as a military commander lies in the fact that he kept his force intact against superior odds. He tied up hundreds of thousands of Allied soldiers for almost six months, thereby delaying the invasion of Italy. He was unwell during this time and was eventually sent back to Germany.

The end of the war in Africa came on 12 May, 1943. A quarter of a million German and Italian prisoners were taken. It was the first victorious Allied campaign and a turning-point in the war.

Operation Barbarossa

Hitler's Aims

> 'The German armed forces must be prepared, even before the conclusion of the war with England, to crush Soviet Russia in a rapid campaign.'
>
> (Hitler, 18 December 1940)

These were the words used by Hitler on 18 December 1940 when he gave orders for an attack on the U.S.S.R. The directive was code-named **'Operation Barbarossa'** (Red Beard) after a medieval German emperor who fought against invaders from the east. Hitler felt that he had secured his position in Western Europe and that he could expect to attack in the east without worrying about a two-front war.

Early in 1941 Hitler said to his generals that 'the communists have never been and never will be our friends'. He called for a 'war of extermination' against

Communism. He would control the inferior Slavs in the east and create *Lebensraum* or living-space for his people (see chapter 16). Germany would also gain supplies of grain from the Ukraine, as well as coal and oil from the Caucasus.

Three-pronged Attack

> We have only to kick in the door and the whole rotten structure will come crashing down.
>
> (Hitler)

Hitler's words indicate his belief that a repeat of the *blitzkrieg* tactics would give the Germans a quick victory. A total of 3 million men equipped with nearly 10,000 tanks and 3,000 aircraft began the attack on 22 June 1941. Army Group North, commanded by **Field Marshal von Leeb** aimed to capture Leningrad. Army Group Centre under **Field Marshal von Bock** hoped to take Moscow, the seat of power and hub of the railway system. Army Group South led by **Field Marshal von Rundstedt** planned to take the Ukraine and then press on to the Caucasian oilfields.

Leningrad

Stalin had not really believed the warnings he had received about the Nazi attack. The Soviet Union was not ready for the lightning strike. In the first fortnight the Russians lost a million men, much of their air-force and thousands of tanks.

The speed of the German advance was such that by September 8, Army Group North had laid siege to **Leningrad**. The city was almost completely surrounded. The only supply route for the inhabitants was across **Lake Ladoga**. The barges which crossed the lake in summer and the lorries which drove over the ice in winter were bombed by the Germans. The result was starvation in the city.

The siege of Leningrad lasted for 900 days. Over half a million citizens died from German shelling or from starvation, but the city refused to surrender. It was not until January 1943 that relief came from Russian forces in the south and the Germans were driven back.

Moscow

Von Bock's Army Group Centre also made rapid progress. At the beginning of December 1941 they were within 70 miles of **Moscow**. By now the German troops were exhausted and were seriously hindered by the sub-zero temperatures and the heavy snow. Communist brigades were formed from among the Russian workers to help the Red Army.

Zhukov's troops prepare a counter-attack against the Nazis, outside Moscow, 1941.

On December 6 Stalin ordered a counter-attack. The Russians were led by **Marshal Zhukov**. They succeeded in pushing the Germans back a little and stalemate followed. In the spring of 1942 the main weight of the German attack was turned south and Moscow was saved.

Stalingrad

Army Group South under von Rundstedt advanced quickly through the Ukraine and besieged **Kiev.** The city fell on 18 September and over half a million Russian troops were captured. In the spring of 1942 Hitler decided to concentrate his main attack on the south. His main target was the **Caucasus** oilfields. Sebastopol and Rostov had been taken by summer 1942. Now Hitler split his forces and sent the Sixth Army under **General von Paulus** to attack Stalingrad. This left a gap in his lines and the oilfields were never taken.

The battle for Stalingrad was a turning-point in the war. Stalin ordered Zhukov to defend the city at all costs. The Germans had gained control of over three-quarters of Stalingrad by November 1942. However they could never capture the Russian artillery across the river Volga. The fighting in the city was horrific. Buildings were bombarded into rubble and troops were reduced to eating rats.

Now the Russian winter intervened again. By the end of November von Paulus found his forces trapped between Russian armies on the Volga and the Don. He wished to withdraw, but Hitler insisted that there should be no retreat. Over 70,000 Germans died in the siege before von Paulus surrendered on 2 February 1943. Most of the 90,000 German prisoners never returned from Stalin's prison camps.

The German army now began the long retreat. In July the Russians and Germans fought a great tank battle at **Kursk** in central Russia. The Russians were victorious and the war was over for Hitler in the east.

Reasons for German Defeat

- Hitler delayed Operation Barbarossa for almost two months in 1941 in order to assist Mussolini's troops who were in difficulties against the Greeks. He also captured Yugoslavia during this time: this was known as the 'Balkan interlude'. The delay meant that the Germans could not achieve victory before winter.
- Hitler's forces were not prepared for the Russian winter. His soldiers did not have proper clothing for the sub-zero temperatures and were not prepared for fighting on the frozen ground covered by snow. German tanks froze and the blitzkrieg tactics came unstuck.
- After he had recovered from the shock of invasion, Stalin proved an inspirational leader. He refused to leave the Kremlin when the Germans advanced on Moscow. He reminded his people of how their ancestors had resisted Napoleon's invasion in 1812. His speeches roused the Russians to suffer terrible hardship in order to resist the enemy in what he called 'The Great Patriotic War'.

- The Russians had dismantled many factories and moved them behind the Urals and into Siberia. These factories produced the vital materials needed for defence and counter-attack.
- The brutality of the Nazi invasion force turned any Russians who might have welcomed the Germans as liberators from Stalin's regime into deadly enemies. This was particularly true in the Ukraine where initial enthusiasm for Hitler's army soon disappeared.
- The Soviets adopted a 'scorched earth' policy of burning and destroying everything as they retreated.

Life in Occupied Europe

Collaboration

Meanwhile the Germans consolidated their position in the countries they had captured. Nazi rule in occupied Europe was dictated by racism. Denmark was allowed retain her own monarch and parliament. In France, **Prime Minister Pétain** and his deputy, **Pierre Laval**, ruled a puppet government from Vichy. They collaborated with the Germans saying that resistance would lead to the destruction of France and the deaths of French citizens. They were right-wingers who thought that their people had become soft and lacked moral fibre. Pétain and Laval cooperated with the Nazis in their persecution of French Jews. After the war, Laval was executed and Pétain was imprisoned.

British children being evacuated to the countryside, to be safe from air-raids

Mussart in Holland and **Quisling** in Norway also led puppet governments. Quisling gave his name to any person who collaborated with an occupying power. Some locals cooperated for reasons of personal advantage. Others believed in the theories of Fascism. However, neither Mussart nor Quisling received widespread support. After the war some collaborators were punished by their countrymen.

The Nazis treated western Europeans well if they showed a spirit of co-operation. They looked on these people as equal, or almost equal, to their own race. The Nazis allowed them a limited form of self-government. This meant that the Germans could devote more time to the 'inferior' people of Eastern Europe. These 'lesser' people were treated harshly and were seen as mere slaves. Millions were transferred to Germany and worked in labour camps.

Resistance

Underground **resistance movements** sprang up in every occupied country. Networks of men and women engaged in spying and sabotage. They derailed trains, blew up bridges, produced underground newspapers, collected information and killed enemy officials. The British set up the **Special Operations Executive** to help resistance movements. Special training was given to volunteers and the B.B.C. broadcast coded messages to them.

The French Forces of the resistance, known as the *Maquis,* helped Allied operations in occupied France. The Polish Home Army rebelled unsuccessfully in Warsaw. In Yugoslavia the *Partisans* under **Josip Broz (Marshal Tito)** were largely responsible for liberating Yugoslavia. The *Chetniks* also operated in that country. Underground fighters in Russia carried on guerrilla warfare and sabotage behind enemy lines. Resistance movements in occupied countries succeeded in tying down large numbers of Nazi soldiers, police and officials. However, the number of people in occupied countries who actively opposed the Nazis was small. Most people kept a low profile. They were not willing to risk the terrible punishments reserved by the Nazis for Resistance workers.

Inside Germany, many people faced a crisis of conscience. Some decided to oppose the Nazis. Socialists and communists engaged in underground resistance. The **Rote Kapelle** (Red Chapel) was involved in espionage. Student opposition also emerged, the best known being the **'White Rose'** group. Franz Halder, Ludwig Beck and some other members of Hitler's staff came to believe that Hitler was leading Germany on the road to ruin and plotted to overthrow him. In July 1944, **Count von Stauffenberg** placed a bomb under Hitler's table at a conference. Hitler escaped with minor injuries in what became known as the **July Plot**. Stauffenberg, and a number of suspects were executed.

The Holocaust

Holocaust means a huge slaughter or destruction. The Nazis regarded all Slavs as inferior people and they were used as cheap labour. It is estimated that a total of 8 million slave labourers was taken from conquered countries to work in German factories.

The outbreak of war and expansion to the east meant that the persecution of the Jews grew worse. **Himmler** and other racist Nazi leaders drew up plans for the **'final solution'** of the 'Jewish problem'.

When the advancing Allied forces entered Germany, Austria and Poland they discovered the full horror and scale of the war crimes. At Auschwitz, Treblinka, Belsen, Buchenwald, Dachau and other concentration camps, they found gas chambers which had been used to put millions of German, Polish, French and Russian Jews to death. Six million Jews are thought to have been killed in this way, one third of them at Auschwitz. It is also estimated that 2 million Russian prisoners of war and a quarter of a million gypsies died in the death camps.

War in the Pacific

Japanese Expansion

Japan needed living space in Asia for her growing population and raw materials for her industries. She seized French Indo-China in 1940. The United States stood against Japanese expansion. President Roosevelt put an embargo on oil and some raw materials being sold to Japan. He also froze all Japanese credit with U.S. banks.

In 1941 **General Tojo**, leader of a group of extremist military officers, became Premier. His representatives continued negotiations with the U.S. Secretary of State, **Cordell Hull**. In November 1941 the Japanese cabinet accepted Tojo's decision to go to war against the United States.

Pearl Harbour

A date that will live in infamy.

(President Roosevelt)

On 7 December 1941 Japanese aircraft attacked and destroyed 340 American planes, eight battleships and 10 other vessels at the American naval base of **Pearl Harbour** in the Hawaiian Islands. The Americans also suffered over 3,000 casualties. The U.S. Pacific Fleet was put out of action but fortunately for the Americans no aircraft carriers, the principal weapon of the war in the Pacific was in the base at the time. The United States immediately declared war on Japan.

After World War I, the United States had reverted to her traditional policy of isolationism. However, America prided herself on her democratic principles and many of her citizens were becoming increasingly concerned at the advance of Nazism. Big business had much to lose if Hitler conquered Europe and industrialists continued to press Roosevelt to declare war on Germany. It seems that even before Pearl Harbour, America was coming close to getting involved in the conflict in

U.S. ships ablaze after Japanese attack, Pearl Harbour, 7 Dec., 1941

Europe. In 1939 President Roosevelt had amended the Neutrality Acts and provided munitions for the Allies on a **'cash and carry'** basis. The following year, fifty destroyers were given to the British Navy. After his re-election in 1940, Roosevelt got Congress approval for his **'Lend-Lease'** programme. He said that the United States must be the 'arsenal of democracy' and lent arms to Britain for the duration of the war.

Hitler believed that it was only a matter of time before America sided with the Allies. He had hoped to complete his conquest of the Soviet Union before this happened. When Roosevelt declared war on Japan, both Germany and Italy declared war on the United States. The world would now witness a truly global conflict.

Japanese Conquests

On Christmas Day 1941 Japanese forces captured **Hong Kong**. In January 1942 they invaded the **Dutch East Indies**. Like the Germans in Europe, Japan also had a stunning series of victories. Her soldiers advanced through the jungles of **Malaya** and captured **Singapore**. Thousands of British and Commonwealth troops were taken prisoner. They had assumed that the attack would come from the sea. General MacArthur's forces were driven from the **Philippines**. Before he left, the general promised that he would return.

By the summer of 1942 Japan controlled a huge Far-Eastern empire, which included Hong Kong, Malaya, Indo-China, Thailand, Burma, the Philippines, Borneo, Java, New Guinea and Dutch East Indies. The Japanese army had proved superior in jungle warfare. The code of the Japanese soldiers was to kill or be killed. This earned them a reputation for cruelty. They preferred death to surrender. Their ***Kamikaze*** pilots flew their aircraft on suicide missions and died by crashing into enemy targets.

WAR IN THE PACIFIC 1941-45

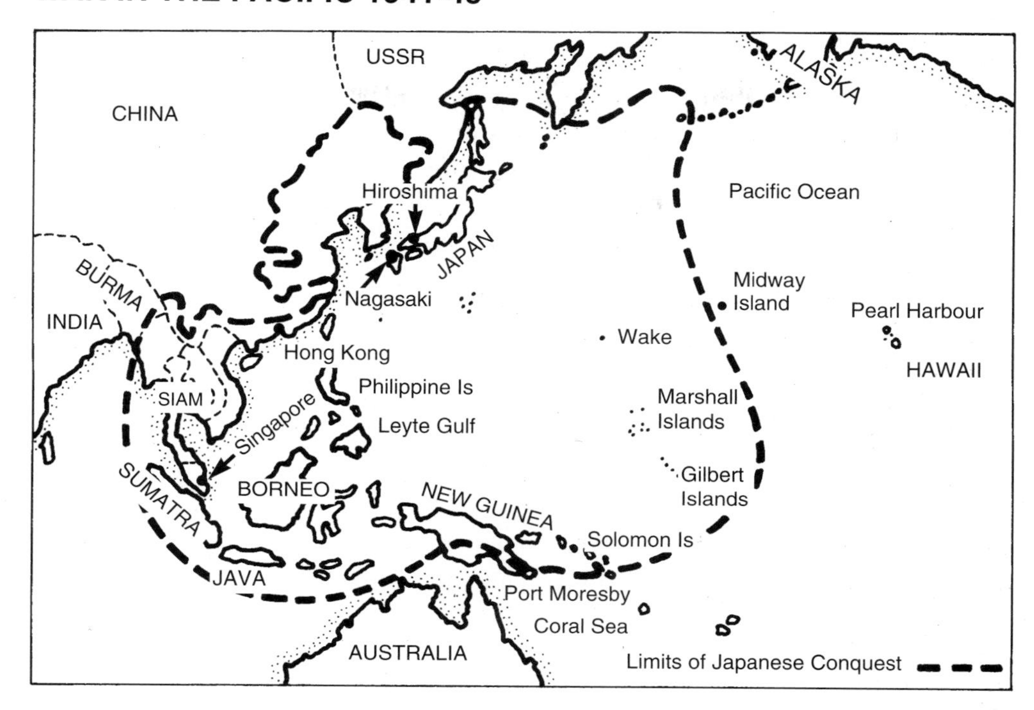

Coral Sea and Midway Island

Admiral Yamamoto of Japan hoped to engage the U.S. Pacific Fleet before the Americans could replace the ships lost at Pearl Harbour. During the early months of 1942, Tokyo and other Japanese cities had been hit by United States bombers operating from aircraft carriers.

Yamamoto's first objective was to capture Port Moresby in New Guinea. This would enable his navy to cut Allied shipping lanes with Australia. On 7 May 1942 carrier planes on both sides fought the **Battle of the Coral Sea**. The Japanese sank the American aircraft carrier *Lexington*. However, two Japanese carriers were put out of action and they were prevented from taking Port Moresby.

On 4 June 1942 the American navy avenged Pearl Harbour in a battle off **Midway Island** in the middle of the Pacific. Four Japanese carriers were sunk and 332 of her planes were destroyed. The Americans had cracked Japan's secret codes and knew about Yamamoto's plans in advance. Defeat at Midway shattered Japan's naval forces and was the turning point of the war in the Pacific.

Island-hopping

After Midway, American forces were on the offensive against Japan. In August 1942 they landed on the island of **Guadalcanal**. This was the first of a number of Pacific islands which were taken by the Americans in the face of bitter Japanese opposition.

The problem now faced by the United States was that of capturing mainland Japan. A number of islands lay in her path. In the next three years the Americans seized key islands and bypassed those which did not have any real strategic importance. After the battle of **Leyte Gulf** in 1944, MacArthur recaptured the Philippines. **Okinawa** and **Iwo Jima** fell in 1945. Meanwhile American, British and Chinese troops hacked their way through the jungles and drove the Japanese out of most of the areas they had occupied. Their way was now open for an attack on the mainland.

Victory for the Allies

Allied Co-operation

The military campaigns which led to victory for the Allies originated in political decisions made at the wartime conferences between the leaders of the United States, Britain and the Soviet Union (see last section of this chapter). Stalin felt that his country was bearing the brunt of the war and accused the other leaders of failing to provide sufficient help for his suffering people. He continued to press for a decision to open up another front in France in order to relieve pressure on the Soviet Union.

President Roosevelt had great sympathy for Stalin's point of view. Churchill did not trust the Soviet leader and was bitterly opposed to Communism. He was reluctant to agree to Stalin's requests. This led to tension at the conferences.

At **Casablanca**, in January 1943, the British and Americans agreed to co-ordinate their bombing raids and to invade Germany through Italy. At **Teheran**, in November of the same year, they finally agreed to Stalin's requests to open up a front in France against the Germans.

Operation Husky

The surrender of the German and Italian forces in North Africa was followed in July 1943 by the invasion of Italy, code-named **'Operation Husky'**. The overall commander of the operation was General Eisenhower of the United States. An Anglo-American force of 160,000 under **General Montgomery** and **General Patton** landed in Sicily. In less than six weeks, the Germans were driven out of Sicily.

Mussolini was deposed by the Fascist Grand Council (see chapter 15) and the new Prime Minister, Marshal Badoglio, surrendered to the Allies. Hitler decided that Italy must be defended at all costs. **General Kesserling** proved a formidable commander of the German troops. Mussolini was rescued and for a time restored by the Germans as leader of a puppet government.

Montgomery's Eighth Army landed at **Reggio** in the south of Italy in September 1943. An Anglo-American force landed further up the coast at **Salerno**. Kesserling established a defensive fortification, the **Gustav Line**, above Naples. It took the Allies five months to capture the monastery of **Monte Cassino** and break through the line. An attempt to get around the line by landing on the beaches at **Anzio** failed and 20,000 Allied soldiers were killed.

The Allies finally took **Rome** in June 1944. Kesserling now set up another defensive barrier, t**he Gothic Line**, in Northern Italy. Allied progress was slow. It was not until April 1945 that the Gothic Line was broken. By that time Germany also faced attack from the east and from the west.

Allied Bombing

> We shall bomb Germany by day as well as by night...making the German people taste and gulp each month a sharper dose of the miseries they have showered upon mankind.
>
> (Churchill)

The Blitz on Britain had virtually ended with the invasion of Russia in 1941. Meanwhile the Allies had developed bombers that could fly long distances. The RAF bombed Germany heavily by night, while the U.S. Air force operated mainly by day. Strategic bombing blasted Axis military and industrial targets. For five days in late February 1944, the Allied air forces bombed Germany's aircraft industries. **General Arnold**, commander of the U.S. Air Force, stated that 'those five days changed the history of the air war'. Allied bombs also destroyed dams in the Ruhr Valley, thus depriving many German industries of their power sources.

The Allies did not adhere strictly to strategic or precision bombing. German industrial and civilian areas were bombed heavily and indiscriminately. The person most closely associated with saturation bombing was **Air Marshal Harris** of the RAF. This saturation bombing led to the destruction of German cities and contributed to the weakening of German morale. Hamburg, Berlin and Dusseldorf were among the cities that suffered. Tens of thousands of German citizens, the majority of them civilians, lost their lives. In 1945 the city of Dresden, which had no wartime industries, was destroyed. It is estimated that about 50,000 of its inhabitants died.

The Germans also pioneered two 'revenge weapons' The V1 was a pilotless flying bomb. The V2 was a rocket that could travel at over 5,000 km per hour. These

Bombs (in tons)	1940	1941	1942	1943	1944	1945
Dropped on Germany	10 000	30 000	40 000	120 000	650 000	500 000
Dropped on Britain	36 844	21 858	3260	2298	9151	761

Air bombardment during the war, 1940-45

weapons brought great terror to London, but caused only limited destruction. In the long term, the advantage gained in the air by the Allies was a major factor in Germany's defeat.

D-Day, 6 June, 1944

As early as May 1943, Britain and the U.S. decided on an invasion of France. This would open up the second front which Stalin had demanded. **General Eisenhower** was placed in command of the proposed invasion, code-named **Operation Overlord**. The Germans had built the 'Atlantic Wall' fortifications stretching from Norway to Spain. Eisenhower insisted on detailed and meticulous planning. Nothing could be left to chance. A huge army was ready in the south of England by June 1944.

Hitler had appointed **Field Marshal von Rundstedt**, assisted by **Rommel**, to defend 'Fortress Europe' from the expected invasion. The Allied Expeditionary Force consisted of almost 3 million troops. It had 5,000 large ships, 4,000 smaller landing craft and more than 11,000 aircraft. Prefabricated harbours, code-named 'Mulberry', had been designed to unload troops and supplies. PLUTO (Pipe-Line Under The Ocean), carried fuel supplies for the invading force.

The D-Day landings, June 1944

On D (Deliverance) Day, 6 June, 1944, Allied troops landed on five **Normandy beaches.** These were code-named Omaha, Utah, Gold, Juno and Sword. The Americans lost over 1,000 men at Omaha. The Germans fought ferociously to prevent the Allies from gaining a foothold on the coast. However, on D-Day, 150,000 Allied troops succeeded in landing in Normandy. About 2,500 of their comrades had been killed. Over half a million Allied soldiers landed in France before the end of June. Their first task was to liberate France. They could then advance on Germany.

The Allies Advance

General Patton led an Allied army which landed in the south of France. He advanced up the Rhone valley and linked up with the Normandy troops which were now moving east. Hitler ordered the destruction of Paris. The German commander of the city, **General Cholitz**, refused to carry out the order. On 25 August 1944 Paris was liberated. **General Charles de Gaulle** had returned to France and insisted on leading the victory parade through the city.

At this stage a dispute occurred between Eisenhower and Montgomery. Eisenhower favoured a cautious advance on a broad front, in order to eliminate pockets of resistance and consolidate gains. Montgomery felt that victory could be achieved more quickly and at less cost. He advocated a direct single attack on a crossing of the Rhine. The British plan **'Operation Market Garden'** was tried in September. British and Polish airborne troops were landed near the Rhine bridge at Arnhem in Holland. They encountered a division of German *Panzers* and suffered heavy losses. After a week, the Allied force was evacuated.

The Battle of the Bulge

On 16 December 1944 **General von Rundstedt** made a last desperate attempt to halt the Allies. He made a surprise counter-attack on a narrow front through the **Ardennes.** The American forces under **General Bradley** were driven back and a 'bulge' appeared in the Allies' lines. Confusion was created by the landing of English-speaking German commandos, dressed in American uniforms, behind the Allies.

General Patton's forces helped to stop the German advance. Allied air superiority was helpful in driving back Runstedt's army. By the end of January 1945, the Germans had retreated. The Allied armies captured Cologne and crossed the Rhine during March. Eisenhower's armies were at the Elbe River, less than 100 kilometres west of Berlin on 11 April 1945.

French villagers welcome U.S. liberation forces, 1944.

Victory in Europe

Between July 1943 and April 1945 Russian armies advanced towards Berlin from the East. They recaptured the Ukraine and the Crimea. They then knocked the four Axis satellites, Romania, Bulgaria, Finland and Hungary out of the war. The Russians crossed into Poland in July 1944. An uprising against the Germans was led by Polish resistance fighters. It is estimated that about 200,000 Poles were slaughtered by the Germans in **Warsaw.** Stalin's troops stood back and allowed the killings take place before they entered the city in January 1945.

From all directions, Allied armies closed in on Berlin. Hitler married **Eva Braun** in an underground bunker in the city. It was there on 30 April 1945 that the *Füehrer* and his wife committed suicide. Two days earlier, Mussolini had been executed by Italian partisans. German forces in Italy surrendered on May 2. On the same day the Russians broke through to Hitler's bunker. On May 4 the German forces in the north of the country surrendered to Montgomery. On May 7 **General Jodl** of the German High Command signed the terms of unconditional surrender on behalf of his government.

On 8 May 1945, V-E (Victory in Europe) Day was celebrated. After five years, eight months and five days, the European phase of World War II had ended. It is estimated that about 40 million people had lost their lives.

The Atomic Bomb

Meanwhile, the war in the Pacific continued and eventually led to the use of the atomic bomb.

> After the missile had been released I sighed and stood back for the shock. When it came, the men aboard with me gasped 'My God', and what had been Hiroshima was a mountain of smoke like a giant mushroom. A thousand feet above the ground there was a great mass of dust, boiling, swirling and extending over most of the city.
>
> (Observer on the *Enola Gay*, quoted in the *Daily Mail*, 8 August, 1945)

Japanese defenders of Iwo Jima and Okinawa had fought to the last man. Over 100,000 Japanese died on Okinawa while the Americans lost 12,000. The Japanese government sent peace-feelers to Stalin. Whether a reasonable peace settlement could have been negotiated has been a matter of controversy ever since. It was against the code of the Japanese warrior to surrender. U.S. military commanders forecast heavy casualties. President Roosevelt had died in April 1945. The new leader of the United States, **President Truman**, demanded that the Japanese surrender unconditionally. When his ultimatum was ignored, he decided to use a new secret weapon, the atomic bomb.

On 6 August, 1945, a B29 Superfortress bomber, the *Enola Gay*, under the command of Colonel Paul Tibbets, released a single bomb attached to parachutes. It descended for five miles and then burst over its target, the city of **Hiroshima**. There

Nagasaki devastated by atomic bomb, 9 August 1945

were over 100,000 casualties, the majority killed outright. Over a generation later, many still suffer from the radiation. Three days later an atomic bomb was dropped on **Nagasaki** and killed over 40,000. The Japanese realised that they were helpless and surrendered on 10 August, 1945.

> The atom bomb was no 'great decision'... It was merely another powerful weapon in the arsenal of righteousness. The dropping of the bombs stopped the war, saved millions of lives.
>
> (President Truman)

The Wartime Conferences

The Atlantic Charter

'That all men in all lands may live out their lives in freedom from fear and want'.

This was the hope expressed by Churchill and Roosevelt in the Atlantic Charter, which showed the sort of world they hoped to see after the war. They met in secret in August 1941, on board a warship anchored off Newfoundland. Their war aims later formed the basis for the objectives of the United Nations.

- All peoples should be free to choose a government of their own liking.
- Countries should have free access to the raw materials they needed.
- There should be economic cooperation in trade and in using the resources of the world.
- Improved labour standards and welfare measures should apply in all free countries.
- All nations of the world should abandon the use of force.
- As a first step the aggressor nations should be disarmed.

The Casablanca Conference

Churchill and Roosevelt met at Casablanca in Morocco in January 1943. They were accompanied by their military advisers. Stalin had been invited to attend but was involved in the defence of his country. The Russian leader had urged the opening up of a 'second front' in mainland Europe in order to relieve the pressure on the Red Army. Churchill felt that the Allies were not ready for such a major undertaking. The conference decided:

- After the defeat of the Germans in Africa, a combined Allied force under the command of General Eisenhower, would attack Germany through Italy in the spring of 1943.
- Bomber raids on Germany would intensify and be coordinated.
- The Germans would have to surrender unconditionally to the Allies.

The Teheran Conference

This conference was held in Teheran, capital of Iran, at the end of November 1943. It was the first to be attended by the three Allied leaders. Churchill proposed an Anglo-American invasion of the Balkans. Stalin opposed this suggestion, as he wanted Eastern Europe to come under the 'protection' of the Soviet Union. Roosevelt sided with the Russian leader. The main decisions were:

- The Soviet Union should declare war on Japan as soon as her war with Germany ended.
- A 'second front' should be opened up in France the following year.
- A world organisation should be founded to take the place of the League of Nations.

The Yalta Conference

In February 1945 Churchill, Roosevelt and Stalin met again at Yalta in the Crimea. By this time the war in Europe was nearly over. Churchill was very suspicious of Stalin's intentions in eastern Europe. Roosevelt did not accept the British leader's reservations. A number of important decisions were made:

- A conference should meet in San Francisco to draw up a charter for the United Nations.
- Germany was to be disarmed, demilitarised and divided into four zones. Berlin would be divided into four sectors. The U.S., U.S.S.R., Britain, and France would each occupy a zone and a sector.
- Germany would pay reparations, half of which would go to the Soviet Union.
- War criminals would face charges when the conflict was over.
- Countries occupied by the Axis powers would choose their own governments through free elections.

Yalta Conference, 1945.
Seated: Churchill, Roosevelt, Stalin.

- The Soviet Union would declare war on Japan within three months of the end of the fighting in Europe. In return, she would receive territory lost to Japan in the war of 1904-1905.
- No firm decision was taken on Poland's borders. A compromise was reached on the formation of a coalition government to rule the country. Stalin had favoured the Communist Lublin Committee. Churchill pressed the claims of the Polish government-in-exile.

The Potsdam Conference

This was the last of the big wartime conferences. It was held at Potsdam near Berlin in July 1945. Roosevelt had died in April and was replaced by President Truman. Churchill was Prime Minister at the beginning of the discussions. The Conservatives were defeated by Labour in the British election and Churchill gave way to Clement Attlee who represented Britain during the second half of the conference. Truman believed that Roosevelt had granted too much to Stalin and he was determined to stand up to the Soviet dictator. The atmosphere was cold while the delegates decided on Germany's future:

- The decision to divide Germany into four zones was confirmed. Austria was also to be split into four zones.
- Germany would be united after she had been denazified, demilitarised and democratised.
- An international court was to be set up to punish war criminals for 'crimes against humanity'.
- A Council of Foreign Ministers was formed to draw up peace treaties with the defeated powers.

- The problem of minorities could be solved by population transfers.
- Reparations by Germany were to be paid in kind. Whole factories could be dismantled and machinery transferred to the occupying powers. However, the Germans were to be allowed a reasonable standard of living.
- No final agreement was reached on Poland. Her borders were pushed temporarily west as far as the Oder-Neisse line. Stalin also won a victory by having the Soviet-backed Lublin Committee recognised as Poland's lawful government.

Conclusion

The outcome of World War II and of the wartime conferences changed the political map of Europe. Even before the end of the war, the Allies were squabbling. Real suspicion existed between the communist east and capitalist west. This suspicion soon deepened into hostility. The clash of ideologies and the division between east and west led to a different type of conflict, known generally as the Cold War.

However, in May 1945, the minds of the politicians in the victorious countries were concerned with peace rather than with war. They hoped that they could avoid future conflicts through international co-operation. An awesome task awaited those who were responsible for setting up the United Nations.

Chapter 24

International Co-operation since World War II

The United Nations

Background

The League of Nations did not disband until April 1940. However, it was already dead when Churchill and Roosevelt drew up the Atlantic Charter in 1941 (see chapter 23). They agreed that there was a need for 'a wider and more permanent system of general security'. In 1942 the ideals expressed in the Atlantic Charter were accepted by all the nations fighting against Germany and Japan. These ideals included self-determination, international co-operation and banning force in settling disputes between nations.

A conference to draw up a charter and decide on the structure of the United Nations met at **Dumbarton Oaks** near Washington in the autumn of 1944. An outline agreement was drawn up and this was later approved at **Yalta** in February 1945. Stalin was worried that the Soviet Union might be outvoted. The question of the veto and representation for Soviet Republics was left over for a later conference.

Only the British Commonwealth, the United States, the Soviet Union and China were involved at Dumbarton Oaks. These were joined by over forty other countries at **San Francisco** in April 1945. They finalised the structure and on 26 June the Charter of the United Nations was signed by 51 nations.

The 'big four' (above) and France were given the power of veto. This meant that they had power to reject a proposed measure. The smaller nations were unhappy with this arrangement but they had to accept it. The United Nations officially came into being on 24 October, 1945. Its headquarters are in Manhattan, **New York**, on a site donated by John D Rockerfeller, Jnr.

Charter of the U.N.

The spirit of the United Nations is expressed in the preamble to its charter which was accepted at San Francisco. The aims expressed in the preamble were:

- to save succeeding generations from the scourge of war
- to reaffirm faith in fundamental human rights, in the equal rights of men and women, and of nations, large and small
- to establish conditions under which justice and respect for international law can be maintained

United Nations Building, New York

- to promote social progress and better standards of life
- to practice tolerance and neighbourly relations among states
- to maintain international peace and security.

The Charter then laid out the structure and rules of the United Nations. The powers and functions of the General Assembly, Security Council, Secretariat and specialised agencies were written into the document.

Organs of the U.N.

The **General Assembly** is the parliament of the U.N. Each member nation has one vote. The Assembly meets once a year and in emergencies. It debates issues and sends proposals to the Security Council for approval. It votes on the budget and elects members to various sub-committees. The powers of the General Assembly are stronger than its counterpart had been in the League of Nations. In emergencies it can recommend actions for the U.N., including the use of armed force.

The **Security Council** has 15 members. The U.S., the U.K., the Soviet Union, China and France are permanent members. The ten other members are elected from the General Assembly for a period of two years. Each of the permanent members has the power of veto. Since 1950 the General Assembly can override such a veto and take action if it can obtain a two-thirds majority on emergency resolutions. The Security Council's main function is to keep the peace between nations. It can meet at any time, in case of emergency.

The **Secretariat** is the U.N.'s civil service. It manages the day-to-day business of the organisation and provides secretarial and specialised services for all the other

U.N. organs. The Secretary General is the most powerful figure in the United Nations and is responsible for its smooth functioning. The first four Secretary Generals of the U.N. were **Trygve Lie** of Norway (1946-53), **Dag Hammarskjold** of Sweden (1953-61), **U Thant** of Burma (1961-71) and **Kurt Waldheim** of Austria (1972-81).

The **International Court of Justice** consists of 15 judges based at The Hague in the Netherlands. It can hear only cases brought by member states, not by individuals. Decisions, by simple majority, are mainly on international disputes and on interpretations of international laws and treaties.

The **Trusteeship Council** administered territories, not yet independent, which were placed under the control of the U.N. at the end of World War II. The Council helped these territories towards independence. By 1980 only the Trust Territory of the Pacific Islands was under United Nations administration.

The **Economic and Social Council** is responsible for supervising the work of the U.N. agencies which specialise in health, education and areas of economic and social concern.

Principal U.N. Agencies

The **WHO** (World Health Organisation) tries to unite all nations in a fight against ill-health. It controls epidemics, monitors drugs and medicines and promotes positive health measures.

The **FAO** (Food and Agricultural Organisation) helps nations improve the produce of farms, forests and fishing waters. It helps farmers, especially those in the Third World, and through agricultural education tries to combat famines and to raise living standards.

The **World Bank** lends money to help countries, especially developing nations, with such projects as irrigation works, communications and power supplies.

British actor, Peter Ustinov, recording for a UNESCO-UN radio series

The **IMF** (The International Monetary Fund) deals with exchange rates and the stability of currencies. It helps with balance of payment difficulties and facilitates international trade.

GATT (The General Agreement on Tariffs and Trade) was founded to promote world trade and settle disputes between trading nations.

UNESCO (The United Nations Educational, Scientific and Cultural Organisation) helps to increase understanding among nations through the exchange of educational, scientific and cultural knowledge. One of its main concerns is the eradication of illiteracy.

UNICEF (The U.N. International Children's Emergency Fund) was founded in 1946 to provide food, clothing and medical supplies for child victims of the war. It continues to provide aid for child development and care.

ILO (The International Labour Organisation) aims to improve working conditions through emphasis on safety standards, good labour relations and good management.

UNICEF Emblem

Keeping the Peace

Despite major difficulties, the U.N. has proved more successful than the League of Nations in its main function of maintaining world peace. The permanent members of the Security Council were unable to agree on the creation of an international peacekeeping force. They have also been only too ready to use the veto when their own vital interests were at stake. However, the United Nations can point to many solid achievements and some valiant efforts in its attempts to keep the peace:

- **Palestine, 1947**: In 1947 the General Assembly approved a plan to divide Palestine into a Jewish state and an Arab state. The Arabs opposed the plan and refused to accept the state of Israel. UN troops have in difficult circumstances tried to keep opposing sides apart in disputed territory in the Sinai Peninsula, the Golan Heights and the Lebanon. In 1974 the General Assembly angered Israel by recognising Palestine's right to nationhood. Hope of a permanent settlement emerged in September 1993 with the mutual recognition of each other's claims by the Israelis and the Palestine Liberation Organisation (PLO).
- **Indonesia 1947-49**: The UN intervened in fighting between Dutch troops and Indonesian Nationalists. As a result the Dutch granted independence to Indonesia in 1949.
- **Kashmir 1949**: The United Nations arranged an armistice which ended fighting in a dispute between Pakistan and India over Kashmir. Although fighting broke out again in 1965, the cease-fire line of 1949 has generally held.
- **Korea, 1950-53**: (see chapter 25) While the Soviets were boycotting the Security Council it voted to intervene against North Korean Communists. The forces were mainly American and repelled the invaders. The intervention proved controversial as it seemed that the U.S. was beginning to dominate the United Nations and to use the organisation for her own purposes.
- **Suez 1956**: (see chapter 27) The UN overwhelmingly condemned the seizure of the Suez Canal by British, French and Israeli forces. A peace-keeping force from ten nations was sent to guard the borders between Egypt and Israel.

Irish contingent of United Nations peace-keeping force in Cyprus, 1964. The Irish Army has also served with the U.N. in the Congo, the Sinai Peninsula and Iraq among other places, and are nowadays in Somalia, Cambodia, Lebanon . . .

- **The Congo 1960-64**: The Congo (now Zaire) in 1960 became an independent republic after 55 years of Belgian rule. A breakaway state was set up in the province of Katanga. The Congolese government requested UN assistance. The Soviet Union vetoed action by the Security Council. The General Assembly sent a large force of 26,000 troops and order was restored. As the Security Council had not approved action, France, the Soviet Union and a number of other countries refused to contribute to the cost of the operation and the UN was soon in serious debt.
- **Cyprus 1964**: Fighting took place between Turks and Greeks on the island of Cyprus. The Secretary General, U Thant, sent a UN peace-keeping force. This force helped prevent conflict on the island during the 1960s and 1970s.
- **The Gulf 1991**: A huge force of about 700,000 UN troops, mostly U.S. marines, took part in the Gulf War. This was in response to UN resolution 678 which authorised the use of force when Iraq failed to withdraw from Kuwait.
- **Somalia 1993**: United Nations forces acted against local warlords to ensure that international food aid could be distributed to starving people.

Assessment

The United Nations experienced many failures, due mainly to the use of the veto by permanent members who claimed that the organisation had no right to get involved in internal matters. It passed votes of censure but was powerless to intervene when

U.N. relief truck brings supplies to the Golan Heights.

the Soviet Union invaded **Hungary** in 1956 or during the **Cuban Missile Crisis** in 1962. This was also the case during the **Vietnam War** from 1956 to 1975 and when the Warsaw Pact nations invaded **Czechoslovakia** in 1968. The UN found it difficult to function effectively because of the Cold War. However, the fact that a Third World War was avoided was due, in no small measure, to the work of the United Nations Organisation.

In 1968 the UN approved a treaty prohibiting powers from giving nuclear weapons to those that did not already have them. A treaty banning the production and use of biological weapons took effect in 1975.

The UN has taken action against the policy of *apartheid* in South Africa. Members have been urged to stop trading with and selling arms to South Africa. Resolutions have also been passed on the peaceful uses of outer space and of the seabed. Some very valuable work has been done by the specialised agencies of the United Nations, especially in the Third World.

The UN, however, has failed to live up to the high hopes expressed in its charter. This has been due to the selfishness of nations who have put their own interests ahead of international welfare. Some countries have expected the United States to always play the leading role and to bear the main financial burden of peace-keeping operations. On the other hand, the question of the U.S. using the U.N. for its own purposes surfaced again in 1991 after the Gulf War, just as it had done after the Korean War.

European Unity

Growth of an idea

> We must re-create the European family in a regional structure called the United States of Europe. France and Germany must take the lead.
>
> (Churchill)

These words from a speech made by Churchill in Zurich in 1946 echoed the thoughts of a number of European statesmen who began to advocate European economic and political cooperation after World War II. European unity was the great goal of two French statesmen, **Jean Monnet** and **Robert Schuman**. Their dream was shared by others, most notably **Paul Henri Spaak**, the Belgian Foreign Minister, and **Alcide de Gasperi**, Italian Prime Minister from 1945 to 1953.

The reasons why these statesmen pursued the idea of European unity can be summed up as follows:

- Europe lay in ruins after the war and there was a determination that such destruction would never happen again. The nationalism of individual states would give way to a European commitment. Many felt that German aggression would be contained within a larger body.

Jean Monnet, pioneer of a 'United States of Europe'

- The individual states of Europe began to realise that they would be too weak to withstand the Soviet threat from the east. They had to cooperate or be in danger of perishing.
- A strong third grouping was needed to provide a balance of power between the two new Superpowers, the USA and the USSR. Europe would have a say over its own destiny and over world peace if it spoke with one strong voice.
- A united approach was needed to help the recovery of Europe's economy. Only through sharing resources and removing trade barriers could real and dramatic growth be achieved.

Benelux Agreement

In September 1944, while their territories were still under occupation, representatives of the Dutch, Belgian and Luxembourg governments met to sign a customs treaty. Tariffs would be gradually reduced so that eventually no duties would be paid on goods passing between the three states. Goods entering from outside would pay a common rate of tariffs.

The Benelux countries gradually developed into a single economic unit. By 1960 free movement of goods, capital and labour had been achieved and a common monetary policy was pursued. The **Benelux Agreement** proved a blueprint for the future of Europe.

The OEEC

In 1947 the American government was anxious that Marshall Aid to Europe should be distributed through a properly structured body. Seventeen countries who received help set up the **OEEC** (Organisation for European Economic Cooperation).

They each contributed an amount towards economic recovery, equal to what they received from the US.

The OEEC later supported the idea of lowering tariffs between member states. They also took part in GATT (see p. 324) talks. The OEEC declined in importance with the ending of Marshall Aid in 1952. It was an important stage in the movement towards greater economic unity in Europe. It later widened out to form the **OECD** (Organisation for Economic Cooperation and Development).

The Council of Europe

In 1948, Churchill, Schuman and de Gasperi were among the politicians who met at the Hague to discuss closer political union in Europe. It was also significant that **Konrad Adenauer**, soon to be first Chancellor of the Federal Republic of West Germany, was present. The **Hague Conference** proposed 'a united Europe, throughout whose area the free movement of persons, ideas and goods is restored'. The delegates expressed a desire that a charter of human rights should be drawn up. They also agreed that a Council of Europe should be formed, consisting of members of national parliaments.

Ten countries signed the statute for the **Council of Europe**. They were Belgium, Britain, Denmark, France, Holland, Ireland, Italy, Luxembourg, Norway and Sweden. The Council held its first meeting at **Strasbourg** in Eastern France in August 1949. The aim of the Council of Europe was to provide social and cultural cooperation between member states and to look after human rights. The **Consultative Assembly** of members of national parliaments discussed issues and made recommendations. **The Committee of Foreign Ministers** could make decisions, but its powers were limited.

The **European Convention on Human Rights** was signed by all member states. While the Council of Europe was dismissed by some as a 'debating society', its structures served as a model for future cooperation.

The ECSC

In 1950 the **Schuman Plan** proposed to make industry more efficient in Western Europe by establishing a common policy on coal and steel. Britain refused to join, as she felt that it would lead to a loss of national sovereignty. Belgium, France, Germany, Italy, Luxembourg and the Netherlands accepted the plan and signed the **Treaty of Paris** in 1951. This created the **ECSC** (European Coal and Steel Community). It was given the task of planning the growth of these industries. Huge increases in production occurred and this led to demands for the further integration of Europe.

In 1953 Jean Monnet addressed a joint meeting of members of the Council of Europe and the ECSC. He said that the ECSC was the start of a 'united Europe':

> Our Community is an open community. We want other countries to join on an equal footing with us... to give up the divisions of the past and, by pooling their

Robert Schuman, French Foreign Minister, who gave his name to the 'Schuman Plan' of 1950

coal and steel production, to ensure the establishment of common bases for economic development as the first stage towards a European federation.

(Jean Monnet)

The European Defence Community (EDC)

After the outbreak of the Korean War in 1950, the United States hoped that West Germany would be allowed to rearm. She could then play her part in the defence of Western Europe against any attack that might come from the Soviet Union. The French Prime Minister, René Pléven, suggested the formation of a European army made up of contingents from each member state, including West Germany.

Britain was not enthusiastic about any form of political or military union and refused to join. The Pléven Plan was welcomed by the West Germans who felt that it might bring to an end Allied Occupation of their country. However, attempts to form a European army led to difficulties between France and Germany. The French National Assembly were worried about any scheme which allowed Germany to rearm and voted against the motion. When the Korean War ended in 1953 the need for such an army eased. The German army was integrated into NATO and the idea of a separate European Defence Union was abandoned. Most countries were not yet ready to surrender control of their armies.

The Spaak Report: EEC Proposed

A meeting of Foreign Ministers of the six ECSC member states in June 1955 issued a statement containing outline plans for a common policy on transport, energy production, economic development and social welfare. A committee under the Belgian Foreign Minister, **Paul Henri Spaak**, was given the task of preparing detailed plans.

Mr Spaak first drew up a report on the creation of **EURATOM** (European Atomic Energy Commission). This body would be concerned with the peaceful uses of atomic energy. Spaak's committee then examined a Dutch proposal for a customs union or common market. Spaak recommended that the six countries with a population of 170 million would eventually form a single economic unit. The Spaak report was accepted as the basis of treaties creating the **European Economic Community** and **EURATOM**.

The Treaty of Rome

The **Treaty of Rome** was signed on 25 March 1957 by the six members of the ECSC. These were Belgium, France, Italy, Luxembourg, the Netherlands and West Germany. The **European Economic Community**, also called the **Common Market**, came into being on 1 January 1958.

The most important articles of the Treaty of Rome were the following:

- All tariffs and trade restrictions between the members were to be abolished. This would take place in stages.
- A common tariff and trade policy towards other countries would be agreed between member states.
- Obstacles to the free movement of people, services, goods and money between members would be removed.
- Common agricultural and transport policies would be drawn up.
- A European Social Fund would help to improve the possibilities of employment and would help raise the standard of living in disadvantaged areas.
- A European Investment Bank would facilitate the economic expansion of the Community.
- Fair and free competition would exist within the E.E.C.
- Members would have the power of veto over proposed new entrants.

The leaders of the six founder members of the EEC sign the 'Treaty of Rome' on 25 March, 1957.

Structures of the E.E.C.

The E.E.C., whose headquarters is in Brussels, has four main institutions:

- The E.E.C. **Commission** administers the affairs of the Community, proposes policies and looks after general community interests rather than those of individual nations. In 1958, the Commission had six members, each appointed by the government of a member state.
- The **Council of Ministers** is made up of Foreign Ministers from member states. They represent the interests of their governments and vote on proposals from the Commission.
- The **European Parliament** meets in Strasbourg. Delegates from member states examine the work of the Commission, debate Community issues and advise the Council of Ministers. Until 1979, the delegates were nominated by their national parliaments. Since then, the members of the European Parliament have been elected by popular vote.
- The **Court of Justice** arbitrates on disputes between member states and interprets the Treaty of Rome.

The European Free Trade Association (EFTA)

Britain had been invited to join the ECSC and the E.E.C. She declined because many of her politicians were not prepared to surrender any powers of the British Parliament to the European Community. They were also worried about the effects on British trade with the Commonwealth and on the special relationship between Britain and the U.S. A number of other countries were also opposed to a supra-national organisation, i.e. a political body representing a number of countries. These decided to establish **EFTA** (The European Free Trade Association) in 1959.

EFTA consisted of Austria, Britain, Denmark, Norway, Portugal, Sweden and Switzerland. They agreed to reduce tariffs on industrial goods by 10 per cent per annum. Agricultural products were not included. Britain's trade with EFTA countries increased significantly, but her trade with the E.E.C. rose even more. Many British politicians began to cast envious eyes on the boom in production and trade in the Community. They saw great advantages in a market which was four times larger than that of EFTA.

The E.E.C. expands

The British government applied for membership of the E.E.C. in 1961. Denmark, Ireland and Norway also sought membership. However, President de Gaulle vetoed British membership (see chapter 26). He claimed that Britain was far more committed to the Commonwealth and to the U.S. than to Europe and that her economy was too weak. When de Gaulle retired in 1969, the E.E.C. invited Britain and the other three states to renew their applications.

THE ECONOMIC INTEGRATION OF WESTERN EUROPE

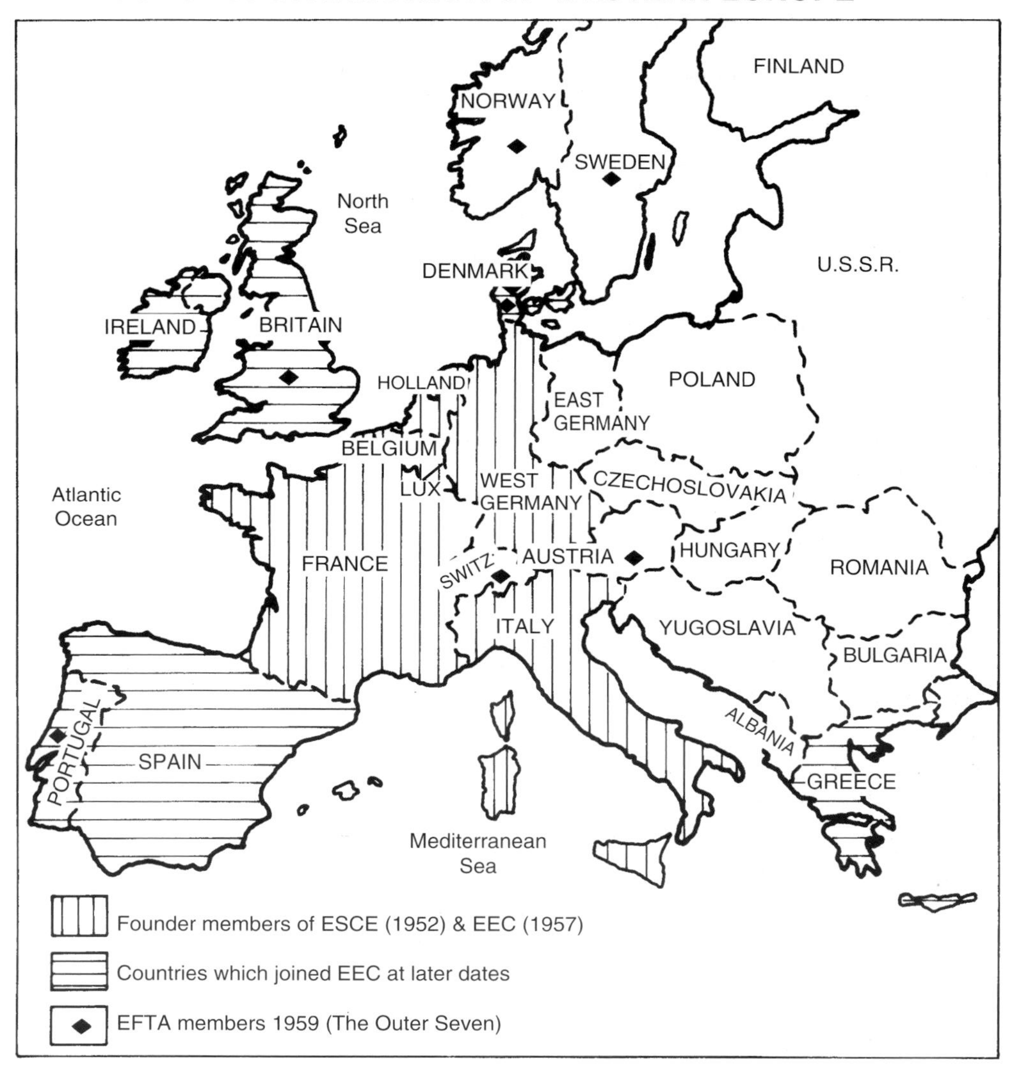

In 1972 the people of Norway voted against joining the Common Market. In 1973 Britain, Denmark and Ireland became members of the E.E.C. and 'the six' became 'the nine'. Greece became the tenth member of the Community in 1981. Portugal and Spain joined in 1986.

The Common Agricultural Policy

> The Common Market shall extend to agriculture and trade in agricultural products.
>
> (Article 38)

Article 38 of the Treaty of Rome meant that there should be no tariffs on farm products sold within the Community. Taxes would be imposed on agricultural products from outside, to stop them undercutting prices to European farmers. In 1962, these farmers were guaranteed minimum prices for their produce. Grants and subsidies were brought in and special schemes helped those in disadvantaged areas. These measures were introduced in order to prevent rural depopulation and to ensure that Europe would not suffer from hunger in the event of war, or from famine.

The **CAP** (Common Agricultural Policy) was successful in protecting European farmers, but it has also created serious problems. The Community began to pile up huge surpluses of food which it had to buy at fixed prices. Huge butter, cheese and grain 'mountains' and large milk and wine 'lakes' accumulated. The Community tried to solve these problems by keeping down guaranteed prices, by selling off surpluses cheaply to the Soviet Union and to Third World countries, and even by destroying produce. In 1993 the E.C. has begun to pay farmers to 'set aside' land, rather than grow unwanted produce.

The CAP still faces problems, especially from the American insistence in GATT talks that grants and subsidies to farmers should be cut. However, for all its faults, it has kept food prices steady and prevented thousands of farmers from going out of business.

Closer Union

The members of the European Parliament have been directly elected by popular vote since 1979. In that year also, the **EMS** (European Monetary System) was established. Through the **ECU** (European Currency Unit) and the **ERM** (Exchange Rate Mechanism), an attempt has been made to stabilise the currencies of member states. This policy has run into difficulties as individual countries have resisted greater monetary union.

In 1986, the twelve members signed the **Single European Act**. All remaining restrictions on free trade were to be removed by 1993. Despite serious reservations in some member states, the **Maastricht Treaty** was signed by all members before the end of 1993. This envisages closer union than agreed in the Treaty of Rome. There will be total freedom of movement for people, goods, services and capital. Members are committed to common health and safety standards, to a social charter and to protection of the environment.

Assessment

The EC (now known as the **EU**, European Union) has brought prosperity and peace to Europe. In some areas, progress has been slow and painful. Unemployment levels have remained high in the Community. However, member nations have committed themselves to complete economic unity and to continued economic growth. Europe can also expect closer cooperation in political and security affairs in the years ahead.

Chapter 25

The Cold War

Origins of the Cold War

Definition

Cold War: A state of tension between countries in which each side adopts policies designed to strengthen itself and weaken the other, but falling short of actual 'hot' war.

(*A Dictionary of Politics*, Penguin, 1984)

The Cold War is the name given to the conflict between the **Eastern Bloc** (communist) and the **Western Bloc** (capitalist) after World War II. The Soviet Union dominated the Eastern Bloc, while the United States adopted the role of leader of the Western world.

No fighting occurred between the U.S.S.R. and the U.S., but they supported opposing sides in a number of local conflicts, such as the Korean War and the Vietnam War. Propaganda, spying, the build-up of arms and the space race were all aspects of the rivalry between East and West. Various crises, such as the Berlin Blockade and the Cuban Missile Crisis, could have turned the Cold War into a third global conflict. Luckily it never became a 'hot' or 'shooting' war between the two power blocs.

The Superpowers

In 1945 much of Europe was devastated. The economies of Britain, France and Germany were in ruins. However, two Great Powers emerged after World War II. Despite the losses sustained in defeating the Nazis, the Soviet Union had the largest army in the world in 1945. America had escaped the ravages of war and experienced a boom in both agriculture and industry. Her economy was now the strongest in the world. In the post-war era, the U.S.S.R. and the U.S. became known as the **Superpowers**.

Reasons for Distrust

Both sides had reasons for distrust in 1945:

- Communism and capitalism are complete opposites. The Cold War was essentially an ideological conflict.

- The West did not trust Stalin, who argued for the creation of a buffer zone of satellite states between the Soviet Union and the West. Stalin spoke of 'a strong defensive perimeter'. The Americans believed that the Soviet leader was merely trying to extend the frontiers of world Communism.
- Stalin also distrusted the Western leaders. Churchill had supported foreign intervention in the Russian Civil War in 1920. Both he and Truman were outspoken opponents of Communism.
- World War II had left a legacy of bitterness. The West felt that in 1939 Stalin had betrayed any principles he might have had, by agreeing to the Nazi-Soviet Pact. The Soviets believed that their Western Allies had deliberately delayed opening up a second front during the war until 1944, in the hope that Hitler would destroy Communism.
- In 1945 The United States had the atom bomb and was unwilling to share her knowledge and technology with the Soviets.

The Conferences

At the conferences during and after the war, serious disagreements emerged between the Soviet Union and the West (see chapter 23). At **Teheran**, Stalin would not agree to Churchill's suggestion that Germany should be invaded through the Balkans. The British Prime Minister concluded that Stalin intended that Eastern Europe would be a Soviet sphere of influence after the war.

Allied leaders, Stalin and Truman: a cooling of relations

At **Yalta** and **Potsdam** the Allied leaders argued fiercely about the question of Poland. Stalin wanted the communist-dominated **Lublin Committee** to form the new Polish government. Churchill spoke in favour of the London-based **Polish government-in-exile**. Serious differences also surfaced at these conferences about the future of Germany. The disagreements at the war-time conferences flared into open hostility during the Cold War.

The Iron Curtain

> From Stettin in the Baltic to Trieste in the Adriatic, an ***iron curtain*** has descended across the continent. Behind that line lie all the capitals of the ancient states of Central and Eastern Europe... all are subject in one form or another not only to Soviet influence, but to a very high and increasing measure of control from Moscow.
>
> (Churchill)

In March 1946, Winston Churchill made a speech at Fulton, Missouri, while visiting the U.S.A. He did not use the language of one Allied statesman describing another ally's actions. He spoke of the 'iron curtain' and made it clear that, within one year of the defeat of Germany, the wartime alliance had broken down.

Churchill had seen communist governments set up all over Eastern Europe, where Stalin's Red Army advanced. This occurred in countries like Poland, Czechoslovakia, Hungary, Romania and Bulgaria (see chapter 28). After March 1946 the term 'Iron Curtain' was used to describe the border between the Soviet-controlled countries and the West. In the popular mind, the term indicated a barrier behind which political and human freedoms were absent and normal democracy did not function.

Winston Churchill's speech at Fulton is sometimes referred to as 'the announcement of the Cold War'. It marked a turning-point in relations between East and West. A week later Stalin reacted angrily:

> There can be no doubt that Mr Churchill's position is a call for war on the U.S.S.R.
>
> (Stalin)

The Cold War begins

Greece and Turkey

British troops helped in the liberation of Greece during the war. Britain tried to secure the election of a non-communist government, but this was opposed by the communist-led **National Liberation Front**. Britain continued to send aid to Greece and a right-wing government took office in 1946. In Turkey, pressure was mounting from the Soviet Union for the return of disputed territories and the right to naval bases.

Britain saw events in Greece and Turkey in terms of Soviet attempts to dominate the Mediterranean region. However, World War II brought Britain to the brink of economic collapse. In February 1947 **Ernest Bevin**, the British Foreign Secretary, informed the United States that Britain would have to cease providing aid to the Greeks and the Turks.

The Truman Doctrine

> I believe that it must be the policy of the United States to support free peoples who are resisting attempted subjugation by armed minorities or by outside pressures.
>
> (President Truman)

The Americans now took a historic step. They had been traditionally isolationist and tried to avoid involvement in European affairs. President Truman asked the United States Congress to allow him to give immediate military and financial aid to Greece and Turkey in order to resist Communism.

The Truman Doctrine went beyond aid to the Greeks and Turks. The President pledged resistance to communist expansion worldwide. The United States provided $400 million in aid to Greece and Turkey. By 1947, the Greek communists were defeated. This was due partly to American aid and partly to a dispute between Yugoslavia and the Soviet Union, which diverted Stalin's attention and Tito's support from Greece.

The Marshall Plan

> Any government that is willing to assist in the task of recovery will find full co-operation, I am sure, on the part of the United States Government. Any governments, political parties or groups which seek to perpetuate human misery in order to profit therefrom, politically or otherwise, will encounter the opposition of the United States.
>
> (General George Marshall, U.S. Secretary of State)

This is an extract from a speech delivered on 5 June 1947 at Harvard University by General George Marshall, the U.S. Secretary of State. **The Marshall Plan**, as it became known, was the economic arm of the Truman Doctrine. America decided to give financial aid to Europe on a massive scale, in order that its shattered economies might be rebuilt and the appeal of Communism lessened.

The economic assistance under the Marshall Plan was offered to all countries of Europe. However, Stalin turned down the offer on behalf of all communist states behind the iron curtain. The Soviet Foreign Minister, **Molotov**, said that it would violate a country's national independence and was evidence of 'encirclement' by a hostile capitalist world. The Russians set up **COMECON** (Council for Mutual Economic Assistance) in 1949, to help the economies of Eastern Europe. Its objective was to challenge 'the world domination of American imperialism'.

George C. Marshall, World War II general and author of the 'Marshall Plan'

Marshall Aid was administered in Europe by the OEEC (see chapter 24). Between 1948 and 1952 it distributed $13,000 million in grants to the European Recovery Programme. The Programme was an instant success. Within two years, industrial production in western Europe rose by twenty-five per cent. The American economy also benefited, as its exports to Europe increased.

Confrontation

The stage was now set for a full-scale confrontation between the Superpowers. Stalin saw that the Soviety Union had been vulnerable to attack from the West. He wanted to create a ring of states which would form a protective barrier around his country. These states would be forced to follow Stalin's hard-line communist policies.

The Americans saw this as an attempt by Stalin to promote global Communism. President Truman was determined to resist this attempt. The first major test of strength between the leaders of the Superpowers took place in Berlin.

Crisis in Berlin

After the war, Germany was divided into American, British, French and Russian occupation zones. The line between the zones became a central part of the Iron Curtain. Berlin's important position as capital meant that it, too, was administered by the four occupying countries, even though it was 150km inside the Russian zone. As a result, access to and from Berlin was across Soviet-controlled territory.

The West believed that Germany would have to be helped with economic recovery before she was ready for democracy. The Russians preferred to keep their former enemy weak. America, Britain and France had merged their zones by 1948. They refused to inform the Russians of their political and economic plans for the future of Germany. Russia walked out of the Allied Control Council. This brought the Allied partnership to an end.

On 20 June 1948, the Western Allies introduced the new Deutschmark into their zones, in order to halt inflation and stimulate economic recovery. The new currency was also used in West Berlin. This was clearly a move towards setting up a separate Germany. Stalin vowed that if this was so, the Western Allies would have to give up their sectors of Berlin. He would not allow the West to have a permanent base well inside the iron curtain.

24 June 1948: The Berlin Blockade

> When Berlin falls, Western Germany will be next. If we mean to hold Germany against Communism, we must not budge.
>
> (General Clay, U.S. Commander in Berlin)

On 24 June 1948 the Russians blocked all road, rail and canal access to Berlin. Electricity and gas supplied to West Berlin from the Soviet zone was cut off. **Berlin was blockaded**. Stalin felt that life would become so difficult for the British, French and Americans, that they would leave the city.

The West responded by deciding to supply West Berlin from the air. Russia had not dared to close the air corridors as she was not ready for war with her former allies. In a great airborne operation, the West kept over two million West Berliners alive. Day and night the planes flew in thousands of tons of fuel and food, to beat the blockade. The **Berlin Airlift** continued for eleven months. By May 1949 Stalin realised that the Western Allies were not going to be beaten. He decided to re-open the transport routes.

Afterwards the three Western Powers created the **German Federal Republic** (West Germany) from their zones. The Russians set up the **German Democratic Republic** (East Germany). Although the Superpowers had avoided war over Berlin, the affair led to increased tension between East and West. It also convinced the West of the need for a military alliance to stand up to the threat of Russian expansion.

NATO and the Warsaw Pact

Background to NATO

In 1948, after Czechoslovakia had been taken over by the communists, Britain, France, Holland, Belgium and Luxembourg signed the **Treaty of Brussels**. They agreed to co-ordinate their forces against the danger of an armed attack.

The Russian blockade of Berlin in June 1948 raised fears that Stalin might use force to gain control of Western Europe. In such a contest, the Soviets would have a major advantage in army and air force numbers. Accordingly, twelve nations signed the **North Atlantic Treaty Organisation** (NATO) in Washington on 4 April, 1949. They were Belgium, Canada, Denmark, France, Great Britain, Iceland, Italy, Luxembourg, The Netherlands, Norway, Portugal and the United States.

Soviet cartoon 1950 depicting American Imperialism

Aims of the Treaty

The Parties undertake, as set forth in the Charter of the United Nations, to settle any international disputes in which they may be involved by peaceful means in such a manner that international peace, security and justice are not endangered, and to refrain in their international relations from the threat or use of force in any manner inconsistent with the purposes of the United Nations.

(Article 1)

The Parties agree that an armed attack against one or more of them in Europe or North American shall be considered an attack against them all;... each of them... will assist the Party or Parties so attacked by taking forthwith, individually and in concert with the other Parties, such action as it deems necessary, including the use of armed force, to restore and maintain the security of the North Atlantic area...

(Article 5)

NATO in action

By signing the treaty, the United States, for the first time in its history, joined a peacetime alliance that committed it to fight in Europe. The U.S. at the same time provided the nations of Western Europe with military weapons and economic aid.

The Soviet threat receded during the 1960s and 1970s. At this time some differences emerged between the Americans and the Europeans. The French felt that the U.S. dominated the organisation. They decided to develop their own nuclear weapons and withdrew their forces from NATO command. However, France remained part of the organisation.

NATO proved successful in preventing the spread of Communism in Western Europe. Greece and Turkey signed the treaty in 1951, West Germany in 1954 and

Spain in 1982. When West Germany formally took its place as a full member of NATO in May 1955, the Soviet Union responded by forming the Warsaw Pact.

The Warsaw Pact

The Warsaw Pact was the treaty which placed the communist nations of Europe under Soviet military command. Albania, Bulgaria, Czechoslovakia, East Germany, Poland, Romania and the Soviet Union signed the Treaty at Warsaw in May 1955. All agreed to the stationing of Soviet forces within their territories and to the appointment of a Russian general as commander of the forces.

It is estimated that, in 1955, the combined military might of the Warsaw Pact countries amounted to six million men. Their charter was almost a copy of the NATO document. Article 1 committed the participants to the peaceful settlement of international disputes. Joint action against an aggressor was also a condition of membership. Marshall Aid to Western Europe also had its counterpart in Eastern Europe's **Council for Mutual Economic Assistance (COMECON).**

The Struggle in the East

Policy of Containment

The struggle between the Superpowers spread to the East as a result of the advance of Communism. China became a communist country in 1949 after a successful revolution, led by **Mao Tse-Tung**. A **Friendship Treaty** was signed between China and the Soviet Union in 1950. Korea and countries in Indo-China such as Vietnam were under threat. The Americans adopted a policy of 'containment'. This meant that Communism should be restricted to its existing boundaries and that all future expansion should be opposed militarily.

Background to the Korean War

Japan had gained control of Korea in 1910 but lost it to the Allies in 1945. It was partitioned along the **38th parallel** (line of latitude 38°N) after the Japanese forces had surrendered to the Red Army in the north and to the Americans in the south. It was intended that the country should be united as an independent state, but as the old alliance broke up, agreement on elections could not be reached between the two occupying powers.

The Americans set up the **Republic of Korea** led by **Syngman Rhee** in their zone in August 1948, after holding free elections. This territory became known as **South Korea**. In September 1948, the Russians formed a communist government in **North Korea** under **Kim II Sung**. The Soviet forces left North Korea in 1948. The U.S. forces pulled out of South Korea in 1949.

THE KOREAN WAR

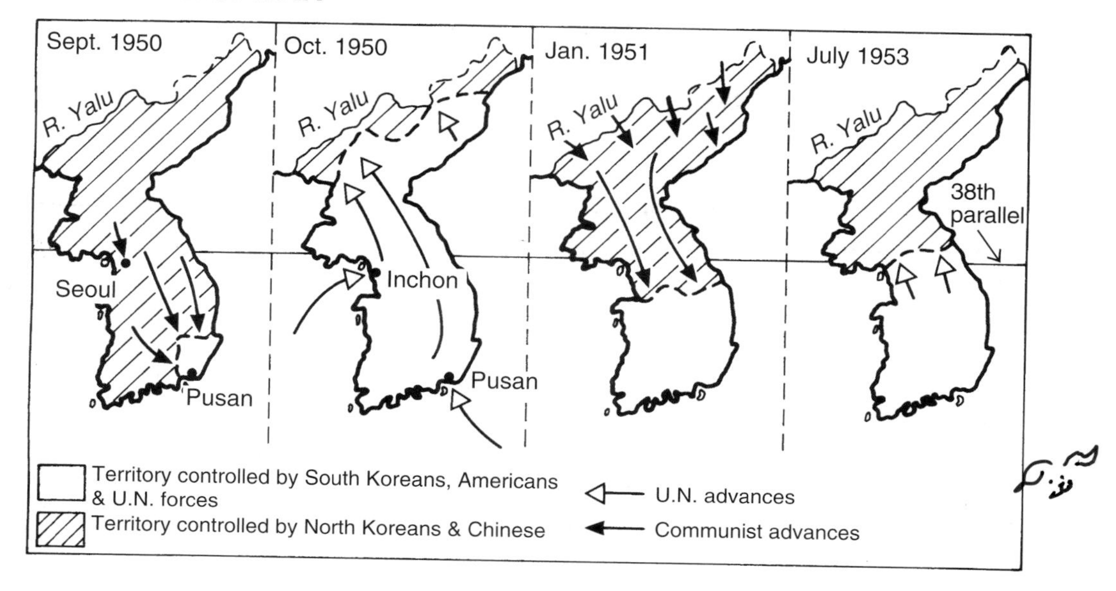

Korean War: Phase One

> The attack upon Korea makes it plain beyond all doubt that Communism has passed beyond the use of subversion to conquer independent nations and will now use armed invasion and war.
>
> (President Truman, 7 June 1950)

On Sunday 25 June 1950 a large North Korean army, supported by tanks, crossed the border and quickly moved south. Within three days this army had captured the capital, **Seoul**, and it soon controlled most of the south. The north Koreans were armed with modern Soviet weapons and the poorly-equipped South collapsed under the attack.

America asked the UN 'to furnish such assistance to the Republic of Korea as may be necessary to repel the armed attack and to restore peace and security in the area'. The Soviet Union was unable to veto the resolution, as she had walked out of the UN a few months earlier when the newly-established Communist State of China was refused admittance by th Americans.

A UN force, with 15 countries represented, was sent to South Korea to repel the invasion. However, most of the troops were American and the force was under the command of the American general, **Douglas MacArthur**.

Korean War: Phase Two

The UN force at first concentrated on gaining a foothold in a narrow area of the south of the country around **Pusan**. Then, on 15 September 1950, MacArthur surprised the invaders by making a daring landing at **Inchon**, near Seoul, 300 km inside their territory. After some fierce fighting, the North Koreans retreated. By October all territory lost by the South had been regained.

Korean War: Phase Three

> It seems strangely difficult for some to realise that here in Asia is where the communist conspirators have elected to make their play for global conquest... that here we fight Europe's war with arms, while the diplomats there still fight it with words, that if we lose the war to Communism in Asia, the fall of Europe is inevitable. There is no substitute for victory.
>
> (General MacArthur)

In October 1950 MacArthur began to pursue the communists northwards. The North Koreans were no match for his forces and by November he had reached the **Yalu River**, the border with China. Two Chinese field armies crossed the Yalu River and by mid-January 1951 the UN forces had retreated south, back across the 38th parallel.

MacArthur began to fight back and pushed the communists into North Korea. President Truman planned a negotiated settlement. However, MacArthur disobeyed orders and again drove forward against the Chinese. He demanded their surrender and even talked about the use of nuclear weapons. Truman was not prepared to run the risk of a third world war and, in April 1951, he dismissed MacArthur for exceeding his authority.

Settlement

Both sides 'dug in' around the 38th parallel and peace talks began in July 1951. South Korea's President, Syngman Rhee, continued to insist on Korea's reunification and fighting resumed.

The new American President, **Eisenhower**, pledged to end the war. Finally an armistice was signed in July 1953. The border remained along the 38th parallel, with a communist government in the North and an anti-communist dictatorial leader in the South. The U.S.S.R. and U.S.A. once again avoided a worldwide conflict. Meanwhile about 3 million people had died in the Korean War.

The Vietnam War

The story of the conflict between the communists of North Vietnam and the non-communists of South Vietnam is dealt with in chapter 26.

In summary, the French colonists were forced out of Vietnam in 1954. A communist government under **Ho Chi Minh** was set up in North Vietnam. The Americans supported the government of South Vietnam against the **Vietcong**, communist guerillas who tried to take power. The United States acted in accordance with the 'domino theory'. This was the belief that a communist victory in South Vietnam would lead to communist rule in all South East Asia.

A Vietnamese woman held at gunpoint

The determination of the American government to save the world from Communism led to a massive build-up of forces. In 1967 there were 470,000 United States troops in South Vietnam. It took a lot of deaths and a long time before America became convinced that she could not win the war against the guerilla tactics of the Vietcong. She gradually withdrew all her troops. Vietnam was united under a communist government in 1975.

Peaceful co-existence

Thaw in the Cold War

Changes in leadership occurred in both East and West in 1953. In March Stalin died and, after a power struggle, **Nikita Khruschev** emerged as leader. In January 1953 **Eisenhower** had succeeded Truman as U.S. President. Relations improved gradually and in general were not as bad as during the Stalin era.

The main reason for the thaw in the Cold War was Khruschev's policy of **peaceful co-existence**. He saw the overthrow of Capitalism throughout the world as a long-term rather than an immediate goal. In the meantime he advocated friendly relations between East and West. His 1956 condemnation of Stalin as a brutal tyrant was warmly welcomed in the capitalist world.

In 1955 both Superpowers withdrew their objections to a number of countries from rival blocs becoming members of the United Nations. Eisenhower and Khruschev held friendly discussions at a summit meeting in Geneva. **The Austrian State Treaty** resulted in the withdrawal of the occupying powers and the declaration of the permanently neutral state of Austria.

The thaw in the Cold War was punctuated by a number of incidents. The West continued to object to the harsh control exerted by the Soviet government in Eastern Europe (see chapter 29). Risings were suppressed in **East Germany** in 1953 and **Poland** in 1956. The most serious rebellion against the Soviets occurred in **Hungary** in 1956. The Hungarian leader **Imre Nagy**, appealed to the United States and Western Europe for help. The revolt was brutally suppressed and Nagy was executed. It was now clear that the West was not going to intervene directly within the Soviet sphere of influence.

The Berlin Wall

> Let every nation know, whether it wish us well or ill, that we shall pay any price, bear any burden, meet any hardship, support any friend, oppose any foe, in order to assure the survival and success of liberty. This much we pledge and more.
>
> (John F Kennedy's Presidential Inaugural Address, 20 January, 1961)

In 1958 Khruschev had tried to get the Western Powers to pull out of West Berlin. Thousands of East Germans were crossing the border into the more prosperous West Germany each year. The East German economy could not afford the loss of so many skilled workers. Since 1945 over 3 million citizens had fled to the West.

The Allies still refused to leave Berlin. The flood of refugees reached 100,000 in the first six months of 1961. In August of that year the East Germans closed the crossing points and erected a wall between East and West Berlin. Western propaganda pointed out that the wall was a symbol of the failure of Communism. The government of East Germany had to imprison its reluctant citizens.

Some East Berliners made daring escapes to the West, but many were killed by East German security guards. The Western powers protested, but were not prepared to use force. In 1963 President Kennedy visited the city and spoke to West Berliners:

> All free men, wherever they live are citizens of Berlin. And therefore as a free man, I take pride in the words *Ich bin ein Berliner* (I am a Berliner).

In the long-term, the Berlin Wall eased tension between East and West, as it defused one of the crisis points between the Superpowers.

The Cuban Missile Crisis

In 1956 **Fidel Castro**, a 30-year-old revolutionary, began a guerilla campaign against the Cuban dictator, **President Batista**. He took power after two years of fighting and introduced valuable social reforms to help the poor people of the island. Some of his policies were directed at American business interests in Cuba. The Americans broke off diplomatic relations and cut trade links with the Cubans. Castro turned to the Soviet Union and signed trade agreements with Khruschev.

Fidel Castro, Cuban revolutionary leader

America also supported an ill-fated invasion of Cuba by Batista's followers at the **Bay of Pigs** in 1961. The invasion was a failure, but it prompted Castro to turn to the Russians for military help. In October 1962 aerial photographs taken by special American spy planes showed that the Russians were building missile-launching sites in Cuba. The missiles when operational would be capable of delivering nuclear warheads to the United States. Other photographs showed that a fleet of Russian cargo ships was heading for Cuba.

Blockade of Cuba

The 'hawks' among President Kennedy's advisers wanted to bomb the sites, but the 'doves' urged caution. The President decided on what he called a 'measured response'. On 22 October he announced a naval blockade of Cuba and demanded that the Russians withdraw their missiles. Tension increased and the worst crisis between the Superpowers began. As the Russian ships approached Cuba, a nuclear exchange looked distinctly possible.

Disaster was averted on 24 October when Khruschev backed down. Twelve Russian ships turned back, before reaching the American blockade. On 28 October Khruschev agreed to dismantle and withdraw the missiles in return for an undertaking from America not to invade Cuba.

The Cuban Missile Crisis showed the Soviet Union that the United States would stand firm on major issues, especially on those within her sphere of influence. However, it also showed the Americans that the Russians could be reasonable and that they did not relish the prospect of a nuclear war. The crisis led to the installation of a direct telephone 'hotline' between the White House in Washington and the Kremlin in Moscow.

Canvas-covered missiles on board a Russian ship bound for Cuba

President Kennedy acted with firmness but with statesmanship. He allowed Khruschev to retreat with dignity:

> Both sides made concessions. We withdrew ballistic rockets and agreed to withdraw IL-28 planes. This gives satisfaction to the Americans. But both Cuba and the Soviet Union received satisfaction: the American invasion of Cuba has been averted, the naval blockade lifted. The situation in the Caribbean is returning to normal.
>
> (Mr Khruschev, 12 December 1962)

The Space Race

Space exploration was another area of rivalry between the Superpowers. In 1957 the Soviet Union launched the first artificial satellite, ***Sputnik 1***. In 1961 the Russians won the race to have the first cosmonaut, when **Yuri Gagarin** orbited the earth. The Americans caught up with the Soviets and had their moment of glory in 1969 when **Neil Armstrong** and **Edwin Aldrin** landed on the moon from the spaceship ***Apollo II***. Since then the 'race' has continued, though at a less hectic pace. The satellite launches have helped to improve our life-styles, especially in the areas of technology and telecommunications. The Cold War might be said to have had at least one beneficial side-effect.

The Cold war continues

The Cuban Missile Crisis brought to a close the most dangerous period of the Cold

COLD WAR DIVISIONS IN EUROPE

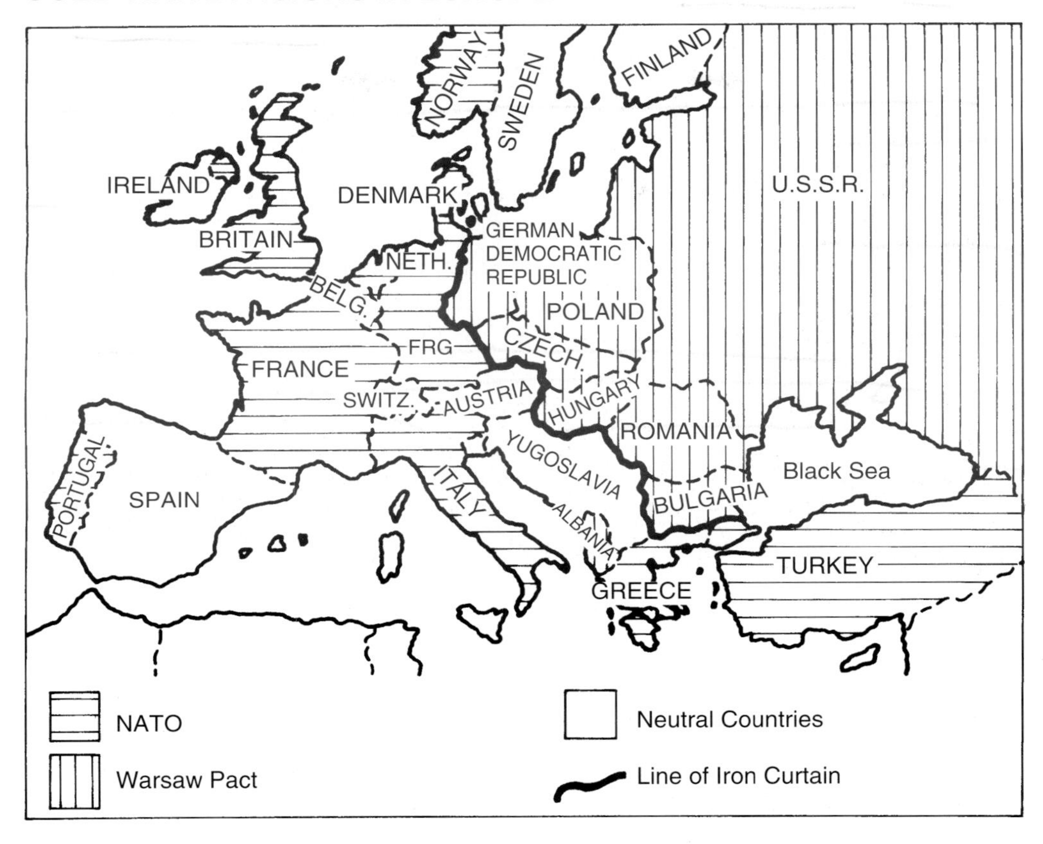

War. However the Superpowers continued to oppose each other in conflicts throughout the world. This mainly involved military and financial support for capitalist or communist regimes in South East Asia, Africa, or South America.

The West objected vigorously when the Soviet Union brought down Alexander **Dubček's** liberal government in Czechoslovakia in 1968, invaded Afghanistan in 1979 and tried to suppress the Polish ***Solidarnosc*** (Solidarity) Movement during the 1980s (see chapter 29). The Soviet Bloc accused the U.S.A. of supporting corrupt and dictatorial rulers in many countries including Iran, the Philippines and various Central and South American states. The West pointed to the denial of human rights and the treatment of dissidents behind the Iron Curtain. Despite these problems, there were signs that the worst of the Cold War was over.

Detente and the end of the Cold War

Detente means an easing of tension and a move towards better relations. During the Cuban Missile Crisis, American missiles had been primed and ready to fire at a second's notice. The Superpowers now became worried about arms reduction.

In 1963 a **Test Ban Treaty** was signed between the U.S. and U.S.S.R. Five years later the two Superpowers along with Britain, agreed to a **Non-Proliferation Treaty**. This was designed to stop the spread of atomic weapons to other countries. In 1972 the **Strategic Arms Limitation Treaty (SALT)** curtailed the actual number of missiles held by both sides. The **SALT 2** agreement, signed in 1979, placed further limits on nuclear weapons.

Relations between the Superpowers also improved as a result of the **Helsinki Agreement, 1975**, which guaranteed respect for human rights such as free speech. The atmosphere between East and West Germany also softened under the guidance of the Mayor of West Berlin, **Willy Brandt**, who became German Chancellor in 1969.

During the 1980s President Gorbachev's policies of ***Glasnost*** (Openness) and ***Perestroika*** (Economic restructuring) also impressed the West. The opening of the Berlin Wall, the reunification of Germany and the collapse of Communism in Eastern Europe brought an end to the Cold War. In the 1990s the two Superpowers have been ready to agree in the United Nations on policy decisions about issues in such places as South Africa, the Middle East and the Gulf. The great ideological debate between Capitalism and Communism and the mighty power struggle between East and West has subsided, at least for the present.

Chapter 26

France, 1945-1970

Liberation of France

Charles de Gaulle: early career

Charles de Gaulle was born in Lille on 22 November 1890. He was interested in pursuing a career in the army and, on leaving school, he entered the Military Academy of Saint-Cyr. In 1916, he was wounded at the Battle of Verdun and was taken prisoner. He was released in 1918 and resumed his military career.

In 1925, de Gaulle was promoted to the staff of the Supreme War Council. During the 1930's he expressed his reservation about France's defence policies in his book, *Vers l'Armée de Metier.* He insisted that France needed a modern mechanised army with specialised armoured divisions and said that the Maginot Line fortification was worthless.

De Gaulle was promoted to the rank of General in 1940 and was given command of a tank brigade. His military theories were now seen as valid and he was appointed Under-Secretary for War.

General de Gaulle leads the Victory Parade down the Champs Elysées, August 1944.

Resistance

General de Gaulle opposed the armistice with Germany in 1940 and fled to London where he set up the **Free French** movement (see chapter 23). However, he failed to get recognition as leader of the government in exile from Churchill or Roosevelt. During the war he became undisputed leader of the French Resistance movement. In 1941, the German invasion of the USSR turned the French Communist Party into militant anti-Nazis. They accepted de Gaulle's call to all French patriots to rally to his leadership. His colleagues in London – **Schuman, Monnet** and **Mendes-France** – later played major roles in post-war French politics.

The Allied landings in North Africa in 1942 gave de Gaulle his opportunity. He arrived there in May 1943 and became co-leader of the **French National Liberation Committee**. His Free French forces liberated Corsica, but de Gaulle still did not gain the trust of the Allies. They thought that he was too abrasive and might attempt to set up a dictatorship in France.

De Gaulle was kept in the dark about the D-Day landings in Normandy in June 1944. He returned to France a week after the landings, determined that his country would not be ruled by the British and Americans. He also tried to ensure that the communists would not get control of local government, as he was a convinced anti-communist.

On 25 August 1944, the Germans surrendered Paris to the French. On the following day Charles de Gaulle had his moment of triumph. He led the victory parade down the Champs Elysées as the acknowledged **Liberator of France**.

Collaborators

In September 1944 de Gaulle announced the setting up of a Provisional Government. It was formally recognised by the Allies in October. He issued orders to the Resistance fighters to disarm and called for a French occupation zone in Germany. De Gaulle was upset when he was not invited to the conferences at Yalta or Potsdam, but France did gain a zone of influence in Germany. He sent the French Army into Germany in 1945.

A French collaborator shot near Rennes soon after the liberation, August 1944

The most immediate problem facing the Provisional Government was what to do with Vichy officials and with collaborators. Special courts were set up and 767 'traitors' were executed after trials. The Vichy Prime Minister, Pierre Laval, was shot by a firing squad in May 1945. He tried to poison himself first and died shouting *'Vive la France'*. The Vichy head of state, Petain, was also sentenced to death. His sentence was commuted to life imprisonment due to his age (89 years) and his record of service in World War I.

At local level some people carried out a campaign of revenge or 'witch-hunt' against collaborators. Communist Resistance groups were most prominent in this savage campaign. It is estimated that about 10,000 collaborators lost their lives during these atrocities.

De Gaulle as Prime Minister

> Aims: 'A more just social order...the subordination of private interests to the general interest'
>
> (French Resistance Charter)

De Gaulle had to include **Socialists** and **Communists** in his Provisional Government. However, he distrusted the Communists and excluded them from key positions. A new Christian Democratic Party, the **MRP** (Movement Républicaine Populaire) was formed under the Resistance leader, **George Bidault**. It supported de Gaulle's leadership.

All three parties were committed to implementing the Resistance Charter. This resulted in the nationalisation of the four largest banks, the coal industry, the Renault car company, Air France and some other major businesses. A comprehensive social insurance scheme was approved.

In a referendum in October 1945, the people rejected the idea of reviving the Third Republic. A Constituent Assembly was elected to draw up a new constitution. De Gaulle became Prime Minister in an Assembly which he considered too left-wing in outlook. The Communists and Socialists were unwilling to support a constitution which would give strong powers to a President. They wanted a strong parliament to counter any possible attempt at dictatorship.

De Gaulle saw that his proposal for a strong Presidency in the constitution would be rejected. He resigned suddenly on 20 January 1946. The new constitution which gave strong powers to parliament was accepted by the people and the Fourth Republic was born.

The Fourth Republic

The Constitution

- Parliament consisted of two houses, a National Assembly and a weaker Council of the Republic.

- The President was elected for a period of seven years. He had very limited powers.
- The Prime Minister had to be able to command the confidence of the Assembly.
- To avoid instability, the constitution favoured the creation of large parties.

The Constitution was accepted by 9 million voters with 8 million voting against. More significant was the fact that over 8 million voters abstained. The Fourth Republic would have to work hard to overcome this apathy.

Weak Government

The Fourth Republic lasted from 1946 until 1958. More than twenty separate governments were formed during this period. These were mostly coalitions of the moderate right (MRP), the centre (the Radicals from the Third Republic) and the moderate left. The system of government became known as *immobilisme* or political immobilism. Governments could not take decisive action or introduce imaginative reforms, for fear of offending any one of the three coalition partners. They also had to guard against the Gaullists on the right and the Communists on the left.

The Gaullists

Charles de Gaulle hoped that his country would quickly recall him in its hour of need. He saw no hope of political stability under a weak constitution. He now wished to be above party politics. He felt that he needed the support of the broad masses of the people.

In 1947 he set up a *Rassemblement du Peuple Francais (Rally of the French People)* which became known as RPF. This was a conservative group to the right of French politics. By 1952 it was the largest group in Parliament. However, it was never large enough to form a single party government, or to bring de Gaulle back to power.

General Charles de Gaulle

De Gaulle continued to attack the Constitution and the Communists. His demand for a referendum on the Constitution was resisted by the Assembly. In 1953 he resigned once again from political life, declaring that only in a national crisis would he return.

The Communists

The Communists became the largest party in the Assembly after the election in 1946. They hoped that their leader, **Maurice Thorez**, would be elected Prime Minister of the Fourth Republic. They failed in their bid and were forced to accept the Socialist **Paul Ramadier** as Premier.

The Communists accepted positions in government but were soon in dispute with Ramadier. They disagreed with his low wages policy and supported anti-government strikes. The United States was not willing to give Marshall Aid to France if Communists continued to hold places in the coalition government. In 1947 the Communists were forced to resign and did not hold any government positions for over 30 years.

The Economy

The French government faced serious post-war economic problems. Limited nationalisation did not provide the hoped-for remedy. Inflation threatened to get completely out of hand and the government tried to overcome its problems by printing more bank notes. As a consequence, the franc lost one third of its value between 1943 and 1947.

The French economy was saved by the infusion of **Marshall Aid** from America. Another factor was the appointment of **Jean Monnet** as head of the Planning Commission. His formula of advance planning, limited nationalisation and strict credit control led to economic recovery, especially in French industry.

Jean Monnet and **Robert Schuman** were two French politicians who were not carried away by the paralysing fear of Germany. They believed that by tying Germany and France together economically and politically, rivalry would end and both countries would prosper. The main credit for the formation of the ECSC and later the European Economic Community (see chapter 24) must go to these two internationalists.

France's steel and electrical industries grew at an unprecedented rate and French cars, planes and railways gained international prestige. Agriculture did not share in the boom. The benefits of the Common Market did not flow until the fifth Republic had come into being.

Relations with the West

During World War II de Gaulle had been upset by the Allies, especially the United States. The French people were also wary of the Germans, having suffered two invasions in the 20th century. However, France soon saw that her place was in a

De Gaulle with the Soviet leader, Mr. Khruschev, who paid an official visit to France, 1960

western alliance against a more dangerous enemy, Soviet-dominated Communism. She signed the NATO charter in 1949 and in the same year agreed to the creation of West Germany. Later René Pléven's plan for a European Defence Community (EDC) failed, but France played its part in the reintegration of Germany into the Western military alliance (see chapter 25). As we have seen, France assumed the leading role in European economic integration through the ECSC (1952) and the E.E.C. (1957).

Indo-China

The twenty-five years after World War II was an era of decolonisation. Countries in Africa and Asia gained their independence, either peacefully or through revolution, from their former European masters. The French had occupied the colonies of Laos, **Cambodia** and an area constituting the modern **Vietnam**, in the late nineteenth century. After World War II, the leader of the Indo-China Communist Party, **Ho Chi Minh,** declared Vietnam independent. The French refused to accept the loss of her colony, or to negotiate. They sent a quarter of a million troops to oppose the Communist forces which were known as the *Viet Minh*.

A bitter war of independence led to the deaths of over 90,000 Frenchmen. The French could not overcome the guerrilla tactics of Ho Chi Minh's forces. The end came in 1954 at the Battle of **Dien Bien Phu**. The French surrendered after a nine-weeks siege.

By the Geneva Agreement of 1954, Laos and Cambodia became independent. Vietnam was divided into Communist North Vietnam and independent non-communist South Vietnam. In 1956 a communist group in South Vietnam called the *Viet Cong*, tried to seize power. The United States believed that the march of

Communism had to be halted. She poured troops into the area and became involved in a long and bloody conflict. The Americans, just like the French, could not defeat the guerrillas and finally withdrew in 1975 after suffering a humiliating defeat.

Algeria

The French were also anxious to retain some hold on their colonies in North Africa. Nationalist forces in **Tunisia, Morocco** and **Algeria** sought independence and engaged in acts of terrorism and guerrilla warfare against the French. The Prime Minister, **Pierre Mendes-France**, was in favour of granting some form of self-rule to the colonies. He could see no end to the loss of French lives and the demands on her resources.

The situation in Algeria was complicated by the attitude of the large white settler population. They knew that their ascendancy depended on maintaining the link with France. Some of these settlers, often known as **Colons**, felt that their very survival was at stake. Tunisia and Morocco were recognised as independent states in March 1956. Over a million Colons resided in Algeria. From 1954 some 350,000 French troops were engaged in a colonial war with **Ben Bella's Algerian Nation Liberation Front (FLN)**. The French army were involved in some very bloody incidents and were frequently accused of torture. The issue of granting independence to Algeria deeply divided public opinion in France.

Return of de Gaulle

The French army felt humiliated by the defeats in World War II and in Indo-China. They could only stand by and see independence conceded to Morocco and Tunisia. France was further discredited in 1956 by her forced withdrawal from Suez (see chapter 27). Now in 1958 the French Prime Minister, **Pflimin**, was ready to negotiate

A French army patrol in Oran, Algeria

with the rebels in Algeria. Some army officers decided that it was time to save France's honour and defend the colonists in Algeria.

On 13 May, 1958 some 40,000 French settlers in Algeria demonstrated against making any deal with the nationalists. The French army in Algeria revolted. Its commander, **General Salan,** declared that 'I have provisionally taken into my hands the destinies of French Algeria'.

The rebels captured the island of **Corsica** and there was a danger that they would invade France itself. They called on de Gaulle to assume control, feeling that he was the only man strong enough to secure the survival of French Algeria.

On 28 May 1958 **President Coty** asked de Gaulle to form a government. The latter had waited for this moment since 1946. Now he accepted, on condition that he would have the power to rule by decree for six months, and that he should draw up a new constitution. The Assembly believed that he was the only person capable of preventing either a military dictatorship or a civil war.

On 1 June 1958 Charles de Gaulle once again became Prime Minister. He drew up a new constitution which was accepted in September by almost 80 per cent of the voters. France now entered into the period of the Fifth Republic.

The Fifth Republic

The Constitution

The Constitution of the Fifth Republic, created by Charles de Gaulle, differed radically from those of the Third and Fourth Republics by strengthening greatly the powers of the President and reducing those of the Assembly. De Gaulle believed that this was necessary in order to avoid the mistakes of the past and provide France with strong and stable government. The President was elected by an electoral college of delegates from local councils, for a seven-year period and had far-reaching powers. The President –

- appointed the Prime Minister who in turn selected a cabinet
- could dissolve Parliament
- could rule by decree in times of emergency
- controlled foreign policy and defence
- could appeal directly to the people by referendum, in order to gain a mandate for his policies.

In December 1958 de Gaulle was elected for a seven-year term to the office of President. In 1962 he used a referendum to amend the Constitution and provide for direct election of the President by the people. In 1965 he was re-elected, this time by direct popular vote. The National Assembly, though its power was reduced, could force a government to resign. The electoral success of the new **Gaullist Party (UNR)** meant that de Gaulle's government did not have to worry unduly about opposition from Parliament.

Solution in Algeria

> I have understood you. I know what has happened here. I see what you have wanted to do. I see that the road you have opened in Algeria is that of renovation and of brotherhood.
>
> (Charles de Gaulle)

When Charles de Gaulle addressed the Algerian white settlers in June 1958, he seemed at first to show approval of their actions. However, it soon became clear that his idea of 'brotherhood' was equal rights for all in Algeria, Europeans and Muslims alike. He sought reconciliation. He proceeded with extreme caution, but acted decisively and firmly when the inevitable confrontation came.

In 1960 the Colons revolted against de Gaulle's plans. The President declared a state of emergency, purged the army of rebels and pursued vigorously a policy of Algerian independence. The **Secret Army Organisation (OAS)**, led by **Salan**, attempted a military coup in Algiers. In an address to the nation, de Gaulle won the support of the people and the rank-and-file soldiers. The coup collapsed.

The OAS adopted terrorist tactics including attempts on the President's life. In 1962 a cease-fire with the FLN ended the 16-year war. Over 90 per cent of the people of France supported the granting of independence to Algeria with equal rights for Colons and Algerians. About a million French colonists returned to France. The solution of the Algerian problem is regarded as de Gaulle's greatest achievement.

Leader of Europe

De Gaulle resented what he considered American domination of NATO. He was determined that France would become independent of the United States and Soviet power blocs. His goal was to make France leader of an independent Europe.

An important aspect of de Gaulle's dream of *grandeur* or greatness for his country was the development of independent nuclear weapons. France exploded her first atomic bomb in 1960. De Gaulle began to suggest that France was capable of defending herself. In 1966 French forces were withdrawn from NATO. The French refused to sign the U.S.-sponsored nuclear Test Ban Treaty in 1963. By 1968 France possessed the hydrogen bomb and a nuclear strike force. De Gaulle had developed French nuclear and military independence.

De Gaulle also vetoed Britain's entry into the E.E.C. He thought that her close links with the United States meant that Britain could never become truly European. Despite his commitment to the E.E.C., de Gaulle was not prepared to give up any of his country's national sovereignty to the organisation. He continued to develop France's friendship with West Germany, confident that the Germans were not yet in a position to challenge his country's position as leader of Europe.

"He says he wants to join – on his own terms . . ."
de Gaulle, Adenauer, Macmillan

De Gaulle travelled the world in order to enhance France's image as a world power. He visited Moscow in 1964 and sought close ties with Eastern European states. However, despite his independent stance, de Gaulle always held France firmly in the Western Bloc in times of crisis and in disputes between East and West.

The Economy

De Gaulle's Fifth Republic ushered in a period of political stability and economic prosperity. Upon taking office he got the French people to subscribe to a 'patriotic loan'. It enabled him to introduce cuts in public spending and stabilise the economy. The stage was set for real economic progress during the 1960s.

France under de Gaulle became a modern industrialised country. She was a leader in new industries such as electronics, chemicals and plastics. Advances in technology were a feature of the decade. French agriculture benefited from the Common Agricultural Policy of the E.E.C. with its guaranteed prices and subsidies to farmers. Economic prosperity led to the development of hospitals, houses and schools.

Decline in Popularity

There were signs in the 1965 election that de Gaulle's popularity was declining. Some felt that he had accumulated too much power and that his personality was too dominant. It was being said that he should be paying more attention to domestic issues than to French *grandeur* abroad. He was 75 in 1965 and younger people wondered if he had lost touch with their needs. In the election he did not achieve an overall majority in the first count.

Student riots, Paris 1968

Student Riots, 1968

> The voice we have just heard is that of dictatorship. The French people will silence that voice.

This was the reply by **Francois Mitterand**, leader of the Left, to de Gaulle's emotional radio broadcast on 30 May 1968. The President had said that France was threatened by an illegal communist dictatorship. He had dissolved the National Assembly and was 'sole custodian' of the state at this time of crisis.

The trouble in 1968 started with student sit-in strikes and demonstrations at the beginning of May 1968. They were critical of student accommodation, the relevance of academic courses and the impossible standards demanded in examinations. Trouble spread and riot police used tear gas against students who put up barricades and threw bricks, paving stones and Molotov cocktails.

By the third week in May, about 10 million workers had joined the students. They had their own grievances: poor state salaries, censorship, discrimination and centralisation. A massive strike threatened to cripple the country, as students and workers called for the President's resignation.

De Gaulle made his famous broadcast, called elections and appealed to the French people to support him against those who tried to bring anarchy to France. Pay settlements gained union support and the President's followers staged counter-demonstrations. The Gaullists gained an extra ninety seats in the elections. The voters had become frightened by the Communist party which, during the strikes, had pledged to 'eliminate the government'.

Resignation

De Gaulle had survived the crisis but his prestige was destroyed. He tried to recover lost ground by bringing in changes. However, it soon became apparent that the reforms were not going to address the most serious grievances. The President decided to put proposed changes in the Constitution to the people in a referendum. He said he would regard the result as a vote of confidence in his Presidency. Eleven million people voted for the changes but twelve million rejected them.

De Gaulle resigned immediately, in April 1969, and retired to write his memoirs. He died in November 1970. While he was 'respected rather than loved' he was honoured in death by all French people for his belief in the future of France in times of crisis and for his untiring devotion to her interests.

Chapter 27

Britain, 1945-1970

Labour in power, 1945-1951

The Beveridge Report

'The Report on Social Insurance and Allied Services... takes Freedom from Want as its aim... The plan for Social Security is designed to secure, by a comprehensive scheme of social insurance, that every individual, on condition of working while he can and contributing from his earnings, shall have an income sufficient for the healthy subsistence of himself and his family, an income to keep him above Want when for any reason he cannot work and earn.'

(Extract from Beveridge Report)

During the Second World War the British government was already planning for better standards of living after the war. In 1941 a committee was set up under **Sir William Beveridge** to identify causes of poverty and to co-ordinate welfare services. **The Beveridge Report** in 1942 identified five 'giant' problems. If these were solved every citizen would have care and security 'from the cradle to the grave':

- Want – to be solved by a complete system of insurance
- Disease – a new health service
- Ignorance – more and better schools
- Squalor – more and better houses
- Idleness – more jobs.

General Election, 1945

A general election was held in July 1945. The Labour Party was unwilling to continue cooperating in a national government which had originated in the crisis of 1940. The Conservatives relied heavily on Churchill's war record during the election campaign. However, the voters associated the Conservatives with the depression of the 1930s and the wartime shortages. Labour offered a programme of full employment and nationalisation. The Labour Party also seemed more enthusiastic about the Beveridge Report and promised a complete system of social security.

In the election Labour won a landslide victory, with 395 MPs out of 600. **Clement Attlee** became Prime Minister. He was quieter and less dominant than Churchill but proved very adept at working with the members of his Cabinet. **Herbert Morrison**, Deputy PM, was in charge of Home Affairs. **Ernest Bevin**, the leader of the trade union movement, was the new Foreign Secretary. **Hugh Dalton**, a well-known lawyer, was appointed as Chancellor of the Exchequer.

Attlee's Labour Government saw its main task as the creation of a welfare state. Its programme was based closely on the Beveridge Report.

National Insurance

> A new Britain, a state which takes over its citizens six months before they are born, providing care and free services for their birth, for their early years, their schooling, sickness, workless days, widowhood and retirement. Finally it helps defray the cost of their departure. All this, with free doctoring, dentistry and medicine, free bathchairs too, if needed, for 4s 11d (under 25p) out of your weekly pay packet.
>
> (*Daily Mail*, 5 July, 1948)

The National Insurance Act and **National Health Servic**e went into operation on that date. All workers except married women had to pay their weekly insurance. Workers were entitled to sickness and unemployment benefits and to retirement and widows' pensions. There was no Means Test. Everyone paid the same contributions and drew the same benefits. The **National Assistance Board** was set up to cover people not covered by these arrangements, people such as the handicapped, unmarried mothers, homeless people and dependents of prisoners.

Aneurin ('Nye') Bevan, Minister for Health, architect of the NHS

The National Health Service (NHS), 1948

Up to 1948 doctors charged fees for seeing patients and prescribing medicines. Many people could not afford these fees. It was also estimated that six million people needed glasses but could not pay for them.

Aneurin (Nye) Bevan, a Welshman of great forcefulness, was Minister for Health. He was a coalminer's son who had seen his father die from silicosis caused by coal dust. He was determined to tackle the 'giant' of disease. His idea was to provide everyone with free medical treatment and supplies, including dental care and glasses. This was to be paid for partly out of taxes and partly from National Insurance contributions.

The British Medical Association objected to the new arrangements. However, Bevan eventually got their agreement. It seemed as if people had saved up their illnesses for the first free day. People rushed to the surgeries and the NHS cost the country a staggering £400 million in the first year. In 1951 the government decided to introduce some charges in order to avoid waste and keep down the cost of the services. Aneurin Bevan resigned from the government in protest. The NHS was a great success and both infant mortality and death rates showed significant decreases.

Education and Housing

The Labour government decided to proceed with implementing the **Butler Education Act of 1944**. This Act had been brought in during the war by the Conservative Minister 'Rab' Butler. The school-leaving age was raised to fifteen. A special examination, the 'eleven plus', divided pupils into three categories for secondary education. The **Grammar Schools** were for those suited to an academic education, the **Technical Schools** for those interested in science and art, and the **Secondary Modern Schools** for those with more practical interests. A rift developed in British education between the different types of schools and during the 1950s and 1960s Labour encouraged the building of **Comprehensive Schools**.

Thousands of houses had been destroyed during the war and many cities contained slum areas. Bevan was in charge of the housing programme. The government succeeded in building nearly 200,000 houses a year, but even this failed to match the demand. Another feature of the period was the building of 'new towns' outside the old cities.

Nationalisation

Labour believed that nationalised industries would run better and that the profits they made would benefit the whole country instead of only the owners or shareholders. These industries were turned into public corporations which were run by boards of directors, with a chairman appointed by the government. The man responsible for the programme of nationalisation was **Herbert Morrison**, Leader of the House of Commons. He concentrated on key industries:

- 1946 – Bank of England nationalised
- 1947 – National Coal Board took over the Coal pits and collieries.
- 1948 – Transport nationalised, British Rail created, London buses and tube trains and long distance road haulage taken over
- 1948 – Electricity
- 1949 – Gas
- 1951 – Iron and Steel.

The Conservatives saved their hostility for the nationalisation of coal and steel. The other areas were public utilities and needed the type of investment that only the government could afford. However, iron and steel was a business which made profits. The Conservatives delayed the measure and returned steel to private ownership when they regained power.

Difficulties for Labour

Labour achieved much between 1945 and 1951. There was no post-war slump, exports grew and unemployment never rose above two per cent. Family incomes increased and the coming of the welfare state brought more comfort and security.

However, all was not well with Britain. The country was broke after the war. Economic disaster was staved off only by Marshall Aid from America. Rationing of foodstuffs continued and life was dreary. People could see no real benefits from nationalisation. The icy cold winter of 1946-47, made more severe by the lack of fuel, caused severe hardship. Despite the building programme, houses were still in short supply and thousands of people set up homes in empty army huts.

Labour fought the 1950 election with the slogan 'fair shares' for all. The party won the election but its majority was cut to six MPs. Taxes were increased to pay for defence spending. The Welfare programme was also proving costly. Bevan and Harold Wilson resigned from the Cabinet in protest against charges in the National Health Service. In 1951 Attlee decided to call another election. The Conservatives asked the electors to vote for freedom – from rationing, high taxation and government controls. They won a narrow overall majority.

Conservative Governments, 1951-1964

Churchill returns to power

Winston Churchill was seventy-seven when he became Prime Minister of a Tory government in 1951. Poor health meant that he left most of the running of the country to his Cabinet, especially the deputy PM, **Anthony Eden.** In office Churchill made it clear that the Welfare state set up by Labour would not be dismantled, because he knew that the measure was very popular with the British people. Only the iron and steel industry and most of the road haulage business returned to private ownership.

This did not mean that the Conservatives were converted to socialism. They believed that free enterprise and controlled capitalism would raise everyone's standard of living and that welfare would eventually become unnecessary. In the meantime they settled for a minimum of nationalisation, cutting out waste and a more efficient and cheaper running of the welfare state. Another result of the principle of free enterprise was the decision made in 1953 that the BBC's monopoly of broadcasting should no longer apply to television. Commercial television, financed by advertising, was to be set up.

'Stop-Go' Economy

The costs of the welfare state and the Korean War meant that the new Chancellor of the Exchequer, **Rab Butler**, inherited a balance of payments deficit of nearly £700 million. He deflated the economy by raising interest rates, restricting credit and reducing imports. At the end of 1952 he achieved a surplus of £300 million.

However, Butler's policies soon led to industrial stagnation. He reduced interest rates and allowed more credit. Spending on consumer goods increased, there was employment for all and the country enjoyed a boom. The old problems surfaced again and the Chancellor was faced with inflation, strikes and falling exports in 1954 and 1955. He reverted to a policy of deflation. Labour said that Butler had lost control and bitterly criticised what they called his **'stop-go'** economic policies.

Housing Success

In December 1952 the Housing Minister, **Harold Macmillan**, was able to tell Parliament that over 300,000 houses had been completed in the previous twelve months. Housing would have a priority second only to defence. Macmillan termed the government's achievement 'a triumph for free enterprise'.

The Eden Government, 1955-1957

Churchill was partly paralysed by a severe stroke in June 1953. He carried on until April 1955 when he handed over to his deputy, **Sir Anthony Eden**. The new Prime Minister immediately called a General Election and the Conservatives increased their majority.

Eden called for the establishment of a property-owning democracy. People were encouraged to buy their homes and become company shareholders. The new Chancellor of the Exchequer, Harold Macmillan, raised extra funds for the government through a Premium Bonds Scheme. The Eden years also saw a major increase in small savings, especially in building societies.

The Government was beset by strikes and a lack of success in its industrial policy. However, the major blow for Eden came at **Suez** (see section on Foreign Policy at the end of this chapter). His health had not been good and he was forced to resign because of 'severe overstrain' brought on by the tensions of the Suez crisis.

Harold Macmillan, 1957-1963

Harold Macmillan became Prime Minister in 1957 at sixty-two. He was a member of the well-known publishing firm. Educated at Eton, he had served as Minister for Housing, Minister for Defence and Chancellor of the Exchequer in successive Conservative governments. At first he supported Britain's aggressive policy at Suez, but later changed his mind. This led Harold Wilson of Labour to say that Macmillan was 'first in and first out' at Suez and benefited by becoming Prime Minister.

Economic Situation

> Let us be frank about it. Most of our people have never had it so good.
>
> (Harold Macmillan)

Macmillan could say this with justification to the British people in a speech in Bedford in 1957. His aim was to improve people's living conditions by bringing prosperity to as many people as possible. This was the new age of the motor car, household gadgets and television. The Prime Minister usually spoke in optimistic terms and was often referred to as 'Supermac' or 'the unflappable Mac'.

In his Bedford speech, Macmillan also sounded a note of caution. He warned about rising prices and said that inflation was pricing Britain out of world markets.

Harold Macmillan, Prime Minister, 1957-63

The 'stop-go' economic policies of deflation and reflation continued during the Macmillan years. It seemed that when the economy was reflated prices rose and the country suffered a balance of payments problem. However, deflation brought an end to growth and increased unemployment figures.

Long-term Planning

There seemed no solution to the underlying problems in Britain's post-war economy. She had lost competitiveness on world markets and had failed to modernise her industry. Macmillan favoured long-term economic and industrial planning:

- The **National Economic Development Council** made up of economists, employers and union leaders advised the government on the long-term development of British industry.
- **Lord Robens** was put in charge of the Coal Board. He closed down unproductive pits.
- **Dr Beeching** closed down 5,000 miles of loss-making railway lines.
- A massive programme for the development of roads and motorways was initiated.
- Technical colleges and universities were expanded and modernised.

Domestic problems

Immigrants from the Commonwealth were allowed unrestricted entry into Britain. Between 1955 and 1961 approximately half a million Commonwealth immigrants arrived in the country, mainly from India, Pakistan and the West Indies. The majority were very poor and found it difficult to integrate. They were not accepted by a section of British society. An **Immigrants Act** in 1962 placed restrictions on immigration. However, the coloured population of Britain continued to grow and racial tensions remained.

In the 1960s the boom began to come to an end. The Chancellor of the Exchequer, **Selwyn Lloyd**, introduced an unpopular pay freeze in 1960. The government continued to lose support. The usually affable Macmillan showed his teeth by sacking Lloyd and six other ministers in what was referred to as the British 'Night of the Long Knives'.

The government was also criticised for its handling of the Vassal, Burgess, Philby and Profumo scandals. **John Vassal**, a Foreign Officer clerk, was convicted of spying for the Soviet Union. **Kim Philby**, a prominent figure in British intelligence, was also a Soviet spy, but he escaped to Russia before he could be arrested. The War Minister, **John Profumo**, lied to Parliament about his involvement in a sex scandal. He was forced to resign. A report on the incident suggested that Macmillan had not dealt firmly with it. The Prime Minister resigned in October, 1963.

The new Prime Minister, **Sir Alec Douglas Home**, could not stop the decline of his party. Economic problems continued and the government was defeated in a general election in October 1964.

Two members of the 'spy-ring': Guy Burgess and Kim Philby

The Wilson Government, 1964-70

Labour in Opposition

Labour had remained in opposition from 1951 to 1964. Clement Attlee had been succeeded as leader of the party by **Hugh Gaitskell**. He was a moderate, middle-of-the-road politician, but he failed to unite the left and right wings of the party. The split surfaced on two main issues:

- **Nationalisation:** Gaitskell favoured a moderate degree of nationalisation. The left wing of the party wanted nationalisation on a broad scale.
- **Defence:** The row over defence centred on the question of whether Britain should have nuclear weapons. The anti-nuclear group was led by Aneurin Bevan. However, at the 1957 Party Conference, Bevan declared that he had changed his mind on the issue and had come to believe that Britain must have a nuclear deterrent. The question continued to divide the party for many years.

Gaitskell died suddenly in 1963. He was succeeded by **Harold Wilson**. The new leader was able to present a united front to the electors in 1964. Labour won the election, but with an overall majority of only four.

Harold Wilson

Harold Wilson graduated from Oxford University in 1937 and lectured in economics there for two years. During World War II he served as an economist for the government. He was elected to Parliament in 1945 and became the youngest cabinet minister in almost two centuries. He resigned in 1951 because of the imposition of health charges. This gained for him the trust of the left-wing. He became party leader in 1963.

In forming his cabinet the new Prime Minister aimed at a balance between the right and left wings of the party. Wilson's style of government was sometimes described as 'presidential'. Some thought that he felt intellectually superior to his ministers. Others argued that he was merely ensuring through strong leadership that the two wings of his party would work together. In 1968, when George Brown resigned as Foreign Secretary, he accused Wilson of running the cabinet in a dictatorial fashion.

The Economy

When Labour came into power in 1964, Wilson forecast a period of economic growth, to be fuelled by 'a white-hot technological revolution'. However the economic problems were acute and were not easily solved. The main problem was the continuing trade deficit with other countries.

Britain was importing much more food, raw materials and manufactured goods than she was exporting. The economy was being kept afloat by borrowings from other governments and foreign bankers. The government had to take a number of measures to satisfy the International Monetary Fund. Otherwise she would not have received loans and the country would have been in a state of bankruptcy. The principal measures were:

- A 15 per cent surcharge on all imports
- Income tax, national insurance contributions and petrol duty were increased
- Restrictions on hire purchase
- A sharp increase in the lending rate to 8 per cent
- Tax on capital gains
- The pound sterling was devalued by 14 per cent in 1967 from $2.80 to $2.40
- A Prices and Incomes Board to curb prices and control wages.

Coloured people, the young, and middle-aged people who have been made redundant wait in a dole queue.

Social Issues

The Labour government fulfilled its election promises by increasing old-age pensions and abolishing prescription charges. These measures helped the government's popularity and it increased its majority in the 1966 election to 96. Among the most important acts passed between 1964 and 1970 were the following:

- Rent Act, 1965, fixed fair rents for private tenants.
- Redundancy Payments Act, 1965, brought in payments for workers who lost their jobs.
- Trades Disputes Act, 1965, gave unions legal protection against employers who claimed strike damages.
- The Race Relations Board was set up in 1965. Three years later a Race Relations Act made discrimination on grounds of race illegal.
- An Ombudsman was appointed in 1967 to investigate cases where a private citizen claimed unfair treatment by a government department.
- The Death Penalty was abolished, 1969.
- In 1969 men and women were given the right to vote at 18.

Conservative Victory, 1970

In 1969 there were 3,000 strikes in Britain, many of them unofficial. The government issued a document *In Place of Strife* which advocated trade union reform. However, the cabinet was split on the matter and was not prepared to bring in sanctions. In the 1970 election the Conservatives, under **Edward Heath**, had an overall majority of 30.

Except for the return of a Labour government from 1974 to 1979, Britain was to be ruled by the Conservatives for over 20 years. **Margaret Thatcher** became the country's first woman Prime Minister in 1979. Her formula for remedy to Britain's ills was control of the unions, privatisation, lower taxes, a curb on welfare spending and more scope for private enterprise.

Decolonisation

Nationalism

> Empires as we have known them must become a thing of the past.
>
> (Ernest Bevin)

The British Foreign Secretary in the post-war Labour government had made his attitude to the colonial problem known during the war. The history of British foreign policy after 1945 is the history of withdrawal from her empire, often referred to as **decolonisation**.

The native peoples had come to realise that occupation of their territories by the

Europeans was not to their advantage. A new spirit of nationalism swept Asia and Africa. This was, ironically, fostered by the education and ideals which originated in Europe. Some of the new leaders had themselves been educated in Britain. The string of victories by the Japanese during World War II had demonstrated that the Europeans were no longer invincible. Ultimately the colonisers could see no long-term economic advantage to maintaining their colonies. Britain hoped that the countries which were granted independence would retain some links by becoming part of the British Commonwealth.

India and Pakistan

Since 1877, the British monarch had been Emperor or Empress of India (see chapter 5). The leader of the demand for Indian independence in the 20th century was **Mahatma ('Great Soul') Gandhi**. He had qualified as a lawyer in London and returned to India in 1915. He organised a campaign of passive resistance to British rule and called for a boycott of British goods.

During World War II Britain crushed an uprising by Indian rebels and banned the Indian Congress Party. In 1946 the Labour government announced that it would grant independence. **Lord Mountbatten**, the last Viceroy, attempted to organise a smooth transition to self-rule for India. However, he was unable to prevent major outbreaks of violence between **Hindus** and **Muslims**. It was decided to partition the country and to grant the Muslims a separate state of Pakistan. Both India and Pakistan remained in the Commonwealth.

Palestine

After World War I Britain had been given Palestine as a mandated territory. In 1917 Britain issued the **'Balfour Declaration'** which promised the establishment of a

Youths stoning a streetcar, Calcutta, 1953

'national home' for the Jews. Nazi persecution before and during World War II resulted in an exodus of Jews to Palestine. This led to tension and violence between Arabs and Jews. The British handed the problem over to the United Nations (see chapter 24). The Jews declared the establishment of the state of Israel in 1948.

Winds of change in Africa

> The most striking of the impressions I have formed since I left London a month ago is of the strength of this African national consciousness. The winds of change are blowing through the continent.
>
> (Harold Macmillan)

This famous speech by the British Prime Minister, Harold Macmillan, in South Africa in 1960, denoted his acceptance of the claims by former British colonies for independence. The following were granted self-rule, or broke away from the British Commonwealth:

> Ghana 1957, Nigeria 1960, Tanganyika and Zanzibar 1961, Uganda 1962, Kenya 1963, Nyasaland and Northern Rhodesia 1964, Bechuanaland 1966, Aden 1967.

Some other countries outside Africa also gained independence: Ceylon and Burma (1948), Malaysia (1957), Cyprus (1960), Malta (1964). The handover of power was not always peaceful. There was serious violence in Kenya, Cyprus and Aden. However, the majority of the new states chose to remain within the British Commonwealth.

Queen Elizabeth II with Commonwealth Prime Ministers, 1962

South Africa and Rhodesia

The British Government was strongly opposed to the South African Government's policy of *apartheid*. The majority of the white population resented British interference in their country. In a referendum in 1961, South Africa voted to leave the Commonwealth and set up a Republic.

Ian Smith, Prime Minister of Southern Rhodesia, advocated white supremacy and refused to accept British insistence on majority rule. He declared independence in 1965 and in 1969 white Rhodesians set up the **Republic of Zimbabwe.**

Foreign Policy

Britain and the Superpowers

During the war Churchill had begun to suspect that Stalin was using the defeat of Germany to spread Communism. He resisted the Soviet leader's demands at the conferences between the Allied leaders in Teheran, Yalta and Potsdam (see chapter 23).

Ernest Bevin was Foreign Secretary in the Labour Government from 1945 to 1951. Together with the Labour Leader, Clement Attlee, he represented Britain at the Second half of the Potsdam Conference, in July 1945. He adopted a policy of close co-operation with the United States in defence of Western Europe against the Soviet Bloc. Bevin came to abhor Stalin and the tactics he used to spread Communism in Eastern Europe. In the ensuing Cold War (see chapter 25), the Labour Party was as strong as the Conservatives in its total opposition to the Soviet Bloc. However,

Ernest Bevin, British Foreign Secretary, signs the NATO Pact, 1949.

Britain was broke after the war and in 1947 informed America that she was unable to fulfil her obligations in Greece and Turkey in the fight against Communism. This led to the Truman Doctrine and Marshall Aid (see chapter 25).

Britain was a founder member of NATO in 1949 and sent troops to help the south in the Korean War. Her opposition to the Soviet Union led to a massive and costly rearmament programme. Unlike France, Britain did not develop an independent nuclear arsenal, but co-operated with the United States. In 1963 Macmillan and John F Kennedy agreed that America would supply nuclear **Polaris** missiles which would be launched from British submarines. The siting of American nuclear missiles in bases in Britain was opposed vehemently by many British citizens.

After the Stalin era, Britain tended to act as middleman in disputes between the United States and the Soviet Union. Macmillan was a strong supporter of Kruschev's 'peaceful co-existence' policy (see chapter 28). He visited both Moscow and Washington and helped in the thawing of the Cold War hostilities.

The Suez Crisis, 1956

> We shall industrialise Egypt and compete with the West. We are marching from strength to strength.
>
> (Nasser)

In 1952 **Colonel Nasser** ousted King Farouk and set up a republican regime in Egypt. He was a strong Arab nationalist and very much opposed to the existence of Israel and to European imperialism. Nasser had been promised American help to finance the construction of the Aswan Dam on the Upper Nile. However, the Americans withheld the help when Nasser established links with the Soviet Union. This meant that the Suez problem became part of Cold War rivalry between East and West. The Egyptian leader nationalised the **Suez Canal** in which Britain and France had major shareholdings.

Invasion of Egypt

> A man with Colonel Nasser's record cannot be allowed to have his thumb at our windpipe.
>
> (Eden)

The British Prime Minister, **Sir Anthony Eden**, saw **Nasser** as a new Hitler who must be stopped by force. Britain, France and Israel plotted his destruction. The Israelis attacked Egypt in the confident expectation of British and French support. The Israelis advanced into the Sinai Peninsula. Britain and France issued an ultimatum: there must be an immediate ceasefire and an Anglo-French force would occupy the **Canal Zone** to keep the combatants apart.

Nasser rejected the ultimatum and Eden sent bombers to attack targets in Egypt. An Anglo-French force was sent in to 'protect the canal' and quickly captured **Port**

Anthony Eden, Prime Minister, broadcasts to the British people during the Suez Crisis, 1956.

Said. It was only a matter of time before they would capture the whole canal. However, they were severely rebuked by the United Nations. Bulganin, the Soviet Foreign Minister, threatened the use of 'modern and terrible weapons' in Britain.

The United States did not wish to alienate the Arab world and expressed strong opposition to the invasion. The UN called for a ceasefire and the Americans withdrew their support for the pound sterling. Eden was faced with hostility abroad and the prospect of devaluation at home. The resolution of the British and French governments quickly weakened. They called off the operation on the day after the capture of Port Said and quickly pulled out their troops.

The Suez episode was a disaster for Britain. She was humiliated and could no longer claim to be a world power. Suez diverted attention from the Soviet occupation of Hungary which occurred at the same time (see chapter 28). Eden had to bear the brunt of the criticism from home and abroad and resigned from his position as Prime Minister.

Britain and Europe

Britain played a leading part in the post-war settlements and in establishing the new state of West Germany. She was a member of the OEEC, which was responsible for the distribution of Marshall Aid. The British government also signed the statute of the Council of Europe but shied away from its aim of forging political union. This was the reason why Britain refused to become part of the European Defence Council. Neither did she become involved in the ECSC.

Britain was invited to become a founder member of the European Economic Community. However, many politicians were unwilling to surrender any part of

Anglo-French Concorde

British sovereignty to European control. The United Kingdom did not sign the Treaty of Rome in 1957. She attempted to increase her trade by becoming a founder member of the European Free Trade Association (EFTA).

In 1961 Britain applied for membership of the E.E.C. Her application was vetoed by President de Gaulle of France. He claimed that Britain was too dependent on the United States and sought unacceptable terms for the British Commonwealth. Britain became part of the European Community in 1973. However, the process of integration into Europe has proved long and painful for British politicians and people alike.

Chapter 28

Post-War Germany

The 'two Germanies'

A Country Destroyed

After the war Germany lay in ruins. Cities like Hamburg, Cologne, Dresden and Berlin had been devastated by bombs. The collapse of the transport system added to the gloom. Not a single major bridge across the Rhine remained intact.

Three million Germans lost their lives in the war, while many were disabled, hospitalised, or in prisoner-of-war camps. The eastern frontier of Germany was redrawn at the **Oder-Neisse Line**. This meant that East Prussia and most of Silesia passed to Poland in compensation for territory taken by the Soviet Union. In the next few years about 10 million refugees from the east, mainly from the lost territory, streamed into an already devastated Germany.

Many Germans suffered terribly after the war, especially during the 'hunger winter' of 1947. The main problem was the lack of fuel and food supplies. Survival techniques included bartering, hoarding, dealing on the black market and theft. There were reports of cannibalism in some areas. The unit of currency on the black market was the American cigarette.

The general population took little note of what was happening on the political front. The fate of Germany was in the hands of the victorious powers. They decided that the country should be de-Nazified, de-militarised and democratised.

War damage, Berlin 1945

De-Nazification: The Nuremberg Trials

> The Allied Powers will pursue them to the uttermost ends of the earth and deliver them to their accusers that justice may be done.
>
> (Churchill)

Thousands of Nazis were arrested after the war. On 24 November 1945 the trials of the leaders of the Nazi regime opened at **Nuremberg**, site of the huge Nazi rallies in the thirties. In the principal trial twenty-two leading Nazis were accused of crimes against peace, war crimes and crimes against humanity. Among those tried were **Göring**, the *Luftwaffe* chief, **Ribbentrop**, the Foreign Minister, **Seyss-Inquart**, the ruthless governor of Austria and **Streicher** who persecuted the Jews. For almost a year their terrible crimes were revealed to a shocked world.

On 1 October 1946 the International Military Court passed sentence. Twelve of the accused were condemned to death. **Rudolph Hess** was one of seven sentenced to imprisonment for long terms. The remaining three were acquitted. Göring committed suicide in his prison cell before he could be hanged.

The trials posed some awkward questions. The Germans were found guilty of crimes against peace, but the Soviets were not charged with attacking Poland or Finland. The defence were not allowed to refer to the Anglo-American destruction of German cities like Dresden. The treatment of German prisoners of war was not deemed relevant. The Military Court derived its authority from the United Nations, but the trials were conducted by the four occupying powers. Many Germans looked on the procedure as an act of revenge and felt that if they had been successful a completely different set of criminals would be in the dock.

De-Nazification: Other Aspects

Perhaps the chief value of the trials was to make the German people aware of the terrible things done under the Nazi regime. An attempt was made, especially by the Americans, to re-educate the German people. They insisted that notions of racial

Göring in the witness box, Nuremberg, April 1946.

superiority be removed from school textbooks. Former Nazis were not permitted to teach and the civil service was purged of the movement's sympathisers.

The German people, in general, tried to get on with their lives and to forget about the past. Even a main munitions manufacturer like Alfred Krupp was freed from prison after three years and reinstated. A former Nazi, **Kurt Kiesinger**, became Chancellor in 1969. The Germans could look at the accusations of guilt with some cynicism when they saw that the victors had helped some top Nazis to escape and enlisted their services. The 'Butcher of Lyons', **Klaus Barbie**, went to work for the American CIA. The Soviets were also prepared to use former Nazis in the Eastern Bloc, provided they obeyed orders.

Occupied Germany

> Germany is part of Europe and recovery in Europe will be slow indeed if Germany with her great resources of coal and iron is turned into a poorhouse.
>
> (American Secretary of State, James Byrnes, September 1946)

After the war, Germany was divided into four separate zones of occupation. Britain, France, the Soviet Union and the United States occupied one zone each. The Allies at first wanted to keep Germany weak because they felt that a weak Germany would ensure a strong and safe Europe. The Soviets immediately began to compensate for

GERMANY AFTER THE SECOND WORLD WAR

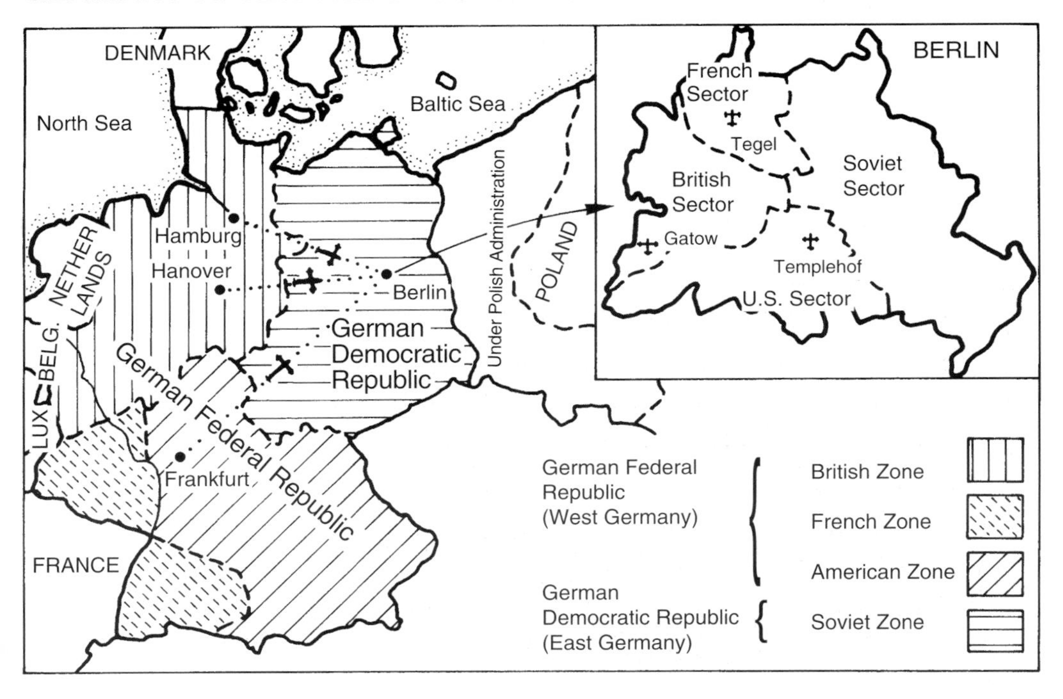

war damage by stripping their zone of machinery, plant and manufactured goods. The French felt very bitter about a second attack in 20 years on their country and also took some equipment.

Differences soon emerged between the Allies and the settlement of Germany was a focal point of the Cold War (see chapter 25). By 1946 the Americans saw the need for a strong Germany as a bulwark against Communism. In January 1947 Britain and the USA combined their occupation zones into one unit, known as **Bizonia.** The French added their zone in June 1948. America granted Marshall Aid (see chapter 25) to these zones and to West Berlin. They decided to stabilise economic conditions by introducing a new currency, the Deutschmark. In response the Soviets blockaded Berlin which was then supplied from the air by the Allies (see chapter 25).

As a result of the Cold War and the Berlin Crisis, the division of Germany lasted for over 40 years. In May 1949 the three western zones combined to form the **Federal Republic of Germany** [FRG]. This became known as **West Germany**. Four months later the Soviets launched their zone as the **Germany Democratic Republic** [GDR] and it was popularly called **East Germany**. The world had to become accustomed to the 'two Germanies'.

The Adenauer years, 1949-1963

The Constitution

In rebuilding Germany, we want to learn from the mistakes of the past.

(Adenauer)

The constitution of West Germany, adopted in 1949, was called the **Basic Law**. This is because it was meant to serve until the country would be reunited. Like the Weimar Republic the new state was founded on liberal democratic principles. However, this time the constitution was much tighter:

- Parliament consisted of two houses. The ***Bundestag*** was elected by all citizens. The ***Bundesrat*** was made up of members of the *Länder* or regional governments.
- A true federal state was created in which the *Länder* were given considerable powers over all areas except defence, foreign affairs and currency control.
- The power of the President was sharply reduced. The role of the holder of the office was not unlike that of the British monarch or Irish President.
- Proportional representation was retained, but a party with less than 5 per cent of the total vote could not enter the *Reichstag*.
- A constitutional court was established to guarantee civil and political liberties and anti-democratic parties were outlawed.

Konrad Adenauer: Profile

Konrad Adenauer was born in **Cologne** in 1876. He joined the **Catholic Centre Party** and served as Mayor of Cologne for twenty years from 1917 until 1937. He opposed the Nazis and was dismissed by them and imprisoned. After the war Adenauer was again appointed Mayor of Cologne. He proved too independent-minded for the British and they dismissed him, alleging that he was inefficient.

In the 1949 elections the **Christian Democratic Union** [CDU] emerged as the largest party. The party sought to represent Catholics, Protestants and Christian Socialists. The second largest party was the **Social Democratic Party** [SPD] led by **Kurt Schumacher,** a socialist who had spent ten years in a concentration camp. Although his party, the CDU, did not have an overall majority, Adenauer was elected Chancellor. Aged 73, and with no experience of national politics, he had all the signs of a temporary leader.

Chancellor-Democracy

Adenauer was often accused of being too autocratic and his period of government has been dubbed 'Chancellor-Democracy'. He took little account of political opposition. Independent-minded ministers were forced to resign while the Chancellor retained the self-confidence of a Churchill or a de Gaulle.

Adenauer was also accused of being 'soft on Nazism', too conservative and too rigidly anti-Communist. He remained in power for fourteen years. During this time he appeared indispensable at home and became a figure of substance abroad. Much of his success was due to the transformation of the economy.

Konrad Adenauer, Chancellor of West Germany, 1949-1963

The Economic Miracle

Ludwig Erhard served as economics minister from 1949 until 1963. The German economy recovered and boomed to such an extent that the recovery was termed an 'economic miracle'. There were a number of reasons for the recovery:

- The currency reform and introduction of the Deutschmark (DM) brought about economic stability.
- the FRG (West Germany) benefited from the Marshall Plan to the tune of 3 billion DM.
- A boom resulted from the Korean War (1950-53), leading to a sharp increase in demand for German goods.
- A restrained wage policy was practised by the West German trade unions.
- The disciplined hard work of the bulk of the population gave the country an advantage.
- The fact that a fresh start had to be made meant that industry was re-launched with the most modern equipment.

In the 1950s the world came to experience the ***Wirtschaftswunder,*** the German 'economic miracle'. It was characterised by high annual growth rates of about 8 per cent, reduction in unemployment and a steady increase in the standard of living. Erhard believed in a market economy. There was little government interference and the economy was free to respond to market demands. The imbalances created by a totally free capitalist system were corrected by a comprehensive social welfare policy. Housing, education, health services and old-age pensions were all promoted.

'Economic Miracle': Volkswagens, Frankfurt, 1952

'Made in Germany' became the trademark of the quality that people now came to expect from the country's products. Volkswagen, Siemens, Mercedes-Benz, Bayer and Krupp became worldwide symbols of the German boom. Cooperation with France and other European countries (see chapter 24) opened up export opportunities. Unemployment disappeared and many immigrants from Eastern Europe, Italy and Turkey flooded into West Germany. These usually filled the unskilled and menial jobs. The social and cultural implications of the arrival of these 'guest workers' was not generally recognised until later.

Foreign Policy: Rearmament

In 1952 the *Bundestag* voted in favour of setting up its own armed forces. The Cold War convinced the Allies that West Germany would have to be allowed rearm, so that she would not be vulnerable to attack from the East and could act as a bulwark against Communism. France was understandably wary of Adenauer's call for an armed Germany and the idea of a **European Defence Community** came to nothing.

The problem of French fears was solved by the acceptance of West Germany's rearmament within NATO. The **Western European Union** was set up to supervise this rearmament. A **Volunteer Act** was passed in Bonn in 1955 and provided for the creation of a Federal army (*Bundeswehr*), air-force (*Luftwaffe*) and navy. These forces were staffed by volunteers only and even by the late 1960s did not exceed half a million. The Germans were treading warily and were determined not to cause anxiety to other European states. The Germans were more interested in economic co-operation and in gaining the trust of their neighbours.

Foreign Policy: Cooperation with the West

Germany's role in European economic cooperation has been examined in detail in chapter 24. Adenauer acted as Foreign Minister from 1951 to 1955 and dominated foreign policy up to 1963. His anti-communist stance meant that he adopted a pro-Western-Europe policy. His role was at least as important as Schuman, Monnet and Spaak in the creation of European institutions. We have seen how he guided the Federal Republic into the **Council of Europe** (1950), the **ECSC** (1951) and the **European Economic Community** (1957) – see chapter 24.

Adenauer has been accused of contributing to the division of Germany. He was not prepared to agree to the idea of German unification if it meant that the country had to become neutral. He accepted the policy of **Walter Hallstein**, one of his foreign affairs advisers. By the **Hallstein Doctrine** the federal Republic claimed to be the legitimate voice of all Germany as it was the only part of the country properly and democratically constituted. Adenauer refused to maintain diplomatic relations with any countries that recognised the German Democratic Republic.

Adenauer was successful in winning back the coal-rich Saarland from France after a referendum in 1957. He developed good relations with Charles de Gaulle and the two countries signed the **Treaty of Franco-German Co-operation** in 1963. Later that

A German poster, 1950, asks for a 'Free pass for the Marshall Plan' throughout the whole of the country.

year he resigned, aged 87, from his position as Chancellor. He may have been very conservative, totally intolerant of Communism and somewhat autocratic. He may have given up power reluctantly. However his country owed him a great debt and his contribution to Europe was enormous:

> Without him no Coal or Steel Pool, no Common Market and no Euratom. Without him the dream of a United Europe would not have become a reality.
>
> (Paul Henri Spaak, Belgian Prime Minister)

The Federal Republic since 1963

The Erhard Years

Adenauer was succeeded as Chancellor by **Ludwig Erhard** in October 1963. In view of Erhard's past success his relative failure as leader was somewhat surprising. He did not exercise the same firm control of his Cabinet as his predecessor. He also suffered from the constant criticism of Adenauer who remained on as a member of the *Bundestag* and chairman of the CDU. The new Chancellor also lacked political experience in foreign affairs.

Erhard did not build on the relationship forged with de Gaulle. Instead he sought more friendly relations with the USA. He helped to finance American forces which were stationed in West Germany under the terms of the NATO Agreement. De Gaulle criticised him for becoming too dependent on the Americans. Erhard adopted a far less hostile approach than Adenauer to East Germany and was accused of being too soft on communism.

The new Chancellor's coming to power coincided with the slow-down in the rate of growth of the German economy. Cheap oil led to the loss of thousands of miners' jobs in the Ruhr. By 1966 the country was in recession. In order to meet welfare benefits Erhard was forced to increase taxes. He lost support in parliament and resigned.

The Grand Coalition, 1966-69

A **'Grand Coalition'** was formed in 1969 between the Christian Democratic Union (CDU) and the Social Democratic Party (SPD). The Chancellor was **Kurt Kiesinger** of the CDU. The 'Grand Coalition' was a successful combination. The economy improved under the Finance Minister, **Franz Joseph Strauss**. The unemployment situation was cured and a fresh influx of 'guest workers' from Turkey, Yugoslavia and other countries entered West Germany. By 1968 industrial growth was back to 6 per cent and inflation was under control.

The SPD played a prominent role in the coalition. This was the first time that the party had returned to government since 1930. It had rid itself of its tarnished image by rejecting its roots in Marxism and class conflict. The leader of the SPD, **Willy Brandt**, accepted the post of Foreign Minister. His name as Foreign Minister and later as Chancellor is identified with the policy of ***Ostpolitik***.

Ostpolitik

> I would be happy if they found that I had done something to make my country a good neighbour in Europe.
>
> (Willy Brandt)

These sentiments expressed by Willy Brandt sum up the principal aims of his political life. Brandt was a former Resistance journalist and later Mayor of Berlin.

Ostpolitik means policy towards the East and its purpose was to establish normal relations with the whole of Eastern Europe. Brandt, as Foreign Minister, established diplomatic relations with Romania and Yugoslavia in 1967. Treaties were signed with Poland and the Soviet Union in 1970 recognising the Oder-Neisse Line, no longer clinging to Adenauer's demand that lands east of the line should be returned to Germany.

Brandt's *Ostpolitik* also found expression in his historic visit to Warsaw in 1970. There he acknowledged Germany's guilt for the death of Polish Jews. In 1971 Willy Brandt received the Nobel Peace Prize. In 1972 a Basic Treaty was signed between the two Germanies which at last **normalised relations** between them.

Extra-Parliamentary Protest

A feature of modern life in Germany has been the growth of protest groups. This first came to the fore with the 1968 left-wing student revolt under **Rudi Dutschke**. This began as a protest against unsatisfactory conditions in the universities, but also reflected wider discontent with society and with the new materialistic culture.

Extra-parliamentary opposition in the 1970s led to a spate of kidnapping and bank robberies by the **Baader-Meinhof** gang. This was quelled by the government but since then the **Red Army Faction** has become totally ruthless and has assassinated a number of public figures.

Another aspect of life and politics in Germany has been the growth of the extremely right-wing **National Democratic Party**. This neo-Nazi group has gained support by calling for a stronger government under a strong president, a turning away from guilt and from concern with the recent past, a return to Germany's 'traditional' values and a strong army to deter possible enemies. In recent times these 'enemies' have come to include over 4 million foreign workers, some of whom have been subjected to vicious attacks.

The German Democratic Republic: East Germany

A Stalinist State

The **German Democratic Republic** which had been born in 1949 quickly accepted a centralised system of government on the Soviet model. The German Socialist Party was forced to unite with the Communists to form the **Socialist Unity Party** (SED) under **Walter Ulbricht**.

Walter Ulbricht, East German Leader, 1950-1971

Elections were still held, but a single list of candidates was drawn up and voters had to accept it *en bloc*. The only method of protest was to reject the list outright and this had to be done in public, a dangerous and futile gesture. Police, education and the management of the economy were all under SED control.

Real power in East Germany lay in the hands of Walter Ulbricht who was General Secretary of the SED from 1950 until 1971. He was a hard-line Stalinist who had fled from Germany when Hitler banned the Communist Party. He lived in the Soviet Union from 1933 until 1945 and was a strong supporter of Stalin.

Ulbricht purged the SED of those who opposed his centralisation and hard-line policies. Parliament met for only a few days each year and then only to rubber-stamp the decisions of Ulbricht and his Politburo. Press, radio and cinema were strictly censored and artists could only produce works of 'socialist realism'.

Economic Recovery

East Germany was in an economic mess when Ulbricht took over, as it had been exploited by the Soviet Union. The 1950s saw an economic recovery which in its own way was as extraordinary as that of West Germany. Ulbricht's planned economy worked along Soviet lines with collectivisation, nationalisation and a concentration on heavy industry the main features. Production of iron and steel increased dramatically and the shipbuilding industry flourished.

At the 1958 Party Congress Ulbricht predicted that within a few years the country would have overtaken its neighbour in terms of per capita income. This, he said, would prove the superiority of the communist system. This was a rash promise, but East Germany gradually closed the gap. In 1963 a more flexible economic policy was introduced, with bonus systems and a recognition of the profit motive. In the long term the East Germans showed that they had the same abilities and will to work as their countrymen in West Germany. They were rewarded with a dramatic increase in the standard of living.

Foreign Policy: Stalin's Attitude

> Communism fits Germany as a saddle fits a cow.
>
> (Stalin)

Stalin must have had his doubts about the suitability of Communism for the German people. He was never really keen on a separate GDR, which after all was not vital for his defensive barrier. In 1952 he offered to establish a unified and neutral Germany on a democratic basis. The West had no trust in Stalin's word at this stage and rejected his offer. Stalin proceeded to tighten his grip on East Germany.

Revolt in 1953

In 1953 trouble started in Berlin when the government demanded extra productivity from the workers. They went on strike, demonstrated through the city and began to

'DO NOT FEAR, HE IS ON A CHAIN.'
Russian cartoon, 1955, after Germany joined NATO

demand free elections and political reforms. The protests turned violent and the government called in Soviet troops and tanks to restore order. This was not achieved without bloodshed and loss of life. The disturbances were significant because this was the first of a number of occasions when Soviet forces intervened in a satellite state.

Key Member of Eastern Bloc

In 1954 reparations to the Soviet Union were ended and the country began to develop as a key member of the Eastern Bloc. East Germany was a founder member of the **Warsaw Pact** (see chapter 25). Despite economic progress at home, people were attracted by the freedom and the higher rewards on offer in West Germany and a flood of emigration began. This led to the construction of the **Berlin Wall** in 1961 (see chapter 25). The East Germans called this the 'Anti-Fascist Protection Wall'.

Relations between East and West Germany were blighted by the wall and by Adenauer's refusal to recognise the Democratic Republic as a legitimate state. However, matters improved after 1966 under Kiesinger and especially after 1969 under Willy Brandt's *Ostpolitik*.

A consequence of *Ostpolitik* was the fall of Ulbricht in 1971. Brandt recognised the frontiers and separate existence of the Democratic Republic. West Germans were allowed to visit family and friends over the border. Ulbricht wanted to drive a harder bargain than the Soviets, who were in any case tired of his constant lectures on ideological purity. He was replaced by **Erich Honecker**, a comparative liberal.

Modern Germany

West Germany

Brandt fell from power in 1974, ironically when one of his closest aides was unmasked as an East German spy. He was succeeded by **Helmut Schmidt** who gave the country stability and overcame many problems. One of these was unemployment caused by the huge rise in world oil prices in 1974. Schmidt dealt firmly with the Baader Meinhof gang and the Red Army Faction. During the 70s concern about the environment, especially pollution of the beloved forests, led to the growth of the Green Party.

Helmut Kohl became Chancellor in 1982. He made further strides in *Ostpolitik* which had become low-key since Brandt's downfall. Massive interest-free loans were given to the GDR. West Germany's economy continued to grow and the country has become the strongest power in the European Community.

East Germany

East Germany got a new constitution in 1974. It stated that the country was 'forever and irrevocably' allied with the USSR. The constitution termed the Democratic Republic a 'developed socialist society' and seemed to allow for inequalities and divisions among the people. A 'social contract' curbed the power of the police and meant that only active dissidents were persecuted.

Sports, especially those with a high profile in the Olympic Games, were increasingly used as a chief means of gaining international prestige. Economic performance became even more impressive and East Germany developed the strongest economy in the Soviet Bloc. Honecker set the seal of legitimacy on the GDR by paying an official visit to the Federal Republic in 1987.

Unification

Despite these advances, citizens in East Germany remained uneasy. The average worker still earned only half the salary of his West German neighbour. People continued to resent lack of freedom. Honecker opposed the reforms of **Mikhail Gorbachev** in the Soviet Union. The East German leader declared that the Berlin Wall would stand for another hundred years if necessary 'to protect our Republic from robbers'.

In 1989 opposition groups began to emerge in East Germany. The country could no longer remain isolated from events in other Eastern Bloc countries (see chapter 29). Parades in Berlin were brutally suppressed by the police. Other demonstrations followed in Leipzig and Dresden and unrest spread across the country.

On 18 October 1989, Honecker was replaced by **Egon Krenz**. However the people were no longer satisfied with the old regime and Krenz gave way to the liberal **Hans Modrow**. The Berlin Wall was opened up on 9 November amid great excitement and celebrations. The two Germanies, separated for so long, rediscovered each other.

Free elections were promised for May 1990. On 19 December 1989 **Helmut Kohl** visited Dresden and declared that his goal was a united Germany. An agreement on economic union came into effect and **Germany was reunited on 3 October 1990.**

Chapter 29

The Soviet Union and the Eastern Bloc after 1945

Stalin's Final Years, 1945-53

War Damage

The Soviet Union emerged victorious in 1945. The Great Patriotic War, as it was called, had strengthened Stalin's position in the Soviet Union as the country had united around its leader to repel the Nazi invasion. It came out of the struggle with the largest force in the world, the Red Army. Stalin depicted the war as a victory for Communism. However, difficulties abounded and in the years ahead it would not be easy to prove that the Soviet system was superior to capitalism.

The cost of the war in terms of human life had been enormous. The real figures may never be known, but it is estimated that the Soviet Union lost up to 20 million of her citizens. The retreating Nazi armies had caused wholesale destruction of property and of agricultural land. The rebuilding of the Soviet economy, along the lines of collectivisation and state planning carried over from the 1930's, was Stalin's first priority after 1945.

Agriculture

The economy was in difficulties after the war. Crops and animals had been destroyed and the country faced food shortages. Stalin decided to forge ahead with his pre-war policy of **collectivisation.** The emphasis now was on larger and in theory more efficient collective farms. The quarter million collectives were merged into bigger units so that by 1953 there were less than 100,000 of them in the Soviet Union.

Stalin called for increased production. The figures for grain increased but meat and dairy goods remained low. Weather conditions in the immediate post-war years led to poor harvests. Stalin blamed the stubborn attitudes of the peasants. The standard of living in rural areas was miserable and many peasants moved to the cities.

Industry

In 1946 Stalin announced a fourth **Five-Year Plan**. The emphasis was again on heavy industry. The Soviet leader linked industry with national security. The key was to be increased production of coal and steel. In these areas the Russians were immensely successful and targets were met. Huge power stations were built, canals were constructed and railways were extended into Siberia and other areas. A feature of the industrial boom was the creation of new towns. The emphasis on heavy industry meant that consumer goods were in short supply or unobtainable. One had to endure long queues in order to buy shoes, clothes and basic household items. The obsession with quantity and the need to meet targets often led to the production of shoddy goods.

The lack of consumer items was in stark contrast to the build-up of military equipment. Conventional weapons and tanks were produced in quantity, but Stalin felt that more was needed so that Russia could match the United States. The foundations of the country's nuclear industry were laid between 1945 and 1950. Scientists were instructed to work on atomic energy and the Soviet Union was able to test her own atomic bomb in 1949.

Cult of Personality

> Stalin is a brilliant leader and teacher of the Party, the great strategist of the Socialist Revolution, military commander and guide of the Soviet State... With the name of Stalin in their hearts, the collective farmers toil devotedly in the fields to supply the Red Army and the cities with food and industry with raw materials. Stalin's name is cherished by the boys and girls of the Socialist land.

Joseph Stalin, who retained power in the Soviet Union until his death in 1953

This is an excerpt from the official biography of Stalin. Its writing by Russian historians was closely supervised by Stalin himself.

Stalin's portrait hung in offices, shops, factories and schools. Most towns had their statues of the leader in bronze or stone and many places bore his name: Stalingrad, Stalinsk, Stalinabad and a host of others. This self-directed cult of Stalin's personality masked an insecure and suspicious leader. In the post-war era he became very isolated from his people, preferring to remain locked in the Kremlin.

Life in Russia

Political life in the Soviet Union between 1939 and 1953 was controlled by Stalin. During that time the Congress of delegates from the Communist Party was never called. Important decisions were taken by Stalin rather than by the Politburo.

The Secret Police under **Beria** carried out Stalin's orders. The persecution of Jews became very severe. Stalin also set out to destroy the culture and nationalist aspirations of non-Russian people. Those who resisted were deported, usually to Siberia. Over half a million people were deported from the Baltic States of Latvia, Lithuania and Estonia which were re-integrated into the Soviet Union after the war.

In 1946 **Zhadnov,** Secretary of the Leningrad Party, began a campaign to force all Soviet writers and other artists to conform exactly to Party thinking. He had Stalin's full support. Artists had to forget about personal feelings and devote their energies and art to the promotion of State, Party and Leader. The censors ensured that no works that did not conform to 'socialist realism' were released. Zhadnov himself compared **Shostakovich's** music to a dentist's drill and advised a lesson in 'communist music'.

Foreign Policy: Cold War

> Mr Churchill now takes the stand of the warmongers, and he is not alone. He has friends not only in Britain but in the U.S. as well. The following circumstances should not be forgotten. The Germans made their invasion of the U.S.S.R. through Finland, Poland, Romania, Bulgaria and Hungary. Governments hostile to the Soviet Union existed in these countries.
>
> As a result of the German invasion the Soviet Union's loss of life has been several times greater than that of Britain and the United States put together. How can anyone, who has not taken leave of his senses, describe the peaceful aspirations of the Soviet Union in Eastern Europe as expansionist tendencies on the part of our state?
>
> (Stalin, 1946)

This was Stalin's reply to Churchill's 'Iron Curtain' speech (see chapter 25). In it he justifies his actions in Eastern Europe and explains the basis of his foreign policy in the post-war era. Stalin's part in the Cold War is examined in detail in chapter 25. His role in Eastern Europe and the emergence of a communist bloc merits further consideration.

Stalin's Motives

Stalin had come to power on a policy of 'socialism in one country'. Now he aimed to create a barrier of 'friendly nations' around the Soviet Union. The leader of a country which had been invaded twice in less than 25 years could certainly pursue this policy with justification. However 'friendly' now meant 'communist' in Stalin's mind. Even this was not enough. These states had to come under Soviet domination and within a few years they were mere satellites of the Soviet Union.

The West explained Stalin's actions as Soviet imperialism and the beginning of world domination by the communists. This was certainly in line with Marxist teachings. However it seems that Stalin was influenced by the hostility of Western leaders, especially Churchill and Truman. The American President's language left little room for compromise:

> Unless Russia is faced with an iron fist and strong language, another war is in the making. Only one language do they understand – 'how many divisions have you?'

Stalin was also aware that the communists would lose out in free elections. He knew that without force the countries of Eastern Europe would fall under the influence of the capitalist West. The Soviet leader was prepared to use all means necessary to create and maintain his barrier.

Political Strategy

At the meeting in Yalta in February 1945, Stalin won from Roosevelt and Churchill the recognition of a Soviet zone of influence in Eastern Europe. He interpreted this as giving him a 'free hand' in the region. His strategy was repeated in various countries and a definite pattern emerged:

- Coalition governments which included communists were at first established.
- Communists insisted on holding the key post of Minister of the Interior. This allowed them to control the police and to intimidate their opponents.
- Communists soon took over all important offices of state.
- Other parties were eventually excluded and the country came under the influence of Moscow.

Economic Strategy

Stalin saw Eastern Europe as a source of economic power for his country. East Germany, Hungary and Romania had to pay reparations to the Soviet Union after World War II. Other East European countries had to trade with Russia on unfavourable terms. However, Stalin soon realised that he could not impoverish the satellite states if they were to stand up to the capitalists. The situation became more critical when Marshall Aid was offered to Eastern Europe. Stalin rejected the offer on behalf of the satellite states but in 1949 founded **COMECON** (Council for Mutual Economic Assistance).

This body brought the economies of the Eastern Bloc under central planning from Moscow. **Collectivisation of land** and **nationalisation of industry** were introduced. The economies of these countries improved, but workers often resented loss of rights and a drop in living standards. There was resistance by many peasants to collectivisation and it was not implemented in all areas, especially in Poland. A feature of the industrialisation of Eastern Europe was the planners' lack of concern for the environment.

Political Cooperation

In 1947 Stalin established **COMINFORM,** the Communist Information Bureau. This was a powerful propaganda weapon and ensured that Eastern Europe spoke with one voice. The organisation also coordinated the work of Communist parties in Europe.

In 1955 the Warsaw Pact was set up as the communist answer to NATO. In practice none of the signatories had to come to another's aid until 1968. However the Pact showed Moscow's determination to build on Stalin's work.

Communist Domination

- In **Czechoslovakia**, a Coalition Government came to power in 1946. It was led by the communist, **Klement Gottwald**. In 1947 Stalin forced the government to refuse Marshall Aid. He became worried about the drift to democracy in Czechoslovakia and in 1948 sent in the Red Army. A communist Government was formed and President Benes, a non-communist, had no option but to resign.
- King Michael of **Romania** was forced into exile in 1947. A Soviet army of occupation intimidated non-communists, especially the Peasant Party. A communist government was formed in 1948.
- A coalition government dominated by communists controlled **Poland** until all opposition parties were banned in 1948. The Peasant Party, supported by those living on or attached to the land and by the Catholic Church, was destroyed. The Socialist Party was forced to merge with the communists. However, discontent was rife and many Poles resented Soviet domination.
- The formation, in 1949, of the **German Democratic Republic** under **Walter Ulbricht** is examined in chapter 28.
- A left-wing Coalition Government in **Hungary** was dominated by the non-Communist Smallholders Party. It introduced reforms which angered Moscow and in 1949 was replaced by a hardline communist government.

Marshal Tito, leader of post-war Yugoslavia

Yugoslavia: A Special Case

I have only to shake my finger and Tito will fall.

(Stalin)

In 1944 the Resistance leader, **Josip Broz (Marshal Tito),** led his forces into Belgrade and went on to liberate Yugoslavia, without assistance from the Red Army. Tito set up a communist government, but acted independently of Moscow. He accepted over $100 million of Marshall Aid and was expelled from COMECON.

Tito went on to establish what he called a Socialist market economy. Competition was allowed between factories and the profit motive was not spurned. Tito showed that there was another road to Communism and was not forgiven by Stalin.

The leadership struggle

Russian Domestic Policy (397 - 405)

Death of Stalin

The Central Committee of the Communist Party of the Soviet Union... notify the misfortune that has overtaken our party and our people, the serious illness of Comrade J.V. Stalin. In the night of March 1-2, while in his Moscow apartment, Comrade Stalin suffered a cerebral haemorrhage affecting the vital areas of his brain... lost consciousness... a serious disturbance in the functioning of the heart and breathing. The best medical brains have been summoned for Comrade Stalin's treatment.

(Moscow Radio, 4 March, 1953)

The Soviet leader died on the following day. The circumstances of his death have remained shrouded in mystery ever since. In January 1953 Stalin had announced the existence of a doctor's plot to kill him. He blamed international Jewish groups. It seemed that this was the signal for another round of purges. Those closest to him on the Politburo feared for their lives. The general theory has been that Beria, Chief of Stalin's Secret Police, or perhaps **Nikita Khruschev**, arranged to have Stalin poisoned. Latest research suggests that he simply suffered a stroke:

> They all knew Stalin was crazy. Khruschev said he would have got them all in the end. They decided to just let him lie there on the floor, hoping he would not recover.
>
> (*The Sunday Times*, 7 November, 1993)

Stalin: Assessment

In twenty-five years Stalin moved the Soviet Union forward from the age of the wooden plough to the atomic era. It has been said that he forced the Russians to do in a lifetime what had taken two hundred years of misery and oppression in other countries. His economic achievements and his victory in World War II made him a world statesman and transformed his country into a major Superpower.

Stalin's achievements can hardly excuse his cruelty and ruthlessness. Strong government was needed but not mass murder. One of the saddest aspects of his rule was that the worst excesses of his terror were practised on his own people. He left behind a country paralysed by fear, dullness and suspicion.

Collective leadership

> What will happen without me? The country will perish because you do not know how to recognise enemies.
>
> (Stalin 1953)

On Stalin's death the two key positions of First Secretary of the Communist Party and Chairman of the Council of Ministers passed to **Georgii Malenkov**, a senior member of the Politburo. However, the top men in the Politburo – Molotov, Khruschev and Bulganin – were determined that never again would too much power rest in one person's hands. On March 14, five days after Stalin's funeral, Malenkov was removed from the position of First Secretary 'at his own request'.

The most feared contender for leadership was **Lavrenti Beria**, Chief of Stalin's Secret Police. He now became political head of all state security. Beria tried to build popular support in order to achieve victory in what he saw as a power struggle with his main rival, Khruschev. He posed as a liberal de-Staliniser, but Khruschev rallied the leadership against him. Beria was arrested and executed in June 1953.

There followed a period of collective leadership. Malenkov lacked charisma and in 1955 was replaced as Prime Minister by **Marshal Bulganin**. Under the guise of collective leadership the power struggle continued until 1958. Bulganin was dismissed and **Khruschev** took over the posts of **First Secretary** and **Prime Minister.** He was the undisputed leader of the Soviet Union until 1964.

The Khruschev era

Nikita Khruschev: Profile

Nikita Khruschev was born in 1894 near Kursk in southern Russia. His father was a poor peasant who in 1908 gave up the struggle to make a living from the land and moved to a nearby coalfield. The younger Khruschev worked as a mechanic in the mines. There he learned to hate the system which exploited the workers.

Khruschev joined the Bolshevik Party in 1918 and served in the Red Army during the Civil War. After the war he became a local party secretary in the Ukraine and took Stalin's side in the power struggle after the death of Lenin. He was rewarded with the post of Secretary of the Moscow branch of the Communist Party. Khruschev took a leading part in the purge of the anti-Stalinists and was given a place on the Politburo.

During the war he served as a political agent in the front lines and was involved in the defence of Stalingrad. After the war he controlled the Ukraine for Stalin until he was recalled in 1949 to serve as head of the Moscow Communist Party. By 1953 he was recognised, along with Beria and Malenkov, as one of the three leading members of the Communist Party. He had studied Stalin's methods well and used his position as first Secretary to build up his power base. He was involved in the execution of Beria and the demotion of Malenkov.

Nikita Khruschev was a man of great energy. He had been in charge of the construction of Moscow's fine metro system. He distrusted intellectuals and tried to identify with the people through his jolly demeanour and homespun philosophy. However, his erratic behaviour in public, especially abroad, was often an embarrassment to his colleagues.

Khruschev genuinely wished to improve the conditions of his people and could be ruthless in the pursuit of his objectives. He had no wish to return to the tyranny of Stalin's rule. He never had dictatorial power, as Stalin had, and needed the backing of the Politburo for his policies.

Khruschev's 'Secret Speech', 1956

> Stalin acted not through persuasion, explanation and patient co-operation with people, but by imposing his concepts and demanding absolute submission to his opinion. Whoever opposed this concept or tried to prove his view-point, or the correctness of his opinion, was doomed .... Mass arrests and deportations of many thousands of people, execution without trial and without normal investigation created conditions of insecurity, fear and even desperation.
>
> (Kruschev)

This is a short excerpt from a remarkable three-hour 'secret speech' made by Khruschev to a closed session of the Twentieth Party Congress on 25 February 1956. The speech was entitled *On the Cult of Personality and its Consequences.* It was an attack on Stalin's 'crimes' and his 'intolerance, brutality and abuse of power'.

Lenin's will which had criticised Stalin in 1924 was read out and Stalin's incompetence at the beginning of World War II was emphasised. The purges of the 1930s were exposed and 'Beria's gang' condemned. Khruschev reserved particular scorn for Stalin's 'cult of personality' and his practice of naming cities, towns, collective farms and prizes after himself.

De-Stalinisation

The 1956 speech was the beginning of a process of **de-Stalinisation**. The Soviet people were at first shocked and confused. However, Khruschev had cleverly broken the link between Stalin and Lenin and distanced the new leadership from the atrocities of the past.

The new openness after 1956 meant reform of the KGB which had taken over from Stalin's Secret Police. Large numbers of prisoners were released from the labour camps. Censorship was reduced in the more liberal atmosphere. **Alexander Solzhenitsyn's** *One Day in the Life of Ivan Denisovich* was published in 1962. It describes vividly life in one of Stalin's ***gulags*** or labour camps. However, **Boris Pasternak** who won the Nobel Prize for literature could not secure publication of *Doctor Zhivago* inside the Soviet Union. It seemed that the communists of Lenin's era were above criticism.

After 1956 Stalin's place in Russian history was gradually downgraded. It was thought that his body should not rest beside that of Lenin in Red Square. It was moved to an unmarked grave inside the walls of the Kremlin.

The Economy: Agriculture

What kind of Communism is it that has no sausage?

(Khruschev)

Khruschev paid particular attention to achieving increased agricultural production and made regular inspections of agricultural areas. He advocated the amalgamation of collectives into huge state farms. This led to more efficient use of machinery but was no more successful than Stalin had been in earning the good-will of the peasants.

Khruschev is closely associated with the **Virgin Lands Project** which began in 1954. This led to the cultivation of previously untilled land in East Russia, mainly in Siberia. The project was initially successful and Khruschev boasted of a record harvest in 1956. However, intense cultivation, lack of fertilisers, soil erosion and dust storms meant that after a few seasons these inhospitable lands gave very poor yields.

The Soviet Premier insisted on a major increase in maize production. During the icy winter of 1962 most of the maize grown in the unsuitable cold areas was lost. Khruschev had increased state investment in agriculture fourfold in ten years, but in 1963 he was forced to introduce bread-rationing and to import large quantities of grain. The food shortages of this time were a major reason for the fall of the Soviet leader.

The Economy: Industry

> We must help people to eat well, dress well and live well. You cannot put theory into your soup or Marxism into your clothes. If, after forty years of Communism, a person cannot have a glass of milk or a pair of shoes, he will not believe Communism is a good thing, no matter what you tell him.
>
> (Kruschev)

This speech by Khruschev reveals a concern for the living conditions of Soviet citizens and his decision to concentrate on consumer goods as well as heavy industry.

The Sixth Five-Year Plan, announced by Khruschev in 1956, decentralised economic planning. However there was little coordination between the 104 regional councils. Some items were over-produced, but others were always scarce or unobtainable. During the Khruschev era targets were not reached, plans were scrapped and at best the industrial programme was only moderately successful.

A feature of the period was the boom in the construction of apartments and flats, although many of these were of poor quality. There was also great progress in the field of technology. The national airline ***Aeroflot*** was expanded and the country

Yuri Gagarin, Russian cosmonaut, first man in space, 1961

opened up to tourism. The launching of the first satellite ***Sputnik***, and of the first man into space (Yuri Gargarin) shows the extent of advances in science and technology (see chapter 25). This was partly due to reforms in the education system. Free second- and third-level education led to a great increase in numbers going beyond first-level. Another notable aspect was the provision of specialist schools for children with particular talents.

The standard of living of the ordinary person went up during Khruschev's leadership. The introduction of shorter working-hours, free medical care and better living conditions were very welcome. However, the provision of full employment and the lack of incentive meant that the communist system remained under stress.

Foreign Policy: Cold War

The Soviet attitude to the West during Khruschev's rule is examined in detail in chapter 25. In summary, there was a 'thaw' in the Cold War between communist East and capitalist West. This was in no small measure due to Khruschev's policy of peaceful co-existence. The major problems during this time were caused by the erection of the **Berlin Wall** in 1961 and the **Cuban Missile Crisis** in 1962.

Khruschev showed that while he could compromise in the interests of world peace, he was determined to keep a hold on Eastern Europe. The existence of COMECON and the Warsaw Pact meant that the satellite states would find it difficult to break out of the Soviet Bloc. Khruschev's speech denouncing Stalin and pointing to 'different roads to socialism' led to expectations of greater freedom in Eastern Europe. After all, their Communism had been imposed by Stalin. Some felt that the Soviet Union would not interfere if countries behind the Iron Curtain tried to overthrow their masters.

May Day 1964: Russian rockets being paraded through Red Square, Moscow

Poland

> We have shed our blood to liberate this country and now you want to hand it over to the Americans.
>
> (Khruschev to Polish leaders, 1966)

Poland was on the verge of an explosion in 1956. Many factory- and shipyard-workers rioted and went on strike. Demonstrators took to the streets. The Polish Communist Party itself revolted and proposed the appointment of **Gomulka** as Party leader in defiance of Moscow.

Khruschev and Molotov flew to Warsaw. They were forced to agree to Gomulka becoming leader. However, he had to accept that Poland would remain within the Communist Bloc. On the other hand, he achieved the right of the Polish government to look after the country's internal affairs.

Gomulka granted a greater degree of freedom to the people of Poland. He allowed the Catholic Church to live in peace. The government did not insist on the collectivisation of agriculture, as it was opposed by the peasants. The Soviet Union was prevailed upon to cancel Poland's debts and the economy improved. However, many Poles looked forward to the day when their country would be completely free.

Hungary

> Today at dawn strong Soviet forces launched an attack against the capital with the obvious purpose of overthrowing the legal Hungarian Government. Our troops are fighting, the Government is at its post. I notify the people of our country and of the entire world of these facts.
>
> (Imre Nagy, Hungary's Prime Minister, Radio Budapest, 4 November, 1956)

Mass riots and strikes in October 1956 showed that Hungary was ready to overthrow its Stalinist government. Hungarian troops joined the rebels. Soviet troops

Stalin's statue was pulled down in Budapest on 2 November 1956 during the Hungarian uprising.

were called in by the government to stop the disturbances. They were in a weak position and after a week withdrew from Budapest.

A moderate communist, **Imre Nagy**, now formed a government. He shocked the Soviets by proposing that Hungary should leave COMECON and the Warsaw Pact and become a neutral independent state. Khruschev ordered the return of Soviet tanks and troops to Budapest on Sunday, 4 November, 1956. Over 30,000 Hungarians were killed in bloody street battles and 200,000 fled westwards through Austria.

A pro-Soviet Government was established under **János Kádár**. Nagy sought refuge in the Yugoslav Embassy, but was arrested and executed in 1958. Although he had been imposed by the Soviets, Kádár established one of the most liberal states in the Eastern Bloc. Many commentators felt that Hungary developed a compromise between capitalism and communism. Relations between the Catholic Church and the Government improved. Cardinal Mindszenty, who had opposed the Soviets in 1956, and was confined to the American Embassy in Budapest for years, was finally pardoned in 1970.

Other States

East Germany became an orthodox member of the Eastern Bloc and a supporter of the Warsaw Pact (see chapter 25).

Romania developed a more independent line. Her leaders fought against Soviet economic domination within COMECON and rejected the presence of Soviet troops on Romanian soil. **Nicolae Ceausescu** became leader in 1965. He established relations with China and with the Western Bloc. This made him popular in America and Britain. However, he proceeded to rule over a rigid communist state.

In Czechoslovakia, strict Communist Party control of the country was maintained until the end of 1967. When the fight for political freedom and economic reform came, Khruschev had lost power.

East German workers build the Berlin wall, 1961.

The most outspoken of all East European leaders was **President Tito** of Yugoslavia. After his 1956 speech, Khruschev tried to heal the rift that had developed between Tito and Stalin. The Yugoslav leader remained cautious and proceeded with a mixture of communism and limited capitalism. In the Cold War clash, he pursued a neutral line and condemned aggression on both sides.

Sharp differences arose between Khruschev and the Chinese. **Mao Tse-Tung** disagreed with the Soviet leader's condemnation of Stalin and with his policy of peaceful co-existence. Khruschev was unhappy with China's refusal to sign a nuclear test ban in 1963. The Sino-Soviet split led to border incidents between the two countries and often looked like developing into outright war.

Fall of Khruschev

Khruschev had to rely on the support of a majority within the Central Committee of the Party. Some pro-Stalinist members opposed his policies and waited for their opportunity. They criticised his humiliation in Cuba in 1962 and his split with the Chinese. Failure to meet economic targets, the collapse of his agricultural policies and scarcity in the shops all contributed to his downfall.

Khruschev was removed from his position by party leaders in October 1964. They announced that he had resigned 'in view of his health and advanced age'. He became a 'non-person' under the pro-Stalinist leadership of Brezhnev and Kosygin. The man who had shown the human face of Communism to the world lived quietly in retirement until his death in September 1971. He was not given a state funeral.

The Soviet Union after Khruschev

The Brezhnev Era

In 1964 **Leonid Brezhnev** was appointed First Secretary of the Communist Party. **Alexei Kosygin** became Chairman of the Council of Ministers. The Central Committee of the Party wanted to ensure that no one person had too much power. In practice Brezhnev emerged as the dominant personality but his policies were tempered by the more moderate Kosygin. The First Secretary retained power until his death in 1982.

Domestic Reforms

Under Brezhnev the Soviet Union was modernised. Agricultural reforms were successful in raising production. Prices were raised and collectives could make a profit. The peasant farmer was rewarded for his contribution to the collective. He could once again use up to an acre of land to raise his own produce and could sell it privately. Social benefits, like old-age pensions and sickness schemes, were extended to the peasant population.

In industry, the regional economic councils set up by Khruschev were scrapped and centralised planning returned to Moscow. However, local factory and farm managers had far more control over production and sales, provided they stayed within the overall plan. These local managers were more aware of the people's needs and production of consumer items, especially household goods, increased. However, the communist system lagged behind its capitalist rival and long queues remained a feature of Soviet life.

Human Rights

At heart Brezhnev was a hard-line politician from the old school. He brought to an end the process of de-Stalinisation and in 1965 praised Stalin's contribution to his country.

Under Brezhnev, a campaign was mounted against writers who criticised the lack of freedom in the Soviet Union. Underground publications, called ***samizdats*** (self-published) were targeted. Those who were found guilty of publishing anti-Soviet literature were sent to labour camps or were exiled within the country.

Writers like **Alexander Solzhenitsyn** could not have their works published in the Soviet Union during the Brezhnev era. He was persecuted and finally allowed to leave the country in 1974. Other major critics who endured persecution were the scientist **Andrei Sakharov** and the poet **Yevgeny Yevtushenko**. The position improved a little after 1975 when the Soviet Union signed the **Helsinki Declaration on Human Rights**. Criticism of the denial of basic freedoms continued to sour relations between East and West until Mikhail Gorbachev took over as Soviet leader in 1985.

Foreign Affairs

The Soviet Union after 1964 continued with the policy of *detente* or lessening of tension. Her negotiators signed an arms **non-Proliferation Treaty, 1963** and two **SALT** agreements during the 1970s (see chapter 25). The Soviet Union continued to supply arms to her allies, such as the Egyptians in 1967 and the Vietcong during the Vietnam War. In summary, it can be said that the Soviet Union continued to support her allies and friends, but was still anxious to avoid a global conflict. However, she still kept as tight a grip as possible on her satellite states in the Eastern Bloc.

Czechoslovakia, 1968

> Yesterday, on 20 August, 1968, about 11 p.m. troops from the Soviet Union, the Polish People's Republic and the Hungarian People's Republic, the German Democratic Republic and the Bulgarian People's Republic crossed the frontiers of the Czechoslovak Socialist Republic.
>
> (Radio Prague, 21 August 1968)

This was the first time that troops from the Warsaw Pact command were used against a member. The occupation was the culmination of a series of events during the previous eight months. On 5 January the hard-line Stalinist leader of

A Czech youth stands defiantly trying to stop a Soviet tank during the invasion of Czechoslovakia, 1968.

Czechoslovakia was replaced by the moderate reformer **Alexander Dubček.** During the **'Prague Spring'** of 1968, Dubček introduced sweeping reforms including freedom of speech and of the press and the right to move freely inside and out of the country. The communist leader said that the Party had no right to rule unless it earned the support of the people.

Czechoslovakian street cartoon, 1968

Czechoslovakia was unable to resist the invasion in 1968. **Dubček** and his followers were removed from office and **Gustav Husák** became Party leader. The reforms of the 'Prague Spring' were cancelled and strict censorship restored. A young student, **Jan Palach,** burned himself to death in January 1969 in protest against the invasion. His sacrifice was remembered when communism fell twenty years later.

The Collapse of Communism

President Mikhail Gorbachev's policy of ***glasnost*** (openness) and ***perestroika*** (re-structuring) were welcomed by most ordinary people in the Soviet Union and the Eastern Bloc countries (see chapter 25). He also decided that the Soviet army would not intervene to keep communist regimes in power. It was as if the lid had been lifted off the can. Movements and groups within the satellite states now sought the election of democratic governments.

In 1989 ***Solidarnosc*** (*Solidarity*), a powerful Polish trade union movement, led by **Lech Walesa,** forced free elections which led to the fall of the communist Government. Like dominoes, the communist regimes in Hungary, Czechoslovakia, Bulgaria, Yugoslavia and Romania gave way to democracies. In Romania the communist hard-line dictator Nicolae Ceausescu and his wife, Elena, were executed by firing squad. On 9 November the Berlin Wall was opened. In October 1990 Germany was re-united under the former West German Chancellor **Helmut Kohl** (see chapter 28).

Mikhail Gorbachev, Russian leader, 1985-1991, who condemned Stalin's excesses during purges of the 1930s

The collapse of Communism has changed the face of Europe and the political stage of the whole world. The Baltic States of Latvia, Estonia and Lithuania have broken away from the Soviet Bloc and re-established their independence. The Soviet Union has become the CIS, the Commonwealth of Independent States. Mikhail Gorbachev has been replaced by **Boris Yeltsin** who became President of Russia in 1991. The establishment of free market economies in former communist states has proved painful. Some Western leaders have had to adjust to life without a traditional enemy and a convenient 'red scare'. The future looks both interesting and exciting.

Questions

ORDINARY LEVEL – D

Answer the following questions briefly. One or two sentences will be enough.

1. Explain why the period of World War II during the winter months of 1939-1940 is referred to as 'the phoney war'.
2. Why did the U.S.A. enter World War II, 1941?
3. Why was the Battle of Stalingrad, 1942-1943, a turning point in World War II?
4. What did Allied leaders at their meeting in Yalta, 1945, decide to do about Germany?
5. Why were the Nuremberg Trials held after World War II?
6. What is the function of the Security Council of the United Nations?
7. Name a specialised agency of the United Nations and briefly describe its work.
8. Why was the North Atlantic Treaty Organisation founded, 1949?
9. Describe one measure taken by Charles de Gaulle, as President of France, to assert his country's independence in foreign affairs.
10. What were the main recommendations of the Beveridge Report published in England, 1942?
11. Explain one major achievement of the Labour Government which took office in Britain in 1945.
12. What is understood by the term 'Cold War' after 1945?
13. Why was there a revolt in Hungary in 1956?
14. What was the significance of the Treaty of Rome, 1957?
15. What was President Kennedy's reaction to the discovery of missile launching sites in Cuba, 1962?
16. Explain one important effect of the Cuban Missile Crisis.

Questions

ORDINARY LEVEL – E

Write a short paragraph on each of the following:

1. Germany's *blitzkrieg* tactics
2. The Battle of Britain
3. The Battle of El Alamein, 1942
4. D-Day, 6 June 1944
5. The setting up of the United Nations Organisation
6. The Treaty of Rome, 1957
7. Student riots in France, 1968
8. The Suez Crisis, 1956
9. The Marshall Plan
10. The Berlin Airlift, 1948-1949
11. The Hungarian Rising, 1956
12. The U.S.S.R. under Nikita Khruschev.

Questions

ORDINARY LEVEL – F

Write a short essay on each of the following:

1 The stages through which the war that began in Europe in 1939 became a 'world war' by 1941.

2 Life in Nazi-occupied Europe during World War II under **three** of the following headings:
(i) The extent of the occupation
(ii) Nazi racialist policies
(iii) Resistance movements
(iv) Economic exploitation of occupied lands.

3 The origins and development of the European Economic Community, 1945-57

4 Europe 1945-1966 under the following headings:
(i) Decolonisation
(ii) Relations with the Superpowers, the U.S.A. and the U.S.S.R.
(iii) Economic recovery.

5 Europe since 1945 under each of the following headings:
(i) N.A.T.O. and the Warsaw Pact
(ii) West Germany under Adenauer
(iii) France under de Gaulle.

Questions

HIGHER LEVEL

Write an essay on each of the following:

1 'Rapid movement was an outstanding feature of the conduct of warfare in World War II.' Discuss. (80)

2 Write an essay on the part played by the United States in World War II. (80)

3 Outline and discuss the stages through which the war in Europe in 1939 became a world war by 1941. (80)

4 Assess the part played by Britain in World War II. (80)

5 The origins, purposes and structure of the United Nations Organisation. (80)

6 Trace the origins and development of the European Economic Community, 1945-c. 1966. (80)

7 Treat of the part played by Charles de Gaulle in the history of France. (80)

8 Treat of the creation of the welfare state in Great Britain by the Labour Government under Clement Attlee. (80)

9 'Britain, in the years after World War II, saw her empire decline but the standard of living of her people improve significantly.' Discuss. (80)

10 'Since 1945, relations between East and West have been dominated by a series of real crises and recurrent tensions.' Discuss. (80)

11 Treat of the part played by the U.S.A. in Europe, 1941-c. 1966. (80)

12 Treat of the relations between the two Superpowers, the U.S.A. and the U.S.S.R., during the period 1945-1966. (80)

13 'By 1948 the domination of the countries of Eastern Europe by the U.S.S.R. had become almost complete.'

(i) Explain briefly how this domination was achieved. (30)

(ii) Discuss the reaction of the countries of Eastern Europe to this domination up to 1966. (50)

14 Treat of economic and political reconstruction in the Federal Republic of Germany under Konrad Adenauer and Ludwig Erhard, 1949-1966. (80)

SECTION FIVE

1870-c. 1970: ASPECTS OF SCIENCE, TECHNOLOGY AND CULTURE

Chapter 30

The Age of Achievement

Introduction

They (the scientists) are the men who are changing the world. Politicians are but the fly on the wheel.

(A.J. Balfour, British Cabinet Minister, 1899)

In 1870 many scientists and educated people believed that the basic laws of nature had been discovered. A century and a half of research had confirmed the general picture of the universe as sketched by **Isaac Newton** (1642-1727). Scientists in 1870 believed in three basic laws: matter consisted of atoms, radiation consisted of waves and both obeyed the principle of the conservation of energy.

A whole series of scientific and technological discoveries in the late nineteenth century and during the twentieth century raised fundamental questions about earlier principles. Advances in biology, chemistry, medicine, physics and technology would make today's world unrecognisable to the scientist of 1870.

Mankind's achievements are also evident in the world of the arts. Artists have expressed their reactions to a changing and sometimes confusing world through literature, painting, sculpture, architecture and the cinema. The historian has much to learn from study of human endeavour in these areas.

Medical Science

Introduction

The story of medicine is one of the most fascinating and most dramatic in the field of science. Discoveries before 1870 had done much to overcome the ignorance and superstition that stood in the way of medical knowledge and proper health-care. Advances on these discoveries since 1870 have revolutionised medical science and made the world a better place to live in.

Evolution and Heredity

In 1859 **Charles Darwin**, an English naturalist, published *On the Origin of the Species*. In this book he explained his theory of natural selection – the fittest individuals of any species survive and pass on their characteristics to their offspring. Darwin elaborated his ideas in *The Descent of Man* which was published in 1871. He showed that man, like other animals, was derived over millions of years from more primitive ancestors. After years of controversy, Darwin's theories were accepted by scientists and by the Christian churches and have influenced scientific thought ever since.

Gregor Mendel, an Austrian monk, performed experiments in his monastery garden involving the crossing of garden peas. He showed that heredity involved pairs of contrasting characteristics, such as tallness and shortness. He explained how children inherited factors from each parent and, by the law of independent assortment, passed certain characteristics on to their own offspring. Mendel's work is regarded as a major influence on genetics, the science of heredity.

Work on heredity continued during the twentieth century and scientists discovered a substance called DNA (deoxyribose nucleic acid). They learned that heredity works through the storing and passing on of genetic information by the DNA molecules which are contained in all living things. The 'genetic code' was cracked in 1953 when **Francis Crick** and **James Watson** discovered the structure of the DNA molecule. Since then scientists have succeeded in isolating genes, and genetics now has a high priority in medical research

Chemistry and Medicine

By the mid-nineteenth century, the use of anaesthetics meant that surgery was no longer a terrifying experience. However, after an operation, patients often died from infection. **Joseph Lister**, an English surgeon at the Glasgow Royal Infirmary, believed that micro-organisms, similar to those which caused food to go bad, might cause infection in wounds. Lister required all surgeons at the hospital to wash their hands and instruments in a carbolic acid before operating. Thanks to this sterilisation, there was a sharp drop in the death-rate following operations. Only gradually did surgeons elsewhere accept Lister's theories and adopt his methods.

The French scientist **Louis Pasteur**, discovered that fermentation in food and alcohol was not caused by chemical reaction, but rather by microscopic living organisms. These findings led to the modern study of bacteriology. They also led Pasteur to the discovery that these micro-organisms or germs could be killed by moderate heat, called pasteurisation. This process is used today to kill germs in milk and other foods.

Pasteur also discovered that a weakened culture of a disease would prevent an animal from developing the disease itself. In 1885 he developed a vaccine against the fatal disease, rabies.

In 1882 a German bacteriologist, **Robert Koch**, isolated the bacillus that caused tuberculosis of the lungs, popularly known as TB or Consumption. TB was incurable at that time, but thanks to Koch's work and modern treatment, patients can now be restored to normal health.

Modern Drugs

Scientists continued to experiment with compounds that would kill bacteria without killing the patient. They succeeded in producing synthetic drugs, as distinct from

Alexander Fleming at work in his laboratory

'home' plant remedies. Progress in the production of these so-called 'magic-bullets' was slow, but a number of major discoveries took place in the twentieth century.

In 1921 **Frederick Banting** and **Charles Best** of Canada discovered the role of insulin in the control of diabetes. In 1928 **Alexander Fleming** showed that harmful bacteria could be killed without injuring normal body cells, through the use of an antibiotic called penicillin. This drug proved useful in the control of a whole range of illnesses. In 1935 scientists developed 'sulfa' drugs for the treatment of pneumonia and meningitis.

Streptomycin has been used against TB since 1944. The 1940s also saw the development of cortisone, which is used to treat rheumatism, burns, arthritis and other illnesses. In 1953 **Joseph Salk** developed a vaccine which was used to control the crippling disease of polio. The world of medicine has continued to be revolutionised by the discovery of new drugs.

Other Medical Advances

In 1895 the German physicist, **Wilhelm Röntgen** discovered X-rays. This had a great effect on diagnosis, that is the identification of a disease through its symptoms. Thus began the science of radiography.

The most notable advance at the beginning of the new century came when **Pierre** and **Marie Curie** extracted the radioactive element radium. This discovery led to the birth of radiotherapy which has proved vital in the treatment of cancer.

The advances of modern medicine are usually the result of the work of teams of researchers rather than of individuals. Immunisation is now available against most diseases. Medical care has been made safer and more comfortable by the development of barbiturates, pain-killers and sedatives.

Nowadays, blood transfusions and kidney machines are commonplace. Skin-grafting was developed during World War II to repair the damage done to airmen

French scientists, Pierre and Marie Curie

whose planes caught fire. Transplant surgery is another feature of modern medicine. In 1967 **Dr. Christian Barnard** of South Africa carried out the first human heart transplant.

Constant progress is being made in overcoming disease and illness. Many challenges remain, including the development of a safe and effective remedy for cancer and for the dreaded modern disease, AIDS. Another challenge is the solution

Doctor Christian Barnard

of moral and ethical problems which have emerged, especially in the fields of transplant medicine and genetic engineering.

Psychology

Psychology is the science which studies the mind. The founder of modern psychology was the Austrian physician, **Sigmund Freud**. He developed his own treatment for mentally-disturbed patients. Freud's method involved the 'free-association' of ideas and drawing memories from the subconscious to the conscious mind. He called his treatment psychoanalysis.

The most famous of Freud's followers were **Alfred Adler** and **Carl Jung**. Their pioneering work with the subconscious processes of the mind has had far-reaching effects on people's attitude to mental illness and on its treatment.

Sigmund Freud, father of psychoanalysis

The Development of Nuclear Power

Nuclear Physics

Henri Becquerel discovered radioactivity accidentally while investigating X-rays in 1896. This marks the beginning of nuclear physics. The important work of **Röntgen** and the **Curies** led other scientists to investigate radioactivity and radiation. **Marie Curie** established that the elements uranium and radium emitted radiation.

A New Zealand physicist, **Ernest Rutherford,** concluded from his experiments that at the centre of the atom was an extremely heavy nucleus or core which was surrounded by electrons. When Rutherford bombarded nitrogen atoms with tiny particles, he produced oxygen and hydrogen. What he had done was to change one chemical element into another.

Nuclear Energy

In 1900 a German scientist, **Max Planck**, published his Quantum Theory. Put simply, this stated that energy is released in small units or 'packages' called quanta. Planck's theory influenced almost every later development, especially those of Einstein and Bohr.

In 1905, the brilliant mathematician and physicist, **Albert Einstein**, a German Jew, published his theory of relativity. He developed a formula ($E = mc^2$) to show that a small amount of mass could be transformed into a tremendous amount of energy. The Danish scientist, **Niels Bohr**, made use of Planck's ideas in his research into atomic structure. The work of Einstein and Bohr led to the development of the atomic bomb and nuclear energy.

Among those whose experiments resulted in the control of nuclear power was the American-based Italian physicist, **Enrico Fermi**. He built a nuclear 'reactor' to produce a controlled chain reaction. Slabs of uranium were surrounded by blocks of graphite which controlled 'fissions' or separations within the uranium. The harnessing of nuclear power has contributed to the growth of modern nuclear reactors which now generate a substantial amount of the world's energy.

The Atomic Bomb

The difference between controlled fission and an atomic bomb is the rate of release of energy. In an atomic explosion the reaction is completed in a split second. On 16 July,

Albert Einstein who developed the Theory of Relativity

1945, at a secret testing ground in New Mexico, the first atomic bomb was tested. It is estimated that the temperature at the centre of the explosion reached about 10 million degrees Centigrade. Three weeks after this test an atomic bomb was dropped on **Hiroshima** in Japan. The bombing of **Nagasaki** three days later brought World War II to an end (see chapter 23).

The horrifying effects of these atomic bombs have raised serious moral questions. Have scientists created a monster, a veritable Frankenstein which they cannot control? Even the peaceful use of nuclear energy has come under scrutiny, especially through the work of Greenpeace and other groups.

The New Technology

Technology defined

Technology: The practice of any or all of the applied sciences that have practical value and/or industrial use

(The Chambers Dictionary, 1993 edition)

The years since 1870 are noteworthy for an abundance of interesting discoveries and inventions. Most of the machines, tools and gadgets which we use today have

Modern technology: BP oil rig in the North Sea

resulted from the application of science and the genius of human beings in the past hundred years. The advances in technology are seen most clearly in the areas of transport, communications and automation.

Transport

In 1887 a German, Gottlieb **Daimler**, produced the first modern motor car, with four wheels and a petrol engine. By 1914 there were almost a million cars on the roads of Europe. The motor car has increased personal freedom and mobility, but has led to congestion in towns and cities and has contributed enormously to the pollution of the environment.

Since the eighteenth century people could fly by using balloons. In 1900 Count Ferdinand **von Zeppelin** attached engines to a balloon and called his invention an airship. In 1903, the brothers **Wilbur** and **Orville Wright** from Ohio in the United States made the first flight in an aeroplane. In 1909 a French pilot, **Louis Blériot,** crossed the English Channel. Aeroplanes had an important role in World Wars I and II (see chapters 11 and 23). In 1919 Captain John **Alcock** and Lieutenant Arthur **Brown** completed the first non-stop transatlantic crossing when they flew from Newfoundland to Ireland. The advances in electronic navigational aids led to the invention of the jet aeroplane by Hans **von Ohm** in Germany in 1939 and **Frank Whittle** in Britain in 1941. The Concorde, a supersonic jet airliner, was designed and constructed during the 1960s, following cooperation between Britain and France.

The first man-made satellite, *Sputnik 1*, was launched by the USSR in 1957. The first manned space flight took place in 1961 when Yuri Gargarin orbited the earth in *Vostok 1*. Soon people walked in space, and in 1969, the American astronaut, Neil Armstrong landed on the moon. *Skylab*, an orbiting space laboratory, was launched by the US in 1973.

Communications

Two inventions which revolutionised communications were the telephone and the wireless. The first telephone was invented in 1876 by Alexander Graham **Bell**, a Scottish-born American. In 1896 an Italian engineer, Gugliemo **Marconi** transmitted messages without using wires. In 1901, he was able to send a signal across the Atlantic.

The new century also saw the development and rapid growth of radio broadcasting. The Scotsman, **John Logie Baird**, succeeded in transmitting an image in 1927 and is credited with the invention of television. The BBC began regular television transmission in 1936. The telecommunications satellite, *Telstar*, was launched in 1962. Since then the 'communications miracle' has continued and we live in a world of electronics, computers, fibre-optics and automation.

Automation

Automatic is a word borrowed from the Greek to describe any process which can carry on without human help. The computer has revolutionised the home and the

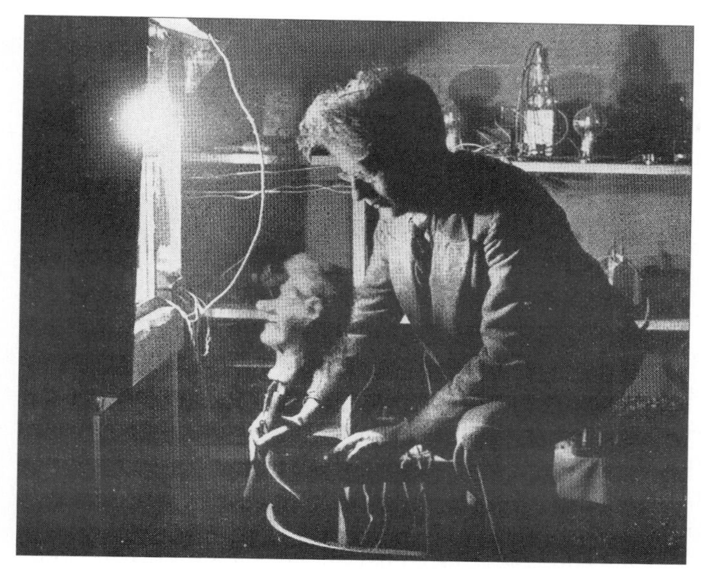

John Logie Baird transmitting an image of a ventriloquist's 'dummy'

work-place. Robots have replaced workers in repetitive production-line industries. Automatic assembly-lines can produce motor-cars 'untouched by human hand'.

The microchip revolution has provided a solution to many problems. However, its most pronounced side-effect has been to increase the number of people unemployed. Many social history students have come to believe that unless the science of living in a community keeps pace with the science of technology, civilisation as we know it may collapse.

The Arts in the Modern World

Literature

The nineteenth century has been called the golden age of the European novel. These novels are valuable sources of social history. The great writers of the era believed that they should place their fictional characters in situations of complete realism. They usually chose contemporary settings which they could describe with the accuracy of an eyewitness.

Charles Dickens, whose best work reveals some of the dreadful social conditions of his age, lived until 1870. By bringing the miseries of child-workers, of the poor, and of the 'hands' in factories to public notice, he did much to remedy their sufferings. Mary Ann Evans, who wrote under the name of **George Eliot**, dealt with the social conditions of farmers and traders in the English Midlands. **Thomas Hardy** lamented the decline of the old country ways of life in his Wessex novels.

Other famous European novelists who tried to depict accurately and objectively detailed aspects of contemporary life were **Gustave Flaubert, Emile Zola, Guy de Maupassant** and **Gerhard Hauptmann**.

Another aspect of nineteenth-century writing is the impact and influence of the Russians. Their work ranged from satire and melodrama to documentary chronicles of Russian life. **Nikolai Gogol**, **Ivan Turgenev**, **Feodor Dostoevsky**, **Leo Tolstoy** and **Anton Chekhov** all became known to European and American readers through translations of their works.

It would be a mistake to imagine that all novelists and writers were totally realistic. Samuel Langhorne Clemens, better known as **Mark Twain**, mixed truth-to-life with a wryly humorous view of it. **Jack London** was well-known for his stories of romance and daring. **Robert Louis Stevenson** wrote some of the best adventure tales of all time. **Rudyard Kipling** achieved fame for his children's stories and for his depiction of life in India under British rule.

Many of the artists mentioned above spanned the nineteenth and twentieth centuries. A group which achieved notoriety in the new century was the **Bloomsbury Group** in England. Famous names were **Bertrand Russell**, **E.M. Forster** and **Virginia Woolf**. This group of philosophers and writers was dedicated to pleasure and to the enjoyment of beautiful objects. The 'War Poets' (see chapter 11) wrote with patriotic idealism, but also gave angry and ironic descriptions of the horror of the battlefront. **Siegfried Sassoon**, **Rupert Brooke**, **Wilfred Owen** and **Francis Ledwidge** became well-known writers of the era.

The twentieth century is also notable for a literary form called 'stream of consciousness'. This intended to portray the internal reality of characters by presenting their thoughts and reactions without direct comment or apparent organisation by the author. **James Joyce**, **Virginia Woolf** and **William Faulkner** used this method which was influenced by Freud's psychological concepts.

The twentieth century French novelists examined how people's consciences work when they encounter difficult moral problems. Among these French writers were **André Gide**, **Marcel Proust**, **Francois Mauriac** and **Colette**, the greatest woman writer of the age. Two other French novelists of note were **Albert Camus**, who examined the reasons for decisions, and **Jean-Paul Sartre** who painted a depressing picture of life. British novelists who were highly regarded were John Galsworthy, Arnold Bennett, D.H. Lawrence, Hilaire Belloc, G.K. Chesterton, George Orwell, Evelyn Waugh, Graham Greene, William Golding, Aldous Huxley and Compton Mackenzie.

American novelists, such as **J.D. Salinger**, **F. Scott Fitzgerald** and **Ernest Hemingway** wrote about the heroic struggles of alienated individuals. **John Steinbeck** was concerned with the fate of the poor, **William Faulkner** was the chronicler of the 'deep south' and **James Thurber** was a brilliant illustrator of his own fantasy stories.

William Butler Yeats and **T.S. Eliot** are probably the best-known poets of the twentieth century. Yeats turned from Irish subjects to explore more universal themes concerning the mysteries of life. His poetry is noted for its symbolism and lyrical quality. T.S. Eliot wrote about the boredom and futility of modern urban society. He

British poets W.H. Auden, Cecil Day Lewis and Stephen Spender, 1949

wrote in free verse and experimented with new techniques. Other leading poets were **W.H. Auden**, **Cecil Day Lewis**, **Stephen Spender**, **Louis MacNeice**, **Edith Sitwell**, and **Dylan Thomas**.

Theatre

The principal influence on European theatre in the nineteenth and twentieth centuries was surely the Norwegian, **Henrik Ibsen**. His plays dealt with actual problems which men and women experience in their lives. Other dramatists whose work has stood the test of time were the Swede, **August Strindberg** and the Russians, **Anton Chekhov** and **Maxim Gorky**.

This was also the age of great Irish playwrights. **Oscar Wilde** was famous for his witty and light-hearted satires. **George Bernard Shaw** made plays into arguments for and against all sorts of topics that interested him. Wilde and Shaw were based in England. In Ireland, the literary theatre dedicated itself to the promotion of plays which drew heavily on Celtic folk history and legend as well as commenting on the contemporary Irish scene. Names associated with the movement were **Lady Gregory**, **W.B. Yeats** and **J.M. Synge**. **Sean O'Casey** chronicled the hardships of the poor in early 20th century Dublin.

Ibsen's ideas of how drama can work were accepted and carried forward by twentieth-century dramatists. **Bertolt Brecht** presented his theatre in episodes rather than acts. He aimed to stimulate the intellect of the audience and to instruct on social and moral problems. The 'theatre of the absurd' in France produced strange fantasies, often without normal conversation. The French dramatists expressed anxiety at man's apparent lack of purpose. The names **Eugene Ionesco**, **Jean Genet** and the French-based Irishman, **Samuel Beckett**, are well known.

A feature of twentieth-century drama is the growth of American theatre. **Eugene**

Portraits: W.B. Yeats and Dylan Thomas

O'Neill emerged as his country's greatest dramatist. **Tennessee Williams** examined neurotic behaviour and highlighted the problems of the 'deep south'. **Arthur Miller** became well-known for his modern tragedies.

British social realism was written from a left-wing standpoint. **John Osborne** was the leader of these so-called 'angry young men' who commented on the social position of the working classes in what was termed kitchen-sink drama. **Alun Owen** was also a well-known British dramatist.

Music

The nineteenth century was the great era of the Romantic Movement in music. Classical music, painting and literature were concerned with form and style. The artists of the Romantic movement considered that what they had to say was more important than the way in which they said it. Romantic music is always concerned with such themes as love, joy, or sorrow. The best-known names of the first half of the nineteenth century were **Beethoven**, **Schubert**, **Chopin**, **Liszt**, **Mendelssohn**, **Schumann**, **Berlioz** and **Rossini**.

The Romantic tradition of the first part of the century continued to dominate after 1850. **Richard Wagner** became famous for his operas. He called these operas, written in German, music-dramas. To obtain both the power and the effects he wanted, Wagner enlarged the orchestra until it reached the size of a modern symphony orchestra. He also revolutionised music by his use of harmony. Another great German of the period was Johannes **Brahms**, famous for his mellow symphonies.

Giuseppe Verdi inherited the tradition of the Italian grand opera which had been established by **Donizetti** and **Rossini**. In his operas he had great settings and a huge cast. Giacomo **Puccini** carried the Italian operatic tradition into the twentieth century. In the same manner **Richard Strauss** and **Gustav Mahler** brought the German tradition forward. These composers also contributed to the great religious musical revival through their masses, oratorios and requiems.

Another aspect of nineteenth- and twentieth-century music is the emergence of nationalism. As countries strove for national independence, they produced composers who wished to write music that belonged especially to their homeland. Students of music note that many composers, including Wagner, based much of their work on folklore and folk melody.

The great composers of the era who were inspired by their native folk songs, dances, rhythms, folk myths and legends are almost too many to mention. Among the great names are: the Russians, **Glinka**, **Borodin** and **Tchaikovsky**; the Scandinavians **Grieg** and **Sibelius**; the Frenchmen, **Debussy**, **Bizet**, **Fauré** and **Ravel**; the English composers **Elgar** and **Delius**. The modern composers, **Igor Stravinsky** and **Dmitri Shostakovich** in the Soviet Union, Arnold **Schoenberg** in Austria, **Bela Bartok** in Hungary and **Vaughan Williams** and **Benjamin Britten** in England deserve special attention.

The twentieth century has also seen the emergence of 'lighter' forms of music. The scene was set in the late nineteenth century by the satirical operettas of W.S. **Gilbert** and Sir Arthur **Sullivan**. The light musical stage in more modern times has been dominated by the Americans, especially through Frederick **Lowe,** Richard **Rodgers**, Oscar **Hammerstein** and Leonard **Bernstein**.

America was also the birth-place of modern jazz. This form of music grew from negro folk songs and was associated with 'singing the blues', that is singing about

The Beatles, Liverpool pop group, 1963

sad themes. Great names in the development of jazz were **Scot Joplin**, **Dizzie Gillespie**, **Charlie Parker**, **Louis Armstrong**, and **Duke Ellington.**

A notable aspect of the twentieth century has been the growth of popular music, often referred to as 'pop'. A greater confidence with new electronic effects and instruments, together with the influence of various folk traditions, resulted in energetic, rhythmic commercial music which gained an international audience. **Elvis Presley**, **Jimi Hendrix**, **Buddy Holly** and **The Beatles** are among the great pioneers of popular music.

Painting

The Romantic movement also dominated art in the nineteenth century. The best-known names from the first half of the century were Eugene **Delacroix**, James **Turner** and the landscape painter, John **Constable**. This movement was maintained in the second half of the century by the Pre-Raphaelite Brotherhood, a secret group dedicated to returning to the simplicity of the mediaeval artists before Raphael. Among these artists were **William Morris**, **Gabriel Rossetti** and **John Millais.**

Impressionism (see chapter 3) marked the beginning of what is termed modern painting. Impressionists were concerned with capturing the immediate visual impression of the changing face of nature. They often painted out of doors and used fresh bright colours to capture the sense of light and atmosphere. The movement was based in France and the most famous Impressionists were Edourd **Manet**, Edgar **Degas,** Claude **Monnet** and Pierre **Renoir**. The Post-Impressionists developed from this group, but practised a more formal style and emphasised the role of the subject. Paul **Cezanne**, Paul **Gaugin** and Vincent **van Gogh** are well-known.

Art Nouveau was a style of decorative art, design and architecture which spread throughout Europe and spanned the old and new centuries. It was characterised by its flowing, curved lines, subtle colours and floral designs. **William Morris** and **Aubrey Beardsley** favoured this style which also influenced the famous American artist, **James Whistler**.

Among the well-known twentieth-century movements was Fauvism. The Fauves (wild beasts) derived their name from their distortion and their use of violent colour. Henry **Matisse** earned a high reputation. However, a more revolutionary style, known as Cubism soon developed. All the different planes shown in a painting looked as if they were composed of little cubes. The Cubists also rejected loud colours and worked mainly in browns and greys. Famous cubists were **Pablo Picasso** and **George Braque**.

Picasso developed his style and became part of the surrealist school of art. The surrealists tried to express the subconscious mind in art. They were influenced by the work of the psychoanalyst, Freud. **Salvador Dali** was another famous surrealist artist. The Expressionist painters tried to express their personal reactions to the confusing world around them.

A logical development of Cubism is what is often referred to as Abstract Art.

'Pop Art': Whaam V, Roy Lichtenstein

Artificial shapes are 'abstracted' or derived from real objects. The Dutchman, **Piet Mondrian**, and the English artist, **Ben Nicholson**, were pioneers of the form. The so-called 'Action Painters' of modern times have also produced abstract, colourful images by spattering paint on canvas. Another feature has been the development of 'pop art'. Artists like **Roy Lichtenstein** have adapted commercial art techniques such as cartoons and advertising posters to reflect the world of mass production.

Architecture

Architectural styles in the nineteenth century were diverse and, in general, traditional. However, in France and America, concrete and steel were seen as the materials of a new architecture.

The French engineer, **Gustave Eiffel**, constructed his famous steel tower between 1887 and 1889 as a centre-piece for the Paris International Exhibition. It was three hundred metres high. The technique of reinforced concrete, that is concrete strengthened by steel rods and wire, was developed in France and used there during the 1890s.

In order to make maximum use of scarce space, 'Skyscrapers' were built in New York during the 1870s and 1880s. In the 1890s the new 'Skyscrapers' were mostly steel-framed. This meant that the frame, not the walls, bore the load of the buildings. The leading figure in this work was **L.H. Sullivan**. His assistant, **Frank Lloyd Wright**, became a great international architect. He was the pioneer of the 'modern' house in his use of a spreading ground plan and extensive windows. He endeavoured to design houses and buildings which would blend into their natural surroundings.

This international modern style was primarily functional and was used extensively in public buildings and in great housing 'tower blocks'. It was characterised by generally undecorated cubist shapes and horizontally-grouped windows. The

Bauhaus school in Germany (see chapter 16) was founded by **Walter Gropius**. It emphasised clarity, straight lines and functional suitability. Other architects who made major contributions to their profession were **Charles Holden** in Britain and the Swiss, **Le Corbusier**. The latter showed great concern for the environment and for ordered town-planning.

Sculpture

The period before 1914 was dominated by **Auguste Rodin**, born in France in 1840. He was influenced by the Impressionists, especially in his use of light. He was well known for his superbly-executed portrait busts.

The main movement in twentieth-century sculpture, as in painting, was towards abstraction. **Jacob Epstein** worked in the language of the Expressionists. Other great abstract sculptors were **Constantin Brancusi** of Romania and the Russian **Nuam Gabo**.

Among the sculptors whose subject was still the human figure were the Italian, **Alberto Giacometti**, and the Englishman **Henry Moore**. Giacometti achieved fame for his elongated wistful forms. Moore did not aim for realism, but sought to express vitality and symbolic power. His great reclining figures were carved from stone or cast in bronze and usually sited in the open. They suggest humanity linked in timeless relationship with nature.

Cinema

The cinema, besides being a popular form of entertainment, may at its best be considered an art. As with most inventions, no one person was responsible for the moving film. The work of **Thomas Edison** in America, the **Lumière** brothers in France and **William Friese-Greene** in Britain led to the showing of the first silent films in the 1890s.

By 1901, **Georges Méliès** was making story films in France. America's first story film was *The Great Train Robbery* in 1903. Two of the early film stars were **Mary Pickford** and the immortal comic, **Charlie Chaplin**. Hollywood, near Los Angeles in California, became the film-centre of the world.

The first great film-director was **D.W. Griffith**, who made *The Birth of a Nation* in 1915. The Russian director, **Eisenstein**, made the masterpiece, *The Battleship Potemkin*, in 1925. The first sound movie, *The Jazz Singer* starring **Al Jolson**, created a sensation in 1927. During the 1930s **Walt Disney's** cartoon characters became world-famous.

The competition from television has made film-makers experiment with improved sound, bigger screens and new spectacles. Now films are truly international. Hollywood has been joined by London, Rome, Paris and Stockholm as world film capitals. Even a small country like Ireland has found out, in recent times, that she can compete with the traditional giants of the cinema world.

Conclusion

> History ... is indeed little more than the register of the crimes, follies and misfortunes of mankind.
>
> (Edward Gibbon, *Decline and Fall of the Roman Empire*)

The study of history since 1870 might tempt us to agree with Gibbon's conclusion. The course of events opened with the Franco-Prussian War. We saw the mistakes made by politicians, the blinkered views of statesmen, the greed of colonists and the problems caused by extreme nationalism.

The twentieth century has been marked by the devastation of war and especially by the horrors of two world wars. The emergence of communism and the reaction of capitalist countries to the problems posed by the communist system have meant that the modern age has been one of tension and conflict. The collapse of communism in Europe has brought its own problems. 'Man's inhumanity to man' seemed to have reached its high-point in the excesses of the regimes of Hitler and Stalin. It has emerged again in the work of latter-day terrorists and in the so-called 'ethnic-cleansing' in south-eastern Europe.

However, an analysis of the advances made since 1870 shows that the picture is not so bleak and gives us some hope for the future. The democratic systems now in place in most countries in Europe contrast favourably with the situation in 1870 when states were ruled by autocrats or by the upper classes, mainly for their own benefit.

Advances in medicine, science and technology reveal the other side of man – the caring genius. A study of culture and the arts helps to restore our faith in humanity. We can only hope that mankind has learned from history and will not repeat the 'crimes' and 'follies' of the past:

> Time present and time past
> Are both perhaps present in time future
> And time future contained in time past.
>
> (T.S. Eliot, *Four Quarters*)

Questions

ORDINARY LEVEL – D

Answer the following questions briefly. One or two sentences will be enough.

1 Impressionism, Expressionism and Cubism are three styles of modern painting. Explain briefly **one** of them and name an artist associated with that style.

2 Say how one of the following contributed to the advancement of medical practice: Louis Pasteur; Robert Koch; Wilhelm Röntgen.

3 Mention **two** forms of popular entertainment in the 1920s.

4 Mention **two** artists working in Paris during the period 1870-1914.

5 With which fields of human endeavour do you associate each of **two** of the following: Alfred Krupp; Gottlieb Daimler; Thomas Mann; Albert Camus; Pablo Picasso?

6 With which fields of human endeavour do you associate **two** of the following: Karl Marx; Karl Benz; Louis Blériot; Sir Alexander Fleming; Mary Pickford?

7 With which fields of human endeavour do you associate each of **two** of the following: Marie Curie, Maria Montessori; Florence Nightingale; Paul Cézanne; Enrico Caruso; Albert Schweitzer?

8 With which fields of human endeavour do you associate each of **three** of the following: Albert Camus; John Maynard Keynes; Henrik Ibsen; Walter Gropius; Werner Von Braun; Jan Sibelius; James Ramsay Macdonald?

9 With which fields of human endeavour do you associate each of **two** of the following: Maria Montessori; Sigmund Freud; Charles Lindberg; John Maynard Keynes; Ingmar Bergman?

10 With which fields of human endeavour do you associate each of **three** of the following: Emile Zola; Gustav Mahler; Emmeline Pankhurst; Louis Blériot; Henri Matisse; Albert Schweitzer?

Questions

ORDINARY LEVEL – E

Write a short paragraph on each of the following:

1. Cultural developments in France, 1870-1914
2. One of the following areas of human endeavour during the period of your course: Science; Art; Music
3. Art in Europe, 1870-1914.

ORDINARY LEVEL – F

Write a short essay on each of the following:

1. Scientific developments and their influence on European civilisation, 1870-1970
2. Trends and movements in the cultural life of Europe, 1870-1970. (You should refer in your answer to trends and movements in some of the following: religion; philosophy; literature; drama; painting; sculpture; architecture; music; the cinema; popular entertainment).
3. Science, literature and the arts in Europe, 1870-1914
4. Technological advances, 1870-1970
5. The influence of scientific developments on European civilisation since 1870.

Questions

HIGHER LEVEL

Write an essay on each of the following:

1 Write on any attempt of European civilisation (art, literature, drama, music etc.) during the period 1870-1970 of which you have made a special study. (80)

2 'The quantity and quality of the cultural production in the fields of literature, painting and music in Europe during the second half of the 19th century were impressive.' Discuss. (80)

3 The importance of economic and technological factors in bringing about change in Europe, 1870-1914 (80)

4 Significant advances in science and technology during the period of your choice (80)

5 The impact of science and technology in the twentieth century on European culture and civilisation. (80)

Select Bibliography

The following books are recommended for further reading:

R.J.Q. Adams	*British Appeasement and the Origins of World War II* (Heath, 1994)
F. Bridgham	*Germany: Unification to Reunification* (Headstart History, 1993)
R. Carr	*The Spanish Tragedy* (Weidenfeld, 1993)
A. Cobban	*A History of Modern France* (Penguin, 1993)
B.J. Elliott	*Hitler and Germany* (Longman, 1984)
B.J. Elliott	*Western Europe after Hitler* (Longman, 1983)
M. Fulbrook	*The Fontana History of Germany, 1918-1990* (Fontana Press, 1991)
M. Fulbrook	*The Two Germanies, 1945-1990* (Macmillan, 1992)
D. Geary	*Hitler and Nazism* (Routledge, 1993)
A. de Grand	*Italian Fascism* (Nebraska Press, 1991)
S. Greenwood	*Britain and European Cooperation since 1945* (Blackwell, 1992)
J. Hite	*Tsarist Russia, 1801-1917* (Causeway Press, 1989)
G. Hoskins	*A History of the Soviet Union, 1917-1991* (Fontana, 1992)
L. Kochan and R. Abraham	*The Making of Modern Russia* (Penguin, 1993)
N. McCord	*British History, 1815-1906* (Oxford University Press, 1991)
T. Pakenham	*The Scramble for Africa* (Abacus, 1991)
K. Randell	*France: The Third Republic* (Edward Arnold, 1986)
J. Robottom	*Modern Russia* (Longman, 1980)
R. Service	*The Russian Revolution, 1900-1927* (Macmillan, 1991)
A. Stiles	*The Unification of Germany, 1815-1890* (Edward Arnold, 1986)
A. Thorpe	*Britain in the era of the Two World Wars 1914-1945* (Longman, 1993)
B. Williams	*The Modern France, 1870-1976* (Longman, 1980)
D.G. Williamson	*Bismarck and Germany, 1862-1890* (Longman, 1986)
Z.A.B. Zoman	*The Making and Breaking of Communist Europe* (Blackwell, 1991)

Index

A HOME RULE; LAND; UNIONISM

B

C GERMANY; FRANCE; RUSSIA (Tzarist)

D FASCISM; FAILURE OF COLLECTIVE SECURITY; COLD WAR

E PROJECT